Study Guide
&
Selected Solutions Manual

Marcia Gillette
Indiana University, Kokomo

Chemistry

FOR CHANGING TIMES

Fourteenth Edition

John W. Hill
Terry W. McCreary

PEARSON

Editor in Chief: Jeanne Zalesky
Acquistions Editor: Chris Hess
Marketing Manager: Will Moore
Team Lead, Project Management Biology, Chemistry,
Environmental Science, and Geo Science: David Zielonka
Program Manager: Lisa Pierce
Project Manager: Mike Reynolds
Operations Specialist: Maura Zaldivar-Garcia
Supplement Cover Designer: Seventeenth Street Studios
Cover Photo Credit: Shutterstock; Getty Images; NASA

Credits and acknowledgments borrowed from other sources and reproduced, with permission, in this textbook appear on the appropriate page within the text.

PEARSON
www.pearsonhighered.com

ISBN 10: 0-13-388904-1 / ISBN 13: 978-0-13-388904-8

1 2 3 4 5 6 7 8 9 10—**V031**— 18 17 16 15

Table of Contents

How to Survive in Chemistry
With a Satisfactory Grade

Handling any challenging and unfamiliar task requires a thoughtful and purposeful strategy. Because chemistry classes often fall in the category of "challenging" and "unfamiliar," it is very important that you adopt a strategy that will allow you to be successful. The first tool you have at your fingertips is your chemistry textbook. Read the appropriate chapter material *before* you attend class so that you'll have a general idea about what the instructor will be covering. After class, review your notes and reread the sections of the textbook that help explain any parts of your notes you don't understand. Fill in your notes with text material to make them more meaningful. Try to relate one topic to another. The different topics should seem integrated to you. Go over the practice problems in the chapter and complete the problems at the end of the chapter. There's a slogan: "CHEMISTRY IS NOT A SPECTATOR SPORT." This means that you have to get off the sidelines and be able to apply yourself to chemistry. Success in chemistry, like success in sports or music, requires practice. If you need help completing the problems, ask for it. Your teacher will be able to help you.

In addition, take good advantage of this study guide. It provides a summary (in outline form) of each chapter, a list of learning objectives, and a self-test, as well as the answers to the odd-numbered end-of-chapter problems. Where appropriate, a discussion of the material in the chapter is also provided. Please note that there isn't a self-test item for every objective or an objective for every test item, but the self-test questions should provide you with an indication of how well you understand the material. For some chapters, additional worked-out examples are provided and additional problems are included. You should work those problems when they, or similar ones from the textbook, are assigned.

Your instructor and the textbook can be of considerable help as you learn chemistry, but your success in the course will depend mainly on the way in which *you* approach the challenge. Before each test be sure that you have completed all assignments and used the self-test to ensure that you understand the material in each chapter. Adopting this approach will help you to understand the principles being covered well enough that you can apply them to new problems and scenarios.

We hope that you find this study guide helpful. Please send us any criticisms and suggestions for improvement. A notation of any errors that you find would be especially helpful.

Study Strategy Suggestions

There are many effective ways to study, but most of them incorporate one or more of the following methods:

1. Preview the material in the text to be covered in the lecture. Just 10 minutes of leafing through the appropriate text pages can make the lecture more meaningful and your note-taking easier.
2. Quickly read through the notes from the last lecture within two hours of that lecture. You forget most of what you hear within two hours. Ten minutes of refreshing your memory shortly after the lecture is a very effective way to retain information.
3. Spend a minimum of three hours of study for each hour of lecture.
4. Move quickly to the problems at the end of the chapter and the self-tests in the study guide after reading the text and studying the lecture notes,
5. Be aware that four ½-hour study periods are better than two 1-hour study periods, which are better than one 2-hour study period.
6. Study with a friend and/or a study group if possible. Each member of an effective study group should prepare individually for the study session. The strengths of each person will help eliminate the weaknesses of the others.
7. Prepare for the exams well in advance. Arrive early and have all materials needed for the test.

CHAPTER

1

Chemistry

A Science for All Seasons

CHAPTER SUMMARY

1.1 Science and Technology: The Roots of Knowledge

Learning Objectives: • Define science, chemistry, technology, and alchemy. • Describe the importance of green chemistry and sustainable chemistry.

A. Science is the primary means by which we obtain new knowledge about nature and our physical world and develop theories to explain that knowledge.
B. Technology is the application of knowledge for practical purposes.
C. Chemistry is that area of knowledge that deals with the behavior of matter.
   1. The ancient Greeks formulated theories about nature.
   2. Alchemists tried unsuccessfully to turn various metals into gold.
   3. Alchemists perfected some chemical techniques such as distillation and extraction.
D. Science developed out of natural philosophy but had its true beginnings when people began to rely on experiments.
E. Green chemistry and sustainability
   1. In response to public concern, chemists developed the field of green chemistry, which places emphasis on reducing pollution and waste of all kinds.
   2. Green chemical concepts led to an emphasis on sustainability or resource utilization, with a concern for the welfare and well-being of future generations.

Answers to Self-Assessment Questions

1. b The ability to change lead or other metals into gold was the goal of alchemists.
2. c The alchemists' approach marks the beginning of modern chemistry.
3. d Rachel Carson's book described the long-term hazardous effects of pesticides.
4. c The goal of green chemistry is to reduce pollution at its source and to conserve resources with an eye to providing for the welfare of future generations.
5. b Sustainable chemistry involves using renewable resources. Hardwood trees are renewable resources.

1.2 Science: Reproducible, Testable, Tentative, Predictive, and Explanatory

Learning Objective: • Define hypothesis, scientific law, scientific theory, and scientific model, and explain their relationships in science.

A. Scientific data must be reproducible.
   1. Scientific work is not fully accepted until it has undergone peer review and has been verified by other scientists.
B. Scientific hypotheses are testable.
   1. Large amounts of scientific data are often summarized in brief verbal or mathematical statements called scientific laws.
C. Scientific theories are tentative and predictive.
   1. A theory represents the best current explanation for a phenomenon.
   2. Theories organize scientific knowledge and are also useful for their predictive value.
   3. Predictions based on theories are tested by further experiments.
D. Scientific models are explanatory.
E. What science is—and what it is not.
   1. Only ideas that have survived experimental testing and peer review are considered valid.
   2. Science is not a democratic process: Majority rule does not determine what constitutes sound science.
   3. Science does not accept notions that are proven false or remain untested by experiment.
F. The limitations of science.
   1. The control of variables can be very difficult.
   2. A variable is something that can change over the course of an experiment.

Answers to Self-Assessment Questions

1. a Hypotheses are tested by experiment.
2. c A law is a summary of a large amount of consistent experimental data.
3. a In order to be useful, a theory must be able to be tested.
4. a This statement cannot be tested.
5. c Scientific research must be experimentally reproducible and peer reviewed in order to be considered valid.
6. a Experimental results are difficult to interpret if all variables cannot be effectively controlled.

1.3 Science and Technology: Risks and Benefits

Learning Objectives: • Define *risk* and *benefit*, and give an example of each. • Estimate a desirability quotient from benefit and risk data.

A. Technology is the sum of the processes by which we modify materials to better serve us.
B. Risk–benefit analysis involves the calculation of a desirability quotient (DQ): DQ = benefits/risks.
C. Benefits and risks often are difficult to quantify and lead to uncertain DQs.

Answers to Self-Assessment Questions

1. d Risk assessment is often based on a social judgment.
2. a According to Table 1.1, of the choices a, the approximate lifetime odds of death from heart disease are 1 in 5, which is higher than even cancer (1 in 7).
3. c Chemistry is central to every aspect of our daily lives.
4. c The U.S. chemical industry is one of the nation's largest industries, accounting for thousands of the products we use in our everyday lives.

1.4 Solving Society's Problems: Scientific Research

Learning Objective: • Distinguish basic research from applied research.

A. Many chemists are involved in applied research.
   1. Applied research is directed toward the solution of a particular problem in industry or in the environment.

B. Many chemists are engaged in basic research.
   1. Basic research is the pursuit of knowledge for its own sake.

Answers to Self-Assessment Questions

1. b Drugs are developed for the benefit of society.
2. a Determining plant responses to drought conditions gives insight into the metabolism of plants, with the promise of future application.
3. a Identification of naturally occurring biologically active compounds provides the basis for inquiry into the relationship of structure and function of substances.
4. b Discovering ways to improve milk production in cows would benefit society in general.

1.5 Chemistry: A Study of Matter and Its Changes

Learning Objective: • Differentiate mass and weight; physical and chemical changes; and physical and chemical properties.

A. Chemistry deals with matter and the changes it undergoes.

B. Matter occupies space and has mass.
   1. Mass measures a quantity of matter that is independent of its relative location.
   2. Weight measures a force, such as the gravitational force or attraction between an object and the earth.

C. Matter is characterized by its properties.
   1. Physical properties can be observed and specified without reference to any other substance.
   2. Physical changes do not change the composition or chemical nature of a substance.
   3. Chemical properties describe and determine how substances combine with other substances.
   4. Chemical changes result in a change in chemical properties.

Answers to Self-Assessment Questions

1. d All but a rainbow have mass and occupy space.
2. b Air pollution is caused by substances that have mass and occupy space.
3. c The identical items will have the same masses but different weights because of differences in gravitational forces on Earth and Mars.
4. b Matter does not change its composition when it undergoes a physical change.
5. d There is no change in composition, so this is a physical change.
6. b This is a physical change as there is no change in the composition.
7. b Copper tarnishes as the result of a chemical reaction.
8. c There is no change in the composition of wool when it is spun into yarn.
9. c There is a change in composition of the tree itself as it absorbs nutrients and grows.

1.6 Classification of Matter

Learning Objective: • Classify matter according to state and as mixture, substance, compound, and/or element.

A. The states of matter.
   1. Solids maintain their shape and their volume.
   2. Liquids maintain their volume but not their shape.
   3. Gases maintain neither their volume nor their shape. They occupy the entire volume of their container

B. Matter: pure substances and mixtures.
   1. Pure substances have fixed compositions that do not vary from one sample of the substance to another.
   2. A mixture is a collection of two or more substances that retain their individual identities and can be mixed in any proportion because they are not chemically bonded to one another.

C. Elements and compounds.
   1. An element is a pure substance that is composed of a single type of atom.
      a. At present there are more than 100 known elements.
      b. Each element is represented by a chemical symbol.
   2. A compound is made up of two or more elements in a fixed ratio.
   3. Elemental symbols are made up of one or two letters derived from the name of the element.
      a. Only the first letter of a chemical symbol is capitalized.
   4. Atoms and molecules.
      a. Atoms are the smallest characteristic part of an element.
      b. Molecules are groups of atoms bound together as a unit

Answers to Self-Assessment Questions

1. c Only liquids have definite volume but vary in shape. Gases vary in volume and shape, while solids have definite volume and definite shape.
2. d Carbon, copper, and silver are elements—soda water is water with carbon dioxide dissolved in it.
3. b Brass is an alloy of copper and tin.
4. a Only a chemical process can separate a compound into its elements.
5. a An atom is the smallest unit of an element.

6. b Na is the symbol for sodium.
7. b Ar is the symbol for argon. Air is a mixture and not an element, Al is the symbol for aluminum, and As is the symbol for arsenic.

1.7 The Measurement of Matter

Learning Objective: • Assign proper units of measurement to observations, and manipulate units in conversions.

A. SI is a modernized version of the metric system.
   1. The SI base unit for Length: meter (m)—slightly more than a yard.
   2. The SI base unit for Mass: kilogram (kg)—(2.2 lb).
   3. The SI base unit for Volume: cubic meter (m^3).
   4. The SI base unit for Time: second (s).
   5. The SI base unit for Temperature: kelvin (K)—0 K is the lowest temperature possible.
   6. The SI base unit for Amount of a substance: mole (mol)—6.022×10^{23} units.
   7. The SI base unit for Electric Current: ampere (A).
   8. The SI base unit for Luminous Intensity: candela (cd).

B. Numbers are often expressed in scientific notation.

C. The most commonly used metric prefixes and their powers of 10.

1. kilo-	(k)	×	1000	(10^3)
2. deci-	(d)	×	0.1	(10^{-1})
3. centi-	(c)	×	0.01	(10^{-2})
4. milli-	(m)	×	0.001	(10^{-3})
5. micro-	(μ)	×	0.000001	(10^{-6})

D. Using equivalencies to accomplish unit conversions is an extremely important skill to use for problem solving.

Answers to Self-Assessment Questions

1. d The meter is the SI unit for length.
2. d The kilogram is the SI unit for mass.
3. a 10^{-2} is the prefix for centi or one one-hundredth.
4. b One inch is equal to 2.54 cm or slightly more than 250 times as big.
5. d 1 cm^3 = 1 mL.
6. c One quart is the equivalent of 0.946 liter.
7. d 5.775 cm = 57.75 mm because there are 10 mm in 1 cm.
8. c This textbook is around 4 kg, an orange is around 125 g, an ant is far less than 1 g, and a peanut is around 1 g.
9. d There are 1000 μg in 1 mg.
10. b 1 cm is roughly 1/250th of an inch. All the objects listed except the knife blade are more than 1 in thick.
11. a 1 cup has a volume of approximately 250 mL.

1.8 Density

Learning Objective: • Calculate the density, mass, or volume of an object given the other two quantities.

A. Density is mass divided by volume: D = m/V.
   1. For liquids and solids, density is usually expressed in units of grams per milliliter (g/mL, for liquids) or grams per cubic centimeter (g/cm^3 for solids).
   2. At 4 °C the density of water is 1.0 g/mL.

Answers to Self-Assessment Questions

1. b In order to float, wood must be less dense than water.
2. b Because they sank, both a pebble and a lead sinker must be more dense than water.
3. a Because it floats on water, the density of ice must be less than that of water (0.998 g/cm^3).
4. b Shaking a pan of mud, gravel, and gold allows the higher-density gold to sink to the bottom of the pan.
5. d Of the choices, only uranium has a density higher than that of mercury, allowing it to sink.
6. c D = m/V; 160.0 g/80.0 cm^3 = 2.00 g/cm^3.

1.9 Energy: Heat and Temperature

Learning Objectives: • Distinguish between heat and temperature. • Explain how the temperature scales are related.

A. Energy is the ability to do work; to change matter either physically or chemically.
B. Definition of heat and temperature.
   1. Heat: energy on the move; the energy that flows from a warmer object to a cooler one.
   2. Temperature is a measure of the average energy of a particle in a given sample; how hot or cold an object is.
   3. Heat flows from more energetic (higher-temperature) to less energetic (lower-temperature) atoms or molecules.
C. Temperature and heat.
   1. The Celsius scale unit is (°C). Water freezes at 0 °C and boils at 100 °C.
   2. The Kelvin scale unit is (K); also known as the absolute scale. 0 K = –273.15 °C.
   3. Conversion between Celsius and Kelvin scales: K = °C + 273.15.
D. Energy: joule (J) or calorie (cal).
   1.0 cal = 4.18 J; 1 Kcal = 1 Calorie (foods) = 1000 cal = 4.18 kJ.

Green Chemistry

A. The enormous technical advances in medicine, manufacturing and chemical processing of the 20th century have led to an increased quality of life in the United States but, at the same time, have taken a significant toll on the environment and human health.
   1. Improper chemical use and disposal and general environmental pollution have significantly degraded the world's ecosystems, threatening the security of future generations.
   2. Sustainable and green chemistry is a different way of thinking about how chemistry and chemical engineering can be done.

a. Different principles have been proposed that can be used when thinking about the design, development, and implementation of chemical products and processes.
b. These principles promote the discovery and development of creative and innovative ways to reduce waste, conserve energy, and replace hazardous substances to protect and benefit the economy, people, and planet.

Answers to Self-Assessment Questions

1. b Heat is energy that flows from a hotter body to a colder one.
2. c Kelvin is the SI temperature unit. A temperature can be expressed in other units using conversion factors or equations.
3. d Water boils at 100 °C (373 K).
4. d We convert temperatures in Kelvins to °C by subtracting 273 from the temperature in Kelvins.
5. c The joule is the SI energy unit.

1.10 Critical Thinking

Learning Objective: • Use critical thinking to evaluate claims and statements.

A. FLaReS: an acronym for critical thinking (ignore the vowels).
1. Falsifiability: Can you conduct an experiment that could prove something wrong? A hypothesis that cannot be tested is not useful.
2. Logic: Does the conclusion follow from the premises; are the premises true?
3. Replicability or Reproducibility: To be valid, experiments must have the ability to be repeated to produce the same evidence.
4. Sufficiency: Evidence provided must be adequate to support the validity of the claim.
   a. The burden of proof is on the claimant.
   b. Extraordinary claims require extraordinary evidence.
   c. Evidence based on authority and/or testimony is not adequate.

LEARNING OBJECTIVES

These objectives are a minimal list. In addition to knowing the material specifically indicated here, you should do the assigned problems in the text. Your instructor may provide additional (or alternative) objectives for any chapter.

You should be able to . . .

1. Define science, chemistry, technology, and alchemy. (1.1)

2. Describe the importance of green chemistry and sustainable chemistry. (1.1)

3. Define *hypothesis*, *scientific law*, *scientific theory*, and *scientific model*, and explain their relationships in science. (1.2)

4. Define *risk* and *benefit*, and give an example of each. (1.3)

5. Estimate a desirability quotient from benefit and risk data. (1.3)

6. Distinguish basic research from applied research. (1.4)
7. Differentiate mass and weight; physical and chemical change; physical and chemical properties. (1.5)
8. Classify matter according to state and as mixture, substance, compound, and/or element. (1.6)
9. Assign proper units of measurement to observations and manipulate units in conversions. (1.7)
10. Calculate the density, mass, or volume of an object given the other two quantities. (1.8)
11. Distinguish between heat and temperature. (1.9)
12. Explain how the temperature scales are related. (1.9)
13. Use critical thinking to evaluate claims and statements. (1.10)
14. Define green chemistry.
15. Describe how *green chemistry* reduces risk and prevents environmental problems.

DISCUSSION

Much of Chapter 1 is intended to place chemistry in both historical and contemporary perspective—to give you a feel for chemistry as it affects society. If, after reading the chapter, you recognize chemistry as something more than just a course that meets certain requirements, you have indeed understood what we were trying to say. In addition to this overview, Chapter 1 introduces several concepts important to our further study of chemistry. These include the international system of measurement, the meaning of terms such as matter and energy, different temperature scales, chemical symbols, and density. Green chemistry represents the recognition of the impacts modern technology has on the environment and on the sustainability of the planet for future generations. The problems at the end of the chapter are meant to check your understanding of this material. The following questions offer another opportunity for you to test yourself on Chapter 1.

Note: Many of these problems are more than tests of your memory. A number of them require preliminary calculations before an answer can be selected. You are expected to know the metric prefixes and units of measure, but consult **Appendix A** in the text if you need help with the metric conversion factors.

EXAMPLE PROBLEM

Challenge: Name something that is not a chemical.

Answer: Everything is a chemical. Energy and matter are equivalent. A vacuum is the absence of matter: It is nothing. Also, a complete vacuum does not exist even in space. Some other possible responses: An idea (Can a thought exist without a brain with neurons and electrons?) Time (Can time exist without something around to mark its passing?) Basically, chemistry is the study of just about everything.

ADDITIONAL PROBLEMS

1. An ad on TV invites you to call for a free "psychic reading." As proof of their legitimacy, psychics offer several testimonials by people claiming that the psychic they called knew facts about the caller that were known by only themselves. Which part of the "FLaReS" test does the ad fail?

   a. Falsifiability
   b. Logic
   c. Replicability
   d. Sufficiency
   e. All of the above

   e. It fails all parts of the test:

   *a. Falsifiability—can't ask testimonial givers questions
   *b. Logic—not applicable
   *c. Replicability—can't reproduce results
   *d. Sufficiency—no evidence given

2. A horoscope printed in the newspaper (*Dayton Daily News*; Joyce Jillson, UPS 1997 ®) for a Leo states: "Creative energy is limitless, so set aside plenty of time for projects. A romantic involvement with a Libra moves to a deeper level of commitment. Make sure that a relative knows how much you are willing to help with a family project."

   Which part of the "FLaReS" test does the horoscope fail?

   a. Falsifiability
   b. Logic
   c. Replicability
   d. Sufficiency
   e. All of the above

   It fails all parts of the test:

   *a. Falsifiability—claims are too vague to be evaluated.
   *b. Logic—not everyone with birthdays July 23–August 22 have the same circumstances.
   *c. Replicability—can't reproduce the experiment.
   *d. Sufficiency—no evidence provided.

ANSWERS TO ODD-NUMBERED REVIEW QUESTIONS AND SOLUTIONS FOR ODD-NUMBERED END-OF-CHAPTER PROBLEMS

Review Questions

1. Science is testable, reproducible, explanatory, predictive, and tentative. Testability best distinguishes science.

3. These problems usually have too many variables to be treated by the scientific method.

5. Risk–benefit analysis compares benefits of an action to risks of that action.

7. DQ, the Desirability Quotient, is benefits divided by risks. A large DQ means that risks are minimal compared to benefits. Often it is hard to quantify risks and benefits.

9. The SI-derived unit for volume is liter (L). Because one liter represents a relatively large quantity, volumes measured in milliliters (mL) and, sometimes, in microliters (μL) are often used in the laboratory.

11. a. applied: This is an example of the application of a technology to a problem with economic consequences.
    b. applied: The Purdue engineer is developing a method for improving operation of automobiles so that their operation will be more environmentally safe. If the method can be applied, automobiles will not require large amounts of fossil fuels and their exhaust will not include greenhouse gases.
    c. basic: Studying atoms under different conditions helps us understand general atomic behavior but does not have a specific application.

Problems

13. The benefits to society are very great. The risk to selected individuals is great but is managed by restricting access to penicillin by prescription and requiring the care of a physician. The DQ is high.

15. The use of isocyanates can cause allergic reactions in some individuals.
    a. When proper safety precautions are taken (proper ventilation and, when necessary, individual masks for workers), the risk of an automotive factory worker being exposed to isocyanates is low and the benefit is high. The DQ is high. The DQ can be increased by ensuring that all safety precautions are observed.
    b. While exposure time is likely to be relatively short, the hobbyist restoring a 1965 Mustang is less likely to be able to work under ideal safety conditions but should know to wear breathing protection and to work in a well-ventilated garage or other area. The DQ is high (but lower than that in scenario a) and can be increased by ensuring that all safety precautions are observed.

17. The DQ for routinely treating sore throats is low, especially if the origin of the problem is unknown. The risk of development of antibiotic-resistant bacteria likely outweighs the possible benefits of the treatment. The flu is a viral problem that does not usually respond to antibiotics, so the DQ for antibiotic use in that instance would likely be considerably lower than for their use to treat a sore throat of bacterial origin.

19. Cellular phones are light but not as light as 100 mg; 100 g is a more likely weight for a cell phone. Laptop computers vary in weight but are likely to weigh several pounds (2.12 lb = 1 kg). The choice of 100 g, 2 kg is the best.

21. Options a and b both seem reasonable. Assuming that the case of bottled water referred to in option d contains 24 16-oz (or 1 lb) bottles, then the mass of those items, converted to kg would be: Mass, kg = (24 lb)(0.454 kg/lb) = 10.9 kg, so the given amount needs to be multiplied by 10. If a case of bottled water weighs approximately 10 kg, then the carpet in option c must weigh a great deal more than 10 kg, and the listed mass needs to be multiplied by 10 at least once and, depending on the room size, perhaps more.

23. An 8-oz tea cup holds about 250 mL.

25. The volume in mile^3 is 3.50×10^8 mile^3. One mile = 1.61 km. Since we are working with a volume with the length cubed, the conversion factor is one mile^3 = 4.17 km^3. (You cube the conversion factor.) 3.50×10^8 mile^3 times 4.17 km^3/1 mile^3 = 1.46×10^9 km^3.

27. One in = 2.54 cm. Without doing a detailed calculation, we know that 0.998 in is nearly 1 in or 2.54 cm (to be precise, it is 0.998 in times (2.54 cm/1 in) or 2.53 cm). The other tube with an inside diameter of 26.3 mm is equivalent to 2.63 cm. Therefore, the 0.998 in tube will fit inside the 26.3 mm tube.

29. a. The composition of glass does not change when it breaks, so glass breaking is a physical property.
    b. The growth of mold on bread involves chemical reactions, so mold formation is a chemical property.
    c. The composition of gold and copper does not change when one or both of them melt. The melting point of a substance is a physical property.
    d. The formation of a custard from eggs, sugar, and milk involves a chemical reaction. Therefore, the fact that this reaction occurs when these three substance are mixed describes a chemical property.

31. a. This is a physical change, assuming that the plastic was just melted and re-formed into a new shape.
    b. This is a chemical change because a chemical reaction is required to transform used cooking oil into biodiesel fuel.
    c. There is no change in the chemical composition: The concentrated orange juice was diluted, which represents a physical change.

33. a. Helium gas is a substance.
    b. Orange juice is a mixture of a variety of substances, including vitamin C, pulp, and water.
    c. Distilled water is water from which all impurities have been removed. It is a substance.
    d. Carbon dioxide is a compound and thus a substance.

35. a. The composition of the copper water pipe is consistent throughout its length and, therefore, a homogeneous mixture of components.
    b. Distilled water is a substance that is homogeneous throughout.
    c. Liquid oxygen is a substance that is homogeneous throughout.
    d. The composition of one spoonful of chicken noodle soup likely differs from the composition of another spoonful. The soup is a heterogeneous mixture.

37. Because glucose has a consistent composition, it is a substance.

39. a. Ir (iridium) is an element.
    b. I (iodine) is an element.
    c. InF is a compound (two elements, indium and fluorine, combined).
    d. IF_7 is a compound form from iodine and fluorine.

41. a. iron
    b. magnesium
    c. helium
    d. nitrogen

43. Observations a–e could apply to an element or a compound. Observation f indicates that baryte is the chemical combination of the particular element that burned and oxygen, making it a compound rather than an element.

45. a. 1.0×10^{-1} L = 1.0 dL
    b. 8.5×10^{-12} g = 8.5 pg
    c. 1.05×10^{-3} m = 1.05 mm

47. a. (5.52×10^4 mL)(1 L/1000 mL) = 55.2 L
    b. (325 mg)(1 g/1000 mg) = 0.325 g

c. (27 cm)(1 m/100 cm) = 0.27 m
d. (27 mm)(1 cm/10 mm) = 2.7 cm
e. (78 μs)(1 millisecond/1000 μs) = 0.078 ms

49. a. 1 cm is larger than 1 mm (1 cm = 0.01 m; 1 mm = 0.001 m)
b. 1 kg is larger than 1 g (1 kg = 1000 g)
c. 1 dL is larger than μL (1 dL = 0.1 L; 1 μL = 0.000001 L)

51. a. 1 cm = 10 mm. (31 cm)(10 mm/cm) = 310 mm.
b. 1 m = 100 cm. (31 cm)(1 m/100 cm) = 0.31 m
c. 1 mm = 0.1 cm. (31 cm)(1 mm/0.10 cm) = 10 mm

53. a. d = m/V = 500.0 g/ 53.6 cm^3 = 9.33 g/cm^3

55. a. d = m/V so m = dV
m = (1.51 g/mL)(49.1 mL) = 74.1 g

57. a. d = m/V so V = m/d
V = 227 g/0.660 g/mL = 344 mL
b. d = m/V so V = m/d
V = 454 g/0.917 g/cm^3 = 495 cm^3

59. Mercury (d = 13.534 g/mL) will be on the bottom, water (d = 0.998 g/mL) in the middle, and hexane (d = 0.660 g/mL) will be on top.

61. d = m/V   37.9 L = 37,900 mL or 37,900 cm^3
m = dV = (1.03 g/cm^3)(37,900 cm^3)
m = 39,037 g
Convert mass of water from grams to pounds: (39,037 g)(1 lb/454 g) = 85.98 lb or 86.0 lb
Mass of aquarium plus seawater = 59.5 lb + 86.0 or 145.5 lb
Yes, the stand will hold the filled aquarium.

63. d = m/V so m = dV
Before calculating volume, change diameter from mm to cm and then calculate the radius:
(148 mm)(1 cm/10 mm) = 14.8 cm/2 = 7.4 cm
$V = 4\pi r^3/3 = [(4)(3.14)(7.4\ cm)^3]/3 = 1.7 \times 10^3\ cm^3$
$m = dV = (3.18\ g/cm^3)(1.7 \times 10^3\ cm^3) = 5.4 \times 10^3\ g$

65. K = °C + 273.15 or rearranging the equation to solve for °C, °C = K – 273.15
°C = 77 K – 273 = –196 °C

67. 1 food Calorie = 1 kilocalorie which equals 4.184 kilojoules
(161 kJ)(1 kcal/4.184 kJ) =38.5 kilocalories

69. 1 μcen = 1×10^{-6} cen
time, min = (1 μcen)(1×10^{-6} cen/ μcen)(100 yr/cen)(365 day/yr)(24 hr/day)(60 min/hr)
= 52.6 min

71. The answer is b. The relationship describes a law. Henry's collected data could be expressed in the form of a mathematical equation that predicts the relationship between concentration of dissolved gas and the partial pressure of that gas in the atmosphere over the liquid.

73. time in years = (200 kw)(1000 w/kw)(15 min/w)(1 hr/60 min)(1 day/24 hr)(1 yr/365 days) = 5.7 yr (6 years to one significant figure).

75. (1) d, repeated observations.
(2) e, a theory that explains the rising power of potassium bromate.
(3) b, a hypothesis or suggested explanation for the observed behavior.
(4) a, an experiment designed to produce some sort of effect.
(5) d, an observation made when the experiment was performed.

77. d = m/V so m = dV
V = (7.6 cm)(7.6 cm)(94 cm) = 5400 cm^3
Mass, g = (0.11 g/cm^3)(5400 cm^3) = 590 g = 5.9×10^2 g

79. Convert all masses to a common unit, for example, kg.
Potatoes: 5 lb of potatoes to kg: (5 lb)(0.454 kg/1 lb) = 2.3 kg
Cabbage: 1.65 kg of cabbage
Sugar: 2500 g of sugar is 2.5 kg.
Cabbage (lightest) < potatoes < sugar (heaviest)

81. d = m/V.
Mass of the metal = (mass of metal + paper) – mass of paper = 18.43 g – 1.21 g = 17.22 g
d = 17.22 g/3.29 cm^3 = 5.23 g/cm^3

83. 5.79 mg = 0.00579 g
Density of gold = 19.3 g/cm^3 (from Table 1.7)
V = m/d = 0.00579 g/19.3 g/cm^3 = 0.000300 cm^3
V = (area)(thickness)
Thickness = V/area
Thickness = 0.000300 cm^3/44.6 cm^2 = 6.73×10^{-6} cm

85. V = 36.1 cm × 36.1 cm × 36.1 = 47,046 cm^3
d = m/V so m = dV
m = 19.3 g/cm^3 × 47,046 cm^3
1 metric ton = 1000 kg; convert mass from g to kg
Mass, kg = 907989 g)(1 kg/1000 g) = 907.989 kg
Mass in metric tons = (908 kg)(1 metric ton/1000 kg) = 0.908 metric tons

87. a. 7.0×10^4 km = 7.0×10^9 cm
Volume of Jupiter = [(4)(3.14)(7.0×10^9 cm)(7.0×10^9 cm) (7.0×10^9 cm)]/3
= 1.44×10^{30} cm^3
Mass of Jupiter = 1.9×10^{30} g
d = m/V = 1.9×10^{30} g/ 1.44×10^{30} cm
= 1.3 g/cm^3

b. 6.4×10^3 km = 6.4×10^8 cm
Volume of Earth = [(4)(3.14)(6.4×10^8 cm)(6.4×10^8 cm)(6.4×10^8 cm)]/3
= 1.10×10^{27} cm^3
Mass of Earth = 5.98×10^{24} kg
d = m/V = 5.98×10^{27} g/1.10×10^{27} cm^3
d = 5.44 g/cm^3

c. 5.82×10^4 km = 5.82×10^9 cm
Volume of Saturn = [(4)((3.14)(5.82×10^9 cm)(5.82×10^9 cm)(5.82×10^9 cm)]/3
= 8.25×10^{29} cm^3

Mass of Saturn = 5.68×10^{29} g
$d = m/V$
$d = 5.68 \times 10^{29}\ \text{g}/\ 8.25 \times 10^{29}\ \text{cm}$
$= 0.688\ \text{g/cm}^3$
The density of water is 1.0 g/cm^3. Saturn would float on water.

89. Green chemistry is the design of chemical products and processes that reduce or eliminate the generation and use of hazardous substances from the beginning to the end of the process.

91. The answer is a. Through the use of less hazardous materials and designing more effective production processes, workers and the environment both benefit.

CHAPTER

2

Atoms

Are They for Real?

CHAPTER SUMMARY

2.1 Atoms: Ideas from the Ancient Greeks

Learning Objective: • Explain the ancient Greeks' ideas about the characteristics of matter.

A. The prevailing view of matter held by Greek philosophers in the fifth century B.C. was that it was endlessly divisible.
B. Democritus, a student of Leucippus, believed that there must be a limit to the divisibility of matter and called his ultimate particles *atomos*, meaning "cannot be cut" or "indivisible."
C. The Greeks believed that there were only four elements: earth, air, fire, and water.

Answers to Self-Assessment Questions

1. a It is impossible to distinguish the individual components in prepared coffee. In the other cases, while the piles or quantities may be large, it is possible to distinguish the individual components (individual leaves, beans, or floor tiles).
2. b People thought at the time that characterizing matter as continuous seemed both logical and reasonable.

2.2 Scientific Laws: Conservation of Mass and Definite Proportions

Learning Objectives: • Describe the significance of the laws of conservation of mass and definite proportions. • Calculate the amounts of elements from the composition of a compound.

A. Boyle (1661) proposed that substances capable of being broken down into simpler substances were compounds, not elements.

B. Lavoisier (1780s) helped establish chemistry as a quantitative science.
   1. Lavoisier summarized the results of many experiments with the law of conservation of mass states that matter is neither created nor destroyed during a chemical change. Matter is conserved.
      a. We make new materials by changing the way atoms are combined.
   2. Lavoisier was the first to use systematic names for elements.
   3. Lavoisier is often called the father of modern chemistry.

C. Proust (1799) concluded from analyses that elements combine in definite proportions to form compounds. He formulated the law of definite proportions (also called the law of definite proportions, the law of constant combination, or the law of constant composition).
   1. J. J. Berzelius, Henry Cavendish, William Nicholson, and Anthony Carlisle further proved the law of definite proportions.
   2. The law of definite proportions is the basis for chemical formulas.
   3. The law of definite proportions also means that compounds have constant properties in addition to constant composition.

Answers to Self-Assessment Questions

1. c The law of conservation of mass tells us that the mass of the reactants must equal the mass of the products in any chemical transformation.
2. c According to the law of conservation of mass, the sum of the mass of the carbon dioxide gas produced must be equal to the sum of the masses of the carbon and oxygen that were decomposed to produce the gas.
3. d The decomposition of water into hydrogen and oxygen indicates that the formula for water contains H and O, so water must be a compound, not an element.
4. c The repeated relationships between the purchase of two gloves with one hat and of two shoes with one purse is an illustration of the law of definite proportions, which indicates that atoms (or items of any kind) combine in definite proportions, in this case two to one.
5. c Water, in whatever quantity, always contains the same percentage by mass of the constituent elements, illustrating the law of definite proportions.
6. c The volume proportions of hydrogen and oxygen in water are constant, illustrating the law of definite proportions.
7. c Only 220.0 g of carbon dioxide can form because once the carbon supply is consumed, there is no carbon left to react with the remaining oxygen and so the reaction stops.

2.3 John Dalton and the Atomic Theory of Matter

Learning Objectives: • Explain why the idea that matter is made of atoms is a theory. • Know how atomic theory explains the laws of multiple proportions and conservation of mass.

A. Dalton extended the ideas of Lavoisier and Proust with the law of multiple proportions.
   1. Elements may combine in more than one set of proportions, with each proportion corresponding to a different compound.

B. Dalton proposed his atomic theory (model) to explain the laws of chemistry.
   1. A chemical law is a statement that summarizes data obtained from experiments.
   2. A theory is a model that consistently explains observations.

C. Dalton's atomic theory.
   1. All matter is composed of small, indestructible, and indivisible atoms.
   2. All atoms of a given element are identical, but the atoms of different elements are different.

3. Compounds are formed by combining elements in fixed proportions.
4. A chemical reaction involves a rearrangement of atoms. No atoms are destroyed or broken apart in the process.

D. Modern modifications of items of Dalton's atomic theory:
   1. Scientists now know that atoms can be divided, as we shall see in the next chapter.
   2. Scientists now know that atoms can have different masses, as we shall see in the next chapter.
   3. Dalton's premise that compounds are formed by combining elements in fixed proportions is valid today as it was when Dalton proposed it.
   4. Scientists today know that chemical reactions involve a rearrangement of atoms. Dalton's premise that no atoms are destroyed or broken apart in a chemical reaction remains valid, although since Dalton's time scientists have learned that atoms do undergo transformations during nuclear reactions.

E. Explanations using atomic theory.
   1. Elements are composed of one kind of atom.
   2. Compounds are composed of two or more kinds of atoms chemically combined in definite proportions.
   3. Matter must be atomic to account for the law of definite proportions.
   4. The rearrangement of atoms explains the law of conservation of mass.
   5. The existence of atoms explains how multiple proportions can exist.

F. Isotopes
   1. Atoms of an element can have different masses. Such atoms are called isotopes.

2.4 The Mole and Molar Mass

Learning Objectives: • Describe what a mole is and how it is used. • Convert between the masses and the moles of a substance.

A. The Mole and Avogadro's Number.
   1. Because the carbon atoms in a diamond are equally spaced at known distances, it is possible to determine the number of carbon atoms in a known mass of diamond.
   2. The number of carbon atoms in a diamond weighing 12.011 g was found to be 6.022×10^{23}.
   3. The quantity 6.022×10^{23} is known as one mole, so one mole of carbon contains 6.022×10^{23} carbon atoms and has a mass of 12.011 g.

B. Molar Mass.
   1. The molar mass of an element is the mass of 6.022×10^{23} atoms of that element, while the molar mass of a compound is the mass of 6.022×10^{23} molecules of that compound. Because carbon-12 is used as a reference, molar masses of all other elements (and compounds) are related to it. Similarly, the atomic masses of all elements are expressed relative to carbon-12, and typically, they are expressed in atomic mass units (amu).
   2. **NOTE**: It is easy to confuse what seem like similar terms: molecule and mole. Molecule refers to an individual unit, while mole refers to collections of 6.022×10^{23} of those units. We measure the mass of molecules in atomic mass units; we measure the mass of moles of those units in grams.

Answers to Self-Assessment Questions

1. d The mass ratio of C to O in CO is 36.0321 g of C to 47.9982 g of O or 0.750697. Carbon dioxide has twice the amount of oxygen as CO, so the C to O mass ratio would be half that of CO, and the mass of oxygen would be (2)(47.9982 g) or 95.9964 g.
2. d Dalton based his theory on the atomic weights of atoms.
3. c Conservation of mass was the key factor for Dalton to propose his theory.
4. c Dalton thought that atoms were conserved and only rearranged to make new materials.
5. a Compounds are composed of two or more elements.
6. d Unlike what is stated in answer d, one mole of carbon contains Avogadro's number of C atoms, which is the amount of carbon in 1 mol of CO_2. This amount of carbon has a mass of 12.0107 g.
7. d Molar mass can be applied to anything: It is the mass of 6.022×10^{23} of those items.
8. b The molecules shown are composed of only one kind of atom, so these represent an element.
9. a The species in the image represent compounds because they are composed of more than one kind of atom. Because all the species in the image are the same, the image represents a single compound.
10 c Image c represents a mixture because it contains two different kinds of species.

2.5 Mendeleev and the Periodic Table

Learning Objective: • Describe how the elements are arranged in the periodic table and why the arrangement is important.

A. By the mid-1800s, 55 elements were known, but no successful way existed to classify them.
   1. Only relative atomic weights could be determined.

B. Emerging Patterns and Periodicity within the Known Elements.
   1. In 1816, Johan Dobereiner found a relationship between the elements based on their atomic masses.
   2. In 1863, John Newland and Béguyer Chancourtois found that atomic masses were separated by regular intervals of eight times the mass of hydrogen.

C. The Periodic Table: Mendeleev and Meyer.
   1. Meyer and Mendeleev linked chemical properties to atomic mass, establishing the periodic law.
   2. Mendeleev's periodic table (1869) grouped 63 elements by increasing atomic weight.
   3. In some instances, heavier elements were placed before lighter elements in order to group similar properties in the same column.
   4. Blank spots, or gaps, were left in the table for elements that were not yet discovered.
   5. Mendeleev predicted the properties of some of the undiscovered elements with remarkable success.

Answers to Self-Assessment Questions

1. a Mendeleev did not announce the discovery of any new elements, although he predicted properties of unknown elements.
2. b Although in general Mendeleev's arrangement of atoms on the table follows an increase in atomic mass, the foundation of the arrangement is based on the chemical behavior and properties of the elements.

2. a The carbon-12 isotope is the basis for all relative masses in the periodic table.
3. b There are a few more than 100 elements known today.

2.6 Atoms and Molecules: Real and Relevant

Learning Objective: • Distinguish atoms from molecules.

A. "Atom" is a concept that is very useful in explaining chemical behavior.
B. Atoms are not destroyed in chemical reactions and thus can be recycled.
C. Materials can be "lost" by scattering their constituent atoms too widely and making it impractical to recover them.
D. A molecule (or formula unit) is the smallest constituent particle of a compound that still retains the properties of that compound.
E. Molecules (or formula units) can be divided into atoms.

Answers to Self-Assessment Questions

1. b Matter cannot be created or destroyed, only transferred from one place to another, which is a form of recycling.
2. c Molecules are atoms of elements combined together chemically.
3. c Atoms are not altered when they are grouped in different combinations as the result of a chemical reaction.

Green Chemistry: It's Elemental

Learning Objectives: • Identify elements that can be classified as hazardous or rare. • Explain how green chemistry can change technologies that rely on hazardous or rare elements.

A. Everything is made up of atoms.
B. It is important to think about the possible impacts of the reactants and products of a chemical process on human health and on the environment.
C. Two important considerations are inherent hazard and natural abundance.
   1. Elements can have different effects depending on whether they are in their elemental form or part of a compound.
   2. Safer alternatives can be designed for materials that involve hazardous or rare materials.
D. Green chemistry supports the design of products where the individual elemental components can be recycled easily.

LEARNING OBJECTIVES

You should be able to . . .

1. Explain the Greeks' ideas about the characteristics of matter. (2.1)

2. Describe the significance of the laws of conservation of mass and definite proportions. (2.2)

3. Calculate the amounts of elements from the composition of a compound. (2.2)

4. Explain why the idea that matter is made of atoms is a theory. (2.3)

5. Know how atomic theory explains the laws of multiple proportions and conservation of mass. (2.3)

6. Describe what a mole is and how it is used. (2.4)

7. Convert between the masses and the moles of a substance. (2.4)

8. Describe how the elements are arranged in the periodic table and why the arrangement is important. (2.5)

9. Distinguish atoms from molecules. (2.6)

10. Identify elements that can be classified as hazardous or rare.

11. Describe how green chemistry can change technologies that rely on hazardous or rare elements.

DISCUSSION

Chapter 2 is a survey of the history of the atomic theory. The Greek philosopher Democritus, who thought that matter was discontinuous, gave us the word "atom." Dalton made the first successful attempt to explain the chemical laws of conservation of matter and definite proportion. We use the atomic theory because it is useful for explaining chemical behavior. You should focus on practicing identifying the interplay between experimenting and theorizing.

The green chemistry emphasis in this chapter relates to the potential impacts of certain elements on human health and on the environment. The green chemistry essay emphasizes issues related to lead and mercury, two particularly hazardous elements, as well as some of the rarer elements. Finding ways to recycle elements is critical to the well-being and sustainability of life on our planet.

EXAMPLE PROBLEMS

1. When 60 g of carbon is burned in air, 220 g of carbon dioxide is formed. How much carbon dioxide is formed when 90 g of carbon is burned?

   Using unit conversions (Appendix C of the textbook), we can multiply the 90 g of carbon by a conversion factor that preserves the original relationship between 60 g of C and 220 g of CO_2 and then apply it to the current question of increasing the amount of carbon to 90 g.

   $$\text{Mass of } CO_2 \text{ formed} = 90 \text{ g C } (220 \text{ g } CO_2/60 \text{ g C}) = 330 \text{ g } CO_2$$

2. When burned in limited air, 6.0 g of carbon forms 14 g of carbon monoxide. How much carbon monoxide is formed when 360 g of carbon is burned?

   $$\text{Mass of CO formed} = (360 \text{ g C})(14 \text{ g CO}/6.0 \text{ g C}) = 840 \text{ g CO}$$

ADDITIONAL PROBLEMS

1. When 12 g of carbon is burned in air, 44 g of carbon dioxide is formed. How much carbon dioxide is formed when 0.060 g of carbon is burned?

2. When burned in air, 8.0 g of sulfur forms 16 g of sulfur dioxide. How much sulfur dioxide is formed when 400 g of sulfur is burned?

3. When electricity is passed through water, it decomposes into hydrogen and oxygen. When 9.0 g of water is electrolyzed, 1.0 g of hydrogen is formed. How much hydrogen is formed when 36 g of water is electrolyzed?

SOLUTIONS TO ADDITIONAL PROBLEMS

1. Mass of CO_2 formed = 0.060 g C(44 g CO_2/12 g C) = 0.22 g CO_2

2. Mass of SO_2 formed = 400 g S (12 g SO_2/8.0 g S) = 600 g SO_2

3. Mass of H_2 formed = 36 g H_2O(1.0 g H_2/9.0 g H_2O) = 4.0 g H_2

ANSWERS TO ODD-NUMBERED REVIEW QUESTIONS AND SOLUTIONS FOR ODD-NUMBERED END-OF-CHAPTER PROBLEMS

Review Questions

1. a. Aristotle argued that matter was continuous (infinitely divisible rather than discrete), while Democritus reasoned that substances (such as water) were composed of small "atoms" of the substance (such as atoms of water).

   b. The modern definition of an element is based on experimentation, while the ancient Greek definition was a philosophical concept based on four elements: earth, air, fire, and water.

3. People, calculators, and M&M candies are discrete at the macroscopic level. Cloth and milk chocolate are continuous, giving the impression that they can be infinitely divided.

5. Cavendish's development of quantitative methods of establishing the proportions of atoms in a substance produced results substantiating the law of definite proportions.

7. The law of definite proportions states that elements combine in the same proportions in any compound regardless of the source of origin of that compound.

9. The existence of these various atomic combinations of nitrogen and oxygen illustrate the law of multiple proportions.

11. Rectangle (c) has 15 oxygen atoms and cannot be representative of the reaction product because there are only 14 reactant oxygen atoms.

13. Mendeleev used the relationship between the chemical properties of the elements and their atomic masses to create his periodic table. His organization caused him to predict the existence of yet-to-be-discovered elements when known elements would otherwise have to be placed in columns with elements whose chemical properties did not "fit."

Problems

15. a. If the container is completely sealed, then its weight after one or two weeks will be the same as what it was at the start of the experiment because all the life processes involving reactions conserve the atoms involved. Some of those atoms will appear in the form of gases, some as solids, and others as liquids, but all will be retained in the container.

    b. If the mouse was housed in a wire container, then any gaseous products of its metabolism would escape through the cage, so the weight after one or two weeks would be less than the original weight of the mouse, its food, and its water.

17. c. The gas that formed as the dissolving pill effervesced, which was likely carbon dioxide, will have escaped into the atmosphere, reducing the combined weight of the resulting mixture and its container.

19. Acetylene contains 2.02 g of H for every 24.02 g of C. The number of g of H that would combine with 78.5 g of C to form acetylene = (78.5 g C)(2.02 g H/24.02 g C) = 6.60 g H.

21. The mass of carbon in 88.20 g of propane is

    Mass of C = (88.20 g C_3H_8)(36.03 g C/44.10 g C_3H_8) = 72.06 g C

    The mass of C in the CO_2 that forms in the combustion cannot exceed 72.06 g according to the law of conservation of mass. The mass of C in 300.06 g of CO_2 as proposed by the student is:

    Mass of C = (300.06 g CO_2)(12.01 g C/44.01 g CO_2) = 81.88 g C, which is more than was available in the combustion.

23. b. The mass of the reactants is 1.00 g + 0.80 g = 1.80 g. The mass of the product, which contains 1.00 g Zn is 1.50 g. Therefore, 0.30 g of the S did not react.

25. a. The ratio of hydrogen to water is 2.02 g to 18.02 g or 0.112. The same ratio applies to 775 g of water or (0.112)(775 g water) = 86.8 g of hydrogen.

    b. The ratio of oxygen to water is 16.00 g to 18.02 g or 0.888. The same ratio applies to 775 g of water or (0.888)(775 g water) = 688 g of oxygen.

27. 3 parts of carbon/11 parts of carbon dioxide = x parts of carbon/14 kg of carbon dioxide or 3.8 kg. (3 parts of carbon plus 8 parts of oxygen gives 11 parts of carbon dioxide.)

29. a. Dalton assumed that different atoms had different masses. This is consistent with Dalton's theory.

    b. Dalton assumed that different atoms had different masses. This is not consistent with Dalton's theory.

31. Dalton's theory states that chemical reactions involved the rearrangement of atoms, so the carbon atoms in the diamond would have combined with oxygen atoms to form a new compound and the hydrogen and oxygen atoms in water would form different combinations when water was electrolyzed.

33. If there are two hydrogen atoms for every one oxygen atom, then the ratio must be answer (d), 16 times.

35. The mass ratio of carbon to oxygen in carbon suboxide is (1.126)(52.96/47.04) = 1.126). A carbon atom weighs 0.750 as much as an oxygen atom (12.01/16.00 = 0.750). In order for the mass ratio of C to O in the compound to be 1.126, the number of C atoms must be 1.126/0.750 or 1.501 for every O atom or 3 C atoms for every 2 O atoms. The formula of carbon suboxide is C_3O_2. This whole number ratio follows the law of multiple proportions.

37. The ratio of Sn to O in the first compound is 0.742 g Sn/0.100 g O. The ratio of Sn to O in the second compound is 0.555 g Sn/0.150 g O, or half the amount of Sn. If the first compound is SnO, then the second compound must be SnO_2.

39. We can calculate the O to Fe mass ratio in each of the compounds, and divide each of the ratios by the smallest ratio (see the model in Table 2.1):

Wustite (FeO): 0.160 g/0.558 g = 0.287	0.287/0.287 = 1.00
Hematite: 0.240 g/0.558 g = 0.430	0.430/0.287 = 1.50 × 2 = 3.00
Magnetite: 1.28 g/3.35 g = 0.382	0.382/0.287 = 1.33 × 3 = 4.00

To convert the values for hematite and magnetite to whole numbers, we multiply the ratios by 2 and 3, respectively. So the formulas for wustite, hematite, and magnetite are FeO, Fe_2O_3, and Fe_3O_4, respectively. This is in agreement with the law of multiple proportions.

41. Number of moles of C = (1.000 g C)(1 mol C/12.01 g C) = 0.08326 mol C

Number of C atoms = (0.08326 mol C)(6.022 x 10^{23} C atoms/mol C) = 5.014 × 10^{22} C atoms

43. Mass of C-12 = (5 mol C-12)(12.0000 g C-12/mol C-12) = 60.0000 g C-12

Mass of C = (5 mol C)(12.0107 g C) = 60.0535 g C

45. Yes, the ratio of carbon to hydrogen in the three samples is 14.90, 14.97, and 14.92. These ratios are in agreement within the limits of experimental error.

Answers and Solutions to Additional Problems

47. A uranium atom weighs 12.5 times as much as a fluorine atom (238/19.0 = 12.5). In order for the mass ratio of U to F in the compound to be 2.09, the number of F atoms must be 12.5/2.09 or 5.99. Therefore, there must be 6 F atoms for every U atom in the compound.

49. We can use the mass ratio of sulfur dioxide to the mass of reacted sulfur (0.623g SO_2/0.312 g S) to solve this problem.

Mass of SO_2 formed, g = (1.305 g S)(0.623 g SO_2/0.312 g S) = 2.61 g SO_2.

51. Divide 0.5836 g by 0.4375 to get 1.334. This gives a ratio of N to O of 1.334 to 1. Multiply both by 3 to get a whole number ratio of 4.002 to 3 or 4 to 3.

53. This experiment illustrates the law of conservation of mass. Although the arrangement of the atoms changed when a gas was formed, all the atoms were retained in the balloon, so their mass did not change.

55. a. According to Figure 2.11, the mass of selenium (Se) should be the average of the masses of sulfur (S) and tellurium (Te): mass of Se = (32 + 125)/2 = 78.6.

    b. Using the masses from the modern periodic table: mass of Se = (32.06 + 127.60)/2 = 79.83.

57. The replacement of lead in paints with titanium dioxide is one example; nontoxic substitutions for lead in plastics and solders are additional examples.

CHAPTER

3

Atomic Structure

Images of the Invisible

CHAPTER SUMMARY

3.1 Electricity and the Atom

Learning Objectives: • Explain the electrical properties of an atom. • Describe how the properties of electricity explain the structure of atoms.

A. Volta invented an electrochemical cell (1800) that was much like a modern battery.
B. Electrolysis.
   1. Davy produced elements from compounds by passing electricity through them (electrolysis).
   2. Faraday continued to work in this field.
   3. Electrolytes are compounds that conduct electricity when melted or dissolved in water.
   4. Electrodes are carbon rods or metal strips that carry electric current when inserted into a molten compound or solution.
   5. The anode is the positively charged electrode. The cathode is the negatively charged electrode.
   6. The anion is a negatively charged ion and is attracted to the anode. A cation is a positively charged ion and is attracted to the cathode.
C. Cathode Ray Tubes.
   1. Crookes (1875) discovered cathode rays by passing electricity through a partially evacuated gas discharge tube.
   2. Cathode rays (beams of current) travel from the cathode to the anode.
D. Thomson's Experiment: Mass-to-Charge Ratio.
   1. Thomson (1897) found that cathode rays were deflected in an electric field and thus must contain charged particles.
   2. These negatively charged particles were called electrons and were found to be the same for all gases used to produce them.
   3. Cathode rays travel in straight lines in the absence of an applied field.
   4. Thomson calculated the ratio of the electron's mass to its charge.
E. Goldstein's Experiment: Positive Particles.
   1. Goldstein (1886) used an apparatus similar to Crookes's tube to study positive atomic particles.
   2. The positively charged particles were found to be more massive than electrons and to vary depending on the type of gas used in the experiment.

3. The lightest positive particle obtained was derived from hydrogen and had a mass 1837 times heavier than that of an electron.

F. Millikan's Oil-Drop Experiment: Electron Charge.
1. Millikan (1909) determined the charge of the electron.
2. The charge on an electron is –1.
3. The mass of the electron is 9.1×10^{-28} g.

Answers to Self-Assessment Questions

1. c Electrolysis is the process of producing a chemical reaction by means of electricity.
2. b Negatively charged ions move toward the positively charged electrode (anode).
3. c Thomson deflected cathode rays in an electrostatic field.
4. d Thomson could only obtain the charge-to-mass ratio, as neither the mass nor the charge was known independently.
5. b Millikan observed the amount of deflection that occurred when he manipulated the charges on two plates.

3.2 Serendipity in Science: X-Rays and Radioactivity

Learning Objective: • Describe the experiments that led to the discovery of X-rays and an explanation of radioactivity.

A. Roentgen: The Discovery of X-Rays.
1. Roentgen (1895) discovered X-rays, a form of electromagnetic radiation.

B. The Discovery of Radioactivity.
1. Becquerel discovered radioactivity while studying fluorescence.
2. Marie and Pierre Curie studied radioactivity.
3. The Curies discovered the radioactive elements radium and polonium.

Answers to Self-Assessment Questions

1. a X-rays are absorbed more by hard, fairly dense materials than by soft, less dense tissues.
2. b Unstable elements become more stable after emitting radiation.

3.3 Three Types of Radioactivity

Learning Objective: • Distinguish the three main kinds of radioactivity: alpha, beta, and gamma.

A. Rutherford classified three types of radioactivity.

Name	Symbol	Mass	Charge
1. Alpha rays	α particles	4 amu	2+
2. Beta rays	β particles	1/1837 amu	1–
3. Gamma rays	γ rays	0	0

Answers to Self-Assessment Questions

1. a An alpha particle is the same as a helium nucleus, which has four times the mass of a hydrogen atom.
2. d An alpha particle with 2 protons has a 2+ charge.
3. b The properties of a beta particle and an electron are the same.
4. b Experiments show that gamma rays are a form of electromagnetic radiation with neither mass nor charge.

3.4 Rutherford's Experiment: The Nuclear Model of the Atom

Learning Objective: Sketch the nuclear model of the atom, and identify its parts.

A. Rutherford's experiment (1911) showed that the positive charge and nearly all the atomic mass are concentrated in the tiny core called the nucleus.
B. Rutherford proposed that negatively charged electrons have almost no mass but occupy nearly all the volume of the atom.

Answers to Self-Assessment Questions

1. b Most of the alpha particles traveled unaffected through empty space. Alpha particles are positively charged, and those that were deflected came under the influence of the combined charge of the protons, which are located in a small volume called the nucleus.
2. c Because only a few alpha particles were deflected, the nucleus had to be a very small positive target.
3. b Most alpha particles missed any obstruction; most of the atom was empty space.
4. b The positively charged alpha particles were repelled by positive nuclei, which contain the protons.

3.5 The Atomic Nucleus

Learning Objectives: • List the particles that make up the nucleus of an atom, and give their relative masses and electric charges. • Identify elements and isotopes from their nuclear particles.

A. Rutherford (1914) proposed that protons constitute the positively charged matter of all atoms, not just that of hydrogen.
B. Chadwick (1932) discovered a nuclear particle, called a neutron, which has about the same mass as a proton but has no charge.
C. Atomic Number.
   1. The number of protons in the nucleus is the atomic number, Z, and determines the identity of the atom.
D. Isotopes.
   1. Isotopes have the same number of protons (that is, the same atomic number) but a different number of neutrons (different atomic mass).

E. Symbols for Isotopes.
   1. The two main nuclear components, protons and neutrons, are called nucleons.
   2. The mass number (A), or nucleon number, is the number of nucleons, or nuclear particles (protons and neutrons).
   3. Atomic symbols are written with the atomic number as a subscript and the mass number as a superscript to the left of the atomic symbol.

Answers to Self-Assessment Questions

1. c Experimental evidence has shown that the atom consists of protons, neutrons, and electrons.
2. d The number of protons, which is given by the atomic number, determines the element's identity.
3. c Experiments deflecting proton beams and electron beams show that electrons are approximately 1800 times lighter than protons.
4. d Experiments have shown that protons and neutrons have about the same mass.
5. b Because the number of protons determines the identity of an element, a change in this number would change the element's identity.
6. d Changing the number of neutrons in an atomic nucleus alters its mass number (the number of protons plus the number of neutrons in the nucleus) but not its atomic number (the number of protons in the nucleus). Isotopes are atoms of the same element with different mass numbers.
7. d The nucleon number, or mass number, is the sum of the number of protons and neutrons.
8. b The atomic number (subscript) is 7, indicating 7 protons; the nucleon number (superscript) is 15, indicating a total of 15 protons and neutrons. Subtracting the atomic number from the nucleon number gives 8 neutrons.
9. b In order for an atom to be neutral, it must have the same number of protons (positively charged particles) as electrons (negatively charged particles): atomic number = 19.
10. d In order to identify a neutral element, we must know its atomic number or the number of electrons it possesses. Knowing only the number of neutrons does not give us enough information to identify the element.
11. c Potassium has atomic number 19; the mass number of this isotope is 40.

3.6 Electron Arrangement: The Bohr Model

Learning Objectives: • Define *quantum*. • Arrange the electrons in a given atom in energy levels (shells).

A. Fireworks and Flame Tests.
   1. Flame tests rely on the color of flames to identify elements.

B. Continuous and Line Spectra.
   1. A prism separates colored light into individual colored lines.
   2. All wavelengths are present in white light, which is a continuous spectrum.
   3. A line spectrum is a pattern of lines, with each line corresponding to a different wavelength or energy emitted by an element.
   4. These different colored lines represent different wavelengths or energies.
   5. Not all the lines in a spectrum are visible. Some have energy that corresponds to parts of the spectrum that are not detectable with an unaided eye such as the ultraviolet and infrared regions.

C. Bohr's Explanation of Line Spectra.
   1. Bohr stated that discrete spectra arise because the energy of the electrons in an atom is quantized, meaning electrons can absorb only discrete values of energy.
      a. Quantum: A tiny unit of energy whose magnitude depends on the frequency of the radiation.
   2. Energy level: Specified energy value for an electron.
   3. Ground state: Electrons in the lowest possible energy levels.
   4. Excited state: Due to added energy, an electron jumps to a higher energy level.
   5. The maximum number of electrons in any given energy level is given by the formula $2n^2$.
   6. Various energy levels are called shells.

D. Ground States and Excited States
   1. Ground state: All electrons are in the lowest possible energy levels closest to the nucleus (this is the most stable configuration).
   2. Excited state: One (or more) electrons is elevated to a higher energy level. Excited states are short-lived because they are unstable.
   3. An atom emits a quantum or particle of light as the electron falls back from an excited state to a ground state. This transition may occur in one step or in multiple steps.
   4. Bohr deduced that a given energy level could contain a limited number of electrons.

F. Building Atoms: Main Shells.
   1. Electrons populate the lowest energy levels possible. These are the levels closest to the nucleus.
   2. Once an energy level is filled, the electrons start filling the next higher level.
   3. An electron configuration (e.g., 2, 6) indicates the number of electrons in each shell or level.

Answers to Self-Assessment Questions

1. a Each element is unique in its electron configuration and in its spectrum.
2. d Electrons release (lose) energy in going from a higher energy level to a lower energy level.
3. a Excited electrons are electrons that are in higher energy levels than when they are in the ground state.
4. a Experiments have shown that the first shell can hold a maximum of 2 electrons.
5. c Experiments have shown that the third shell can hold no more than 18 electrons.
6. b The atomic number of Al is 13. and thus it has 13 electrons—2 in the first shell, 8t in the second shell, and 3 in the third shell.
7. b The atomic number of sulfur is 16 and thus has 16 electrons—2 in the first shell, 8 in the second shell, and 6 in the third shell.

3.7 Electron Arrangement: The Quantum Model

Learning Objective: • Write an electron configuration (in subshell notation) for a given atom.

A. de Broglie suggested the wavelike properties of electrons.

B. Schrödinger developed equations to describe the behavior of electrons in atoms.

C. Electrons move in specifically shaped volumes of space, called orbitals.
   1. These different shapes are indicated by the letters *s*, *p*, *d*, and *f*.
   2. The *s* sublevel has 1 orbital.
      The *p* sublevel has 3 orbitals.
      The *d* sublevel has 5 orbitals.
      The *f* sublevel has 7 orbitals.
   3. An orbital can hold a maximum of 2 electrons.

D. Building Atoms by Orbital Filling.
   1. Each electron orbital can contain a maximum of 2 electrons, and some shells can contain more than 1 orbital.
   2. Orbitals in the same shell that have the same letter designation make up a subshell (sublevel).
   3. Electron configurations are expressed by numbers that indicate the main energy level (shell) and letters that indicate the sublevel(s). For example, the notation $2p^5$ indicates that there are 5 electrons in the *p* orbitals of the second energy level.
   4. Valence electrons are the electrons in the outermost main shell.
   5. An order-of-filling chart is extremely useful for determining the electron configuration of an atom (see Figure 3.16 in the text). Note: It is helpful to learn this diagram.

Answers to Self-Assessment Questions

1. c Very complex mathematical calculations show that the 1*s* orbital has a spherical shape.
2. c Very complex mathematical calculations show that an orbital can hold 2 electrons.
3. c The 3*p* orbital is higher than the 3*s* orbital (see the order-of-filling chart).
4. c The third main shell is the first one that is large enough to accommodate *d* orbitals (see the order-of-filling chart).
5. b Nitrogen with 7 electrons has 3 *p* electrons, which means that the *p* subshell is half-filled.
6. b The *p* subshell can accommodate a total of 6 electrons in 3 orbitals.
7. b For a 3*d* subshell the number 3 indicates the shell number.
8. c All *p* subshells can hold a maximum of 6 electrons.

3.8 Electron Configurations and the Periodic Table

Learning Objective: • Describe how an element's electron configuration relates to its location in the periodic table.

A. The modern periodic table is arranged in order of increasing atomic number, and grouped according to electronic structure.
   1. Vertical columns are called groups or families.
   2. Horizontal rows are called periods, and indicate how many main electron energy levels an atom has.
   3. Elements in one group have similar chemical properties.
      a. Elements in A groups are called main-group elements.
      b. Elements in B groups are called transition elements.

B. Family Features: Outer Electron Configurations.
   1. Electrons in the outer shell of an atom are called valence electrons
      a. The groups (families) have similar outer electron (valence electron) configurations.
      b. Valence electrons determine most of the chemistry of an atom.
   2. Groups were once designated by a Roman numeral and letter A or B.
      a. The new IUPAC system renumbers groups 1 to 18 and does not use A or B designations.

C. Family Groups.
   1. Group 1A—Alkali Metals (highly reactive). ns^1
   2. Group 2A—Alkaline Earth Metals (moderately reactive). ns^2
   3. Group 7A—Halogens (highly reactive nonmetals). ns^2np^5
   4. Group 8A—Noble Gases (nonreactive nonmetals). ns^2np^6

D. Metals and Nonmetals.
   1. Elements in the periodic table are divided into two main types: metals and nonmetals. A heavy, stairstep-like line separates the two classes.
      a. Metals—elements to the left of the line.
      b. Nonmetals—elements to the right of the line.
      c. Metalloids (or semimetals)—elements bordering the line.
   2. The characteristics of metals are luster, good conductors of heat and electricity, solid at room temperature (except mercury, a liquid), malleable, and ductile.
   3. The characteristics of nonmetals: lack metallic properties.
      a. Most are solids or gases at room temperature.
      b. Bromine is a liquid.
   4. The characteristics of metalloids: They possess properties of both metals and nonmetals.
   5. The group number for the main-group elements (A groups) gives the number of electrons in the outer energy level (valence electrons).

D. Which model to use: We use the model that is most helpful in understanding a particular concept.

Answers to Self-Assessment Questions

1. b Cl and Br are in the same group, 7A, so their chemical properties are similar.
2. c The period number is determined by the highest main shell number that the electrons are filling.
3. c Group 7A elements (the halogens) have seven valence electrons.
4. a Group 1A are the alkali metal elements.
5. b Bromine is in group 7A and has 7 valence electrons.
6. c Li, Na, and K are all in group 1A, so their chemical properties are similar.
7. b The period number gives the number of main shells that are populated (but not necessarily filled) with electrons.

Green Chemistry: Clean Energy from Solar Fuels

A. Green plants use photosynthesis to make carbohydrate fuels.

B. Using sunlight and "artificial photosynthesis" to split water into hydrogen and oxygen is one of the big challenges of chemistry.
   1. Hydrogen is a carbon-free fuel that, when combined with oxygen, releases the energy required when it was originally formed from the reaction of water.

LEARNING OBJECTIVES

You should be able to . . .

1. Explain the electrical properties of an atom. (3.1)

2. Describe how the properties of electricity explain the structure of atoms. (3.1)

3. Describe the experiments that led to the discovery of X-rays and an explanation of radioactivity. (3.2)

4. Distinguish the three main kinds of radioactivity: alpha, beta, and gamma. (3.3)

5. Sketch the nuclear model of the atom and identify its parts. (3.4)

6. List the particles that make up the nucleus of an atom and give their relative masses and electric charges (3.5)

7. Identify elements and isotopes from their nuclear particles. (3.5)

8. Define *quantum*. (3.6)

9. Arrange electrons in a given atom in energy levels (shells). (3.6)

10. Relate the idea of a quantum of energy to an orbital. (3.7)

11. Write an electron configuration (in subshell notation) for a given atom. (3.7)

12. Describe how an element's electron configuration relates to its location in the periodic table. (3.8)

13. Distinguish the conversion of solar energy into electrical energy in a solar cell from the conversion of solar energy into the chemical-bond energy of a solar fuel.

14. Explain why splitting water into the elements hydrogen and oxygen requires an energy input and why producing water by the reaction of hydrogen and oxygen releases energy.

DISCUSSION

We can see the images of atoms even though they are too small to be seen by visible light. The experiments that led to the modern atomic theory started with the beginnings of electricity and radioactivity and progressed through spectroscopy. The descriptions of the arrangement of electrons around the nucleus and quantum mechanics are the result of many years of the proposal of hypotheses, experimentation, and observation. Finally, we organize the periodic table to maximize our understanding of atomic structure. Applications of our understanding allow us to devise new technologies, such as solar fuels, to replace some of the nonrenewable resources and reduce pollution.

ANSWERS TO ODD-NUMBERED REVIEW QUESTIONS AND SOLUTIONS FOR ODD-NUMBERED END-OF-CHAPTER PROBLEMS

Review Questions

1. Many scientists contributed to our knowledge of the atom.

   a. Crookes invented the gas discharge tube. He also investigated the cathode rays and showed them to be negatively charged and light mass.

   b. Goldstein discovered positive particles (protons) in experiments similar to those of Crookes.

   c. Faraday showed that matter is electrical in nature.

d. Thomson determined the mass-to-charge ratio for the electron.

e. Rutherford recognized and named the nuclear particles now known as alpha and beta particles. He also named gamma radiation. He also recognized that protons are the positively charged matter in all atoms.

f. Chadwick discovered neutrons, which are particles with almost the same mass as protons but with no electrical charge.

3. Both X-rays and gamma rays are forms of energy. Gamma rays are more penetrating than X-rays.

5. Rutherford postulated that all the positive charge and nearly all the mass of an atom (later known to be contributed by the protons and neutrons) are concentrated in a tiny nucleus. The negatively charged electrons have almost no mass, yet they occupy nearly all the volume of an atom.

7. Atoms A and B are not isotopes because they have different numbers of protons and are thus different elements. Atoms A and C are not isotopes; they are different atoms because they have different numbers of protons. Atoms A and D are isotopes because they differ only in the number of neutrons. Atoms B and C are isotopes because they differ only in the number of neutrons. Atoms C and D are not isotopes because they have different numbers of protons.

9. Dalton assumed that atoms were indivisible and that all atoms of a particular element were identical (see Chapter 2, Section 2.3). The nuclear model indicates that atoms are divisible in the sense that they are made up of protons, neutrons, and electrons, and that all the atoms of a particular element are not necessarily identical. While the number of protons in all atoms of a particular element is the same, the number of neutrons can vary, producing different isotopes of that element.

11. Ruthenium is the element whose atoms contain 44 protons in their nuclei. The atomic number of ruthenium is 44, and its atomic symbol is Ru.

13. More energy is required to excite an electron from the first to the third shell than is required to excite an electron from the second to the third shell.

15. The quantum (or quantum-mechanical) model, derived from the mathematical quantum mechanics approach, described electrons as having wave properties as well as particle properties.

17. Because atoms are electrically neutral, the number of electrons surrounding the nucleus of an atom is the same as the number of protons in its nucleus.

a. Beryllium (Be) nuclei have 4 protons, so Be atoms have 4 electrons.

b. Sodium (Na) nuclei have 11 protons, so Na atoms have 11 electrons.

c. Sulfur (S) nuclei have 16 protons, so S atoms have 16 electrons.

d. Selenium (Se) nuclei have 34 protons, so Se atoms have 34 electrons.

e. Copper (Cu) nuclei have 29 protons, so Cu atoms have 29 electrons.

f. Uranium (U) nuclei have 92 protons, so U atoms have 92 electrons.

Problems

19. Alpha particles, which are positively charged, have the largest mass. Beta particles, which are equivalent to electrons, have a negative charge and a much smaller mass than alpha particles. Gamma radiation has no charge and no mass.

21. The answer is d. Radioactive elements spontaneously emit particles and/or rays as a means of stabilizing their nuclei.

23. A nuclear symbol has the atomic number to the left as a subscript to the atomic symbol and the nucleon number to the left as a superscript. The atomic number determines the element and indicates the number of protons. The nucleon number is determined by the sum of the number of protons and neutrons.

    a. $^{37}_{17}Cl$ or chlorine-37 or Cl-37

    b. $^{55}_{25}Mn$ or manganese-55 or Mn-55

25. Four different elements are listed: two isotopes of sodium (Na, atomic number 11) and one atom each of neon (Ne, atomic number 10), boron (B, atomic number 5), and magnesium (Mg, atomic number 12).

27. The most likely mass number of this element is the whole number closest in magnitude to the atomic mass shown on the periodic table. Thus, for the element with atomic number 14, which is silicon, this mass number would be 28.

29. Atoms are electrically neutral, which means that they have the same number of electrons and protons. When an atom loses 2 electrons, the number of protons is two greater than the number of electrons, giving the resulting species a charge of 2+.

31. The number of electrons is given by the sum of the superscripts of the electron structures; this is the atomic number (assuming neutral atoms). The atomic number identifies the element.

    a. 8 electrons or atomic number 8

    b. 14 electrons or atomic number 14

    c. 28 electrons or atomic number 28

33. The order-of-filling chart tells us in which order the orbitals fill in ground-state atoms. Any pattern that violates this order represents an atom in an excited state.

    a. Possible excited state: The 1*s* orbital holds 2 electrons. In this case, one of those electrons appears to be excited into the next higher, or 2*s*, orbital

    b. Incorrect. The maximum number of electrons that can populate the three *p* orbitals is six.

    c. Possible excited state: Two electrons have been raised from the 2*s* to the 2*p* subshell.

    d. This atom is in its ground state.

35. A fluorine atom (F) has 9 electrons. Adding one electron would give 10 electrons or atomic number 10. This is the element neon, Ne.

37. Silicon (Si) and germanium (Ge) are both in group 4A. Each has four electrons in its outermost electron shell designated as ns^2np^2. Silicon's 4 valence electrons are in the third main shell, while those of germanium are in the fourth main shell, so a germanium atom is larger than a silicon atom.

39. In general, metals are to the left on the table and nonmetals are to the right.

    a. Manganese is a metal.

    b. Strontium is a metal.

    c. Cesium is a metal.

    d. Argon is a nonmetal.

41. Ne and Kr are noble gases.

43. Four: Ne, P, Kr, and N are nonmetals.

45. a. The atom has a total of 15 electrons, so it must have 15 protons. Its atomic number is 15.

    b. Phosphorus.

    c. Because atoms are electrically neutral, an atom of phosphorus has 15 electrons.

    d. The electron configuration for phosphorus is $1s^22s^22p^63s^23p^3$. The atom has 6 *s* electrons.

    e. According to the filling pattern (see d), this atom has no electrons in *d* orbitals.

47. a. The fourth period noble gas is Kr.

    b. The third period alkali metal is Na.

    c. The fourth period halogen is Br.

    d. The metal in the fourth period and group 3B is Sc.

49. a. Fe (atomic number 26): $1s^22s^22p^63s^23p^63d^64s^2$ (Note: This configuration can also be correctly written to follow the filling pattern in Figure 3.16: $1s^22s^22p^63s^23p^64s^23d^6$).

    b. Sn (atomic number 50): $1s^22s^22p^63s^23p^63d^{10}4s^24p^64d^{10}5s^25p^2$ (or $1s^22s^22p^63s^23p^64s^23d^{10}4p^65s^24d^{10}5p^2$).

    c. Pb (atomic number 82): $1s^22s^22p^63s^23p^63d^{10}4s^24p^64d^{10}4f^{14}5s^25p^65d^{10}6s^26p^2$ (or $1s^22s^22p^63s^23p^64s^23d^{10}4p^65s^24d^{10}5p^66s^24f^{14}5d^{10}6p^2$).

51. Both atoms are in the fifth period. Atom L must be strontium (Sr) with an electron configuration of $1s^22s^22p^63s^23p^63d^{10}4s^24p^65s^2$. Atom M, which is adjacent to atom L, must be yttrium (Y), with an electron configuration of $1s^22s^22p^63s^23p^63d^{10}4s^24p^64d^15s^2$.

53. b. Energy is required to break the bonds that hold the two H atoms to the O atom in water.

55. d. Scientists are looking for electrodes and other components that can be fabricated from more abundant materials than the platinum that is typically used.

CHAPTER

4

Chemical Bonds

The Ties That Bind

CHAPTER SUMMARY

Introduction

A. Chemical bonds hold atoms together in molecules and ions together in ionic crystals.
   1. The nature of bonding within the molecule determines the forces between atoms in molecules and between ions in ionic compounds.
   2. Bonds determine the shape and properties of the molecule.

4.1 The Art of Deduction: Stable Electron Configurations

Learning Objective: • Determine the number of electrons in an ion.

A. Atoms can gain or lose electrons to form electron configurations that are isoelectronic with (have the same electronic configuration as) the noble gases.
B. When atoms gain or lose electrons and acquire a charge, they are called ions.

Answers to Self-Assessment Questions

1. c Elements in group 8A all have a completely filled outermost shell.
2. b Sodium loses one electron to achieve the same electron configuration as neon.
3 a S^{2-} and Ar both have 18 electrons and the same electron configuration.
4. d S^{2-} and Ca^{2+} both have 18 electrons like Ar and the same electron configuration.

4.2 Lewis (Electron-Dot) Structures

Learning Objective: • Distinguish between an ion and an atom.

A. Electron-dot symbols, or Lewis symbols, are a useful way to represent atoms or ions.
   1. The chemical symbol represents the core (nucleus plus inner electrons).
   2. Valence electrons are represented by dots.

B. Chemical Symbolism
   1. Electron-dot symbols are most convenient to draw for main-group elements where, in most cases, the group number is equivalent to the number of valence electrons. (Na, in group 1A has one valence electron; Cl, in group 7A has seven valence electrons.)

Answers to Self-Assessment Questions

1. b Three valence electrons because B is in group 3A.
2. b Eight valence electrons because while Br is in group 7A and so the neutral atom has seven valence electrons, the negative charge indicates that the atom has gained one more electron for a total of 8.
3. c Li is in group 1A and only has one valence electron.
4. a As is in group 5A and needs 5 valence electrons.

4.3 The Reaction of Sodium and Chlorine

Learning Objectives: • Distinguish between an ion and an atom. • Describe the nature of the attraction that leads to formation of an ionic bond.

A. Sodium is a very soft, reactive metal. Chlorine is a poisonous greenish-yellow reactive gas.
   a. When sodium metal (Na) is added to chlorine gas (Cl_2), a violent reaction takes place producing a white, stable, water-soluble solid known as sodium chloride (table salt).
   b. Sodium reacts with chlorine by giving an electron to each of the Cl atoms.
   c. The ions produced are stable because both Na^+ and Cl^- ions are isoelectronic with noble gases (Na^+ is isoelectronic with Ne and Cl^- is isoelectronic with Ar).
   d. Positively charged ions are called cations. Negatively charged ions are called anions.

B. Ionic Bonds.
   a. Oppositely charged ions are strongly attracted to one another.
   b. The ions assemble in regular patterns called crystals.
   c. The forces holding the ions together in the crystal are called ionic bonds.

C. Atoms and ions are distinctively different.
   a. The properties and reactions of neutral atoms (for example, Na and Cl_2) are very different from the properties and reactions of the ions they form (for example, Na^+ and Cl^- ions).

Answers to Self-Assessment Questions

1. a Both have 10 electrons and the same electron configuration.
2. d The Cl^- ion has 8 valence electrons and a 1− charge.
3. b The Li^+ and F^- ions form an ionic bond.
4. c Atoms in group 6A have 6 valence electrons. To become isoelectronic with the nearest noble gas, these atoms must gain 2 electrons, forming 2– ions.

4.4 Using Lewis Symbols for Ionic Compounds

Learning Objectives: • Write symbols for common ions, and determine their charges.
• Describe the relationship between the octet rule and the charge on an ion periodic table.

A. Potassium (in the same family as sodium) reacts with chlorine the same way sodium does.
B. Potassium reacts with bromine (same family as chlorine) the same way it reacts with chlorine.
C. Magnesium (group 2A) reacts with oxygen (group 6A) by giving up two electrons.
D. Metallic elements of group 1A, 2A, and 3A react with nonmetallic elements in group 5A, 6A, and 7A to form stable crystalline solids.
   1. Metals give up electrons, forming (+) ions.
   2. Nonmetals gain electrons, forming (–) ions.
   3. The attraction of (+) ions for (–) ions is the basis of ionic bonds.
E. The Octet Rule.
   1. All noble gases except helium have an octet (8) of valence electrons. Helium has two valence electrons.
   2. Main-group metals react with main-group nonmetals to form ionic compounds. These atoms lose or gain electrons to follow the octet rule according to their group number:
      Group 1A metal atoms each lose one electron to form 1+ cations.
      Group 2A metal atoms each lose two electrons to form 2+ cations.
      Group 6A nonmetal atoms each gain two electrons to form 2– anions.
      Group 7A nonmetal atoms each gain one electron to form 1– anions.

Answers to Self-Assessment Questions

1. c Generally, ionic compounds form between metals, which lose electrons, and nonmetals, which gain electrons to achieve stability. Cs is in group 1A and Se is in group 6A. Cs can easily lose 1 electron and Se can easily gain 2 electrons to achieve an octet.
2. b O is in group 6A and can easily gain 2 electrons to produce an anion with a 2– charge.
3. b Sr is in group 2A and can easily lose 2 electrons to produce a cation with a 2+ charge.
4. b There are 5 valence electrons in a neutral P atom. When 3 electrons are added, there are 8 valence electrons in the P^{3-} anion.
5. b Li loses 1 electron to become Li^{+}, and N gains 2 electrons to become N^{3-}. Three Li^{+} ions are needed to balance the 3– charge on the N^{3-}.
6. c Mg loses 2 electrons to become Mg^{2+}, and Br gains 1 electron to become Br^{-}. Two Br^{-} ions are needed to balance the 2+ charge on the Mg^{2+} ion.
7. b. Br is in group 7A and will easily gain 1 electron to become Br^{-}.

4.5 Formulas and Names of Binary Ionic Compounds

Learning Objective: • Name and write formulas for binary ionic compounds.

A. Binary compounds are composed of two different elements.
B. Names and symbols for simple ions.
   1. To name simple cations, add "ion" to the name of the parent element.
      a. Example: Na^{+}, sodium ion
   2. To name simple anions, change the ending of the name of the element to "-ide" and add "ion."
      a. Example: Cl^{-}, chloride ion.
   3. Elements in the B groups on the periodic table can form more than one ion.

B. Formulas and names for binary ionic compounds.
   1. The "crossover" method is used to determine the number of cations and anions needed for the correct formula of an ionic compound. For aluminum sulfide, $Al^{3+} + S^{2-} \rightarrow Al_2S_3$. Make certain that the subscripts are reduced to the lowest possible multiple. For example, calcium oxide is formed from Ca^{2+} and O^{2-} ions. Note: Using the crossover method might imply that the formula of calcium oxide is Ca_2O_2. These subscripts must be divided by 2 to reduce them to the lowest possible ratio, so the formula of calcium oxide is CaO.

Answers to Self-Assessment Questions

1. a Ba in group 2A loses 2 electrons to become Ba^{2+}, and S in group 6A gains 2 electrons to become S^{2-}. The formula of barium sulfide is BaS.
2. c Zn is a transition metal that loses 2 electrons to become Zn^{2+} (see Table 4.2 and Figure 4.4), and Cl is in group 7A, so it gains 1 electron to become Cl^-. Two Cl^- ions are needed to balance the 2+ charge on Zn^{2+}. The formula is $ZnCl_2$.
3. b Al in group IIIA loses 1 electron to become Al^{3+}, and P in group VA gains 3 electrons to become P^{3-}. The formula is AlP.
4. c K^+, the cation, is named first and keeps the original element's name, potassium. Cl^-, the anion, has the suffix -ide added to the root, chlor-. The correct name for KCl is potassium chloride.
5. a Al^{3+}, the cation, is named first and keeps the original element's name, aluminum. S^{2-}, the anion, has the suffix -ide added to the root, sulf-. The correct name for Al_2S_3 is aluminum sulfide.
6. d Li^+, the cation, is named first and keeps the original element's name, lithium. Br^-, the anion, has the suffix -ide added to the root, brom-. The correct name for LiBr is lithium bromide.
7. b Zn^{2+}, the cation, is named first and keeps the original element's name, zinc. I^-, the anion, has the suffix -ide added to the root, iod-. The correct name for ZnI_2 is zinc iodide.

4.6 Covalent Bonds: Shared Electron Pairs

Learning Objectives: • Explain the difference between a covalent bond and an ionic bond. Name and write formulas for covalent compounds.

A. Elements that are unable to transfer electrons completely between them achieve electronic configurations by sharing pairs of electrons.
B. A bond formed when atoms share one (or more) pairs of electrons is called a covalent bond. Covalently bonded atoms, except hydrogen, seek an arrangement that surrounds them with eight valence electrons (octet rule).
   1. If one electron pair is shared, the bond formed is a single bond.
   2. The two shared electrons forming the bond are called a bonding pair.
   3. The electrons that are not shared are called nonbonding electron pairs or lone pairs.
   4. Covalent bonding can be symbolized in several ways, most commonly with a pair of dots or a single line to represent a shared electron pair.
C. Multiple covalent bonds.
   1. Atoms can share more than one pair of electrons, resulting in:
      a. Double bonds (two pairs of electrons shared between two atoms).
      b. Triple bonds (three pairs of electrons shared between two atoms). Three is the maximum number of shared electron pairs that can exist between any two atoms.

Answers to Self-Assessment Questions

1. a Covalent bonds form as the result of two atoms sharing one or more pairs of electrons. Bonds involving just one shared electron pair are called single covalent bonds.
2. a Atoms in group 7A each have 7 valence electrons, one of which is unpaired. Sharing the unpaired electron of one 7A atom with the unpaired electron of a second 7A atom completes the octet for both atoms through the formation of a single covalent bond.
3. c Oxygen atoms are in group 6 and have 6 valence electrons. In order to complete their octets, each of the two oxygen atoms shares one pair of its valence electrons, with the other oxygen forming a double covalent bond. This leaves two nonbonding electron pairs on each of the oxygen atoms in the resulting O_2 molecule.

4.7 Unequal Sharing: Polar Covalent Bonds

Learning Objectives: Classify a covalent bond as polar or nonpolar. Use electronegativities of elements to determine bond polarity.

A. In a polar covalent bond, the electron pair is unequally shared between the two bonded atoms.
   1. The bonding pair of electrons is more strongly attracted to the element that is the most electronegative.
   2. The chlorine atom has a partial negative charge because it attracts the shared pair of electrons more strongly than the hydrogen does. The H atom has a partial positive charge.
      a. δ^+ δ^-
H—Cl
B. Electronegativity is a measure of the attraction of an atom in a molecule for a shared electron pair.
   1. Nonmetals are more electronegative than metals. Fluorine (upper-right corner of the periodic table) is the most electronegative element, and cesium (lower left) is the least electronegative.
C. When the electronegativity difference is < 0.5, the bond is considered to be nonpolar covalent; when the electronegativity difference is large (> 2.0), complete electron transfer occurs producing ions, the attraction of which to each other results in an ionic bond; when the electronegativity difference is between 0.5 and 2.0, the bond is polar covalent.
D. Names of Covalent (or Molecular) Compounds.
   1. When two nonmetals form a covalently bonded compound, prefixes are used to indicate the number of atoms of each element in the molecule.

Prefix	Number
mono-	1
di-	2
tri-	3
tetra-	4

   2. The prefix mono- is omitted; SO_2 is sulfur dioxide (not monosulfur dioxide).

Answers to Self-Assessment Questions

1. c Both Br atoms in Br_2 can complete their octets by sharing their unpaired electrons, forming a single covalent bond. This bond is nonpolar because the electronegativities of both bonding atoms are the same.

2. d. The only way to distribute 10 valence electrons (4 contributed by the C atom and 6 by the O atom) so that both C and O have octets of electrons is with a triple bond between the two atoms and a nonbonded electron pair on each of the atoms. Note that CO is an exception to the general "rules" that C atoms share 4 pairs of electrons and O atoms share 2 electron pairs. The C-O bond is polar covalent because of the magnitude of the difference in the electronegativities of C (EN = 2.5) and O (EN = 3.5).
3. c The H and Cl atoms each have one unpaired electron. Those electrons are shared, producing the covalent bond between the two atoms.
4. c The difference in electronegativity is greatest in a C-F bond (1.5 based on Figure 4.5), so it is the most polar of those shown.
5. b The difference in electronegativity is greatest for C and F, and the electronegativity of F is higher than that of C. Therefore, the C atom has a smaller share of electrons in the bonding electron pair in the C-F bond than it does in a C-C, C-O, or C-N bond.
6. c The atom with the greater electronegativity has the stronger attraction for electrons in a bond, so the electrons (negatively charged) are closer to that atom, as designated by the $\delta-$ symbol.
7. c The accepted guideline for distinguishing between covalent polar and ionic bonding is an electronegativity difference of 2.0 or more.
8. c The name *phosphorus trichloride* indicates a substance composed of one phosphorus and three chlorine atoms, so the correct formula for the compound is PCl_3.
9 c SF_6 is composed of one sulfur and six fluorine atoms, and the prefix for "6" is hexa. The correct name for SF_6 is sulfur hexafluoride.
10. b The prefix for two N atoms is di- and the prefix for four S atoms is tetra-, so the correct name for N_2S_4 is dinitrogen tetrasulfide.
11. a The prefix for two I atoms is di- and the prefix for five O atoms is penta-, so the correct name for I_2O_5 is diiodine pentaoxide.

4.8 Polyatomic Molecules: Water, Ammonia, and Methane

Learning Objective: • Predict the number of bonds formed by common nonmetals (the HONC rules).

A. To calculate the number of covalent bonds, a nonmetallic element (groups 6A–7A) will almost always form, subtract the group number from 8. This generalization is summarized by the HONC rules below.
   1. **H**ydrogen forms one bond.
   2. **O**xygen in group 6A: 8 – 6 = 2 bonds.
   3. **N**itrogen in group 5A: 8 – 5 = 3 bonds.
   4. **C**arbon in group 4A: 8 – 4 = 4 bonds.

B. Water.
   1. The molecular formula for water is H_2O.
   2. Because the electronegativity difference between H (2.1) and O (3.5) is 1.4, the bonds in a water molecule are polar covalent.

C. Ammonia.
   1. The molecular formula for ammonia is NH_3.
   2. The electronegativities of N and H are 3.0 and 2.1, respectively, so the electronegativity difference is 0.9 and the bonds in NH_3 are polar covalent.

D. Methane.
   1. The molecular formula for methane is CH_4.
   2. The electronegativities of C and H are 2.1 and 2.5, respectively. Because the difference between these electronegativities is less than 0.5, the C-H bonds are nonpolar covalent.

Answers to Self-Assessment Questions

1. d The electronegativity difference between O and H is 1.4, so the H-O bond is polar covalent.
2. d C shares electrons with 4 H atoms to obtain a share in 8 valence electrons.
3. d H follows the duet rule (a share in two valence electrons) in all three molecules, H_2, H_2O, and CH_4.

4.9 Polyatomic Ions

Learning Objective: • Recognize common polyatomic ions and be able to use them in naming and writing formulas for ionic compounds.

A. Polyatomic ions are charged particles containing two or more covalently bonded atoms.
B. A number of polyatomic ions are so common that they have been given names.
   1. Examples

Name	Formula	Name	Formula
ammonium ion	NH_4^+	nitrate ion	NO_3^-
carbonate ion	CO_3^{2-}	phosphate ion	PO_4^{3-}
hydroxide ion	OH^-	sulfate ion	SO_4^{2-}

C. When more than one polyatomic ion is present in a substance the formula of the polyatomic ion is enclosed in parentheses. A subscript outside the parentheses surrounding a polyatomic ion indicates the number of those polyatomic ions required to balance the charge of the counter ion. For example, the formula of aluminum hydroxide is $Al(OH)_3$, and the formula of ammonium sulfate is $(NH_4)_2SO_4$. The subscript(s) written outside the parentheses apply to all atoms enclosed by the parentheses. Note: When just one polyatomic ion is part of the structure, parentheses are not used. For example, the formula of ammonium chloride is NH_4Cl, and the formula of calcium sulfate is $CaSO_4$. If a second level of parentheses is necessary, square brackets, [and], are used.

Answers to Self-Assessment Questions

1. c SO_3^{2-}, the sulfite ion, is polyatomic.
2. d The formula of the ammonium ion is NH_4^+, and the formula of the phosphate ion is PO_4^{3-}. Three NH_4^+ ions are required to counterbalance the 3– charge of the phosphate ion.
3. a Sodium ions have the formula Na^+, and the formula for the hydrogen carbonate ion is HCO_3^-.
4. a Cu(I) is Cu^+, and the hydrogen sulfate ion is HSO_4^-.
5. c The formula of the ammonium ion is NH_4^+, and the formula of the dichromate ion is $Cr_2O_7^{2-}$. Therefore, it takes two ammonium ions to counterbalance the 2– charge on the dichromate ion.
6. c The formulas for the ammonium and hydrogen carbonate ions are NH_4^+ and HCO_3^-, respectively.
7. d All three substances, NaCN, NaOH, and $Na[Au(CN)_2]$, contain polyatomic ions.
8. c NaCN contains a sodium ion (Na^+) and a cyanide ion (CN^-).

4.10 Rules for Writing Lewis Formulas

Learning Objectives: • Write Lewis formulas for simple molecules and polyatomic ions. • Identify free radicals.

A. Construct a skeletal structure that shows the order in which atoms are attached to each other.
   1. H atoms form only one bond, so they cannot be in the middle of a skeletal structure.
      a. H atoms are often bonded to C, N, or O.

2. In polyatomic molecules, the central atom is almost always surrounded by more electronegative elements (H_2O is an exception).

B. Calculate the total number of valence electrons by summing the number of valence electrons contributed by each atom.
   1. For anions, add the number of negative charges to the valence electron total.
   2. For cations, subtract the number of positive charges from the valence electron total.

C. Draw single lines to represent bonds between atoms in the skeletal structure.
   1. Subtract two electrons from the total number of valence electrons for each bond drawn.

D. Starting with the most electronegative atoms, distribute the remaining valence electrons so that each atom is surrounded by eight electrons in keeping with the octet rule (except H and He, which are surrounded by two electrons, duet rule).
   1. If there are not enough valence electrons to satisfy the central atom, move one or more lone pairs from an outer atom to the space between atoms to form a double or triple bond between atoms.
   2. Count the electrons around each atom to make certain every atom except H (or He) has an octet of electrons and that the correct number of valence electrons is used in the structure.

E. Odd-electron molecules: free radicals.
   1. A free radical is an atom or molecule with an unpaired electron.
   2. Many free radicals are highly reactive.
   3. Some nitrogen oxides, major components of smog, are examples of free radicals, as are chlorine atoms that form from the breakdown of chlorofluorocarbons in the stratosphere.

Answers to Self-Assessment Questions

1. b Each F atom has 3 lone pairs of electrons. Four F atoms have a total of 12 lone pairs.
2. b Each F atom and each Cl atom has 3 lone pairs of electrons. Taken together, there are 12 lone pairs in the structure.
3. b Each O atom has 2 unshared electron pairs for a total of 4 unshared electron pairs or 8 unshared electrons.
4. a All the electrons are paired in SCl_2.
5. d Each Cl atom in the Lewis structure has 3 unshared electron pairs, or 6 unshared electrons. The total number of unshared electrons is 24.
6. d There are 16 valence electrons that must be distributed between the central carbon atom and the two oxygen atoms. We can give each of the three atoms an octet of electrons by drawing two double bonds, one between the carbon atom and each of the oxygen atoms. (Can you think of two other ways of correctly drawing this structure?)
7. d Cl_2O is an exception to the rule that usually the least electronegative element is the central atom in a molecule or in a polyatomic molecule. The number of valence electrons available for the structure is ($7 + 7 + 6 = 20$). Note that the structure in answer b has only 18. The chemical formula of the compound shown in answer c is incorrect (ClO_2).
8. b There are 18 valence electrons to distribute between the central S atom and each of the O atoms. This is done by creating a double bond between one O atom and the central S atom, and a single bond between the other O atom and that S atom. To complete the octet on S, there must also be one nonbonded electron pair on the S atom.
9. d All the molecules listed except O_2 have triple bonds in their structure. The structure for HCN is H−C≡N: ; the structure of N_2 is :N≡N: , and the structure of CO is :C≡O: .There is a double bond between the two oxygen atoms in O_2.
10. d There should be 26 valence electrons in the correct structure [$(6 \times 4) + 2$], and answers a and b each have 24 valence electrons. Answer c has 26 valence electrons, but the central S atom is surrounded by 10 electrons and not the desired 8.

4.11 Molecular Shape: The VSEPR Theory

Learning Objective: Predict the shapes of simple molecules from their Lewis structures.

A. VSEPR is used to predict the three-dimensional arrangement of atoms around the central atom.
   1. Electron pairs are arranged around a central atom to minimize repulsion.
      a. The term *electron set* refers to either a lone pair on a central atom or a bond, whether single, double, or triple, between the central atom and another atom.
      b. Two electron sets (bonding or nonbonding) are arranged so that they are on opposite sides of the central atom at an angle of 180°.
      c. Three electron sets assume a triangular arrangement about the central atom forming angles of 120°.
      d. Four electron sets form a tetrahedral array around the central atom, giving angles of separation of 109.5°.

B. To determine molecular shape:
   1. Draw a Lewis structure in which a shared electron pair (bonding pair) is indicated by a line. Indicate unshared electron pairs with dots.
   2. Count the number of electron sets around the central atom. Note: Multiple bonds count as only one electron set.
   3. Draw the shape as if all bonds were bonding pairs, placing them as far as possible from each other.
   4. To visualize the shape of the molecule, erase nonbonding pairs because the molecular shape refers to the arrangement of bonded atoms only.

Answers to Self-Assessment Questions

1. a The bond angles in a tetrahedral arrangement are 109.5°, so it is not possible to create a structure in which the central atom has four electron sets with a 180° angle that would represent a linear structure.
2. b A central atom surrounded by three electron sets will have a trigonal planar structure. If one of those electron sets is a nonbonded electron pair, the molecular structure will be bent.
3. b The central S atom in SO_2 and in SO_3 both have three electron sets, so their structures are trigonal planar with 120° angles. The central S atom in SO_3^{2-} has four electron sets, so its structure will be tetrahedral with 109.5° bond angles.
4. b The central C atom is attached with double bonds to each of the S atoms, giving it two electron sets to separate and a linear shape.
5. a OF_2 is bent because the central O atom is surrounded by four electron sets, two of which are bonding pairs and two are nonbonding pairs.
6. d The central Si atom in SiF4 is surrounded by four electron sets, resulting in a tetrahedral shape.
7. c NH_3 is trigonal pyramidal because the central N atom is surrounded by four electrons sets, one of which is a nonbonded electron pair.

4.12 Shapes and Properties: Polar and Nonpolar Molecules

Learning Objective: • Classify a simple molecule as polar or nonpolar from its shape and the polarity of its bonds.

A. To be polar, a molecule must have polar bonds AND separate centers of positive and negative charge. Therefore, both the polarities of bonds and the molecular geometry must be considered when deciding whether or not a molecule is polar.

1. Diatomic molecules.
    a. A molecule is polar if its bonds are polar.
    b. A molecule is nonpolar if its bonds are nonpolar.
2. Methane: a tetrahedral molecule.
    a. VSEPR predicts a tetrahedral shape. The four bonds form angles of 109.5°.
    b. Very slight bond polarities are canceled by the symmetrical shape of the molecule.
    c. The molecule is nonpolar.
3. Ammonia: a pyramidal molecule.
    a. VSEPR predicts a tetrahedral shape, but the nonbonding pair on the central nitrogen atom pushes the bonds closer together, forming an angle of 107° rather than 109.5°.
    b. The shape is pyramidal and the molecule is polar because the center of partial negative charge is located on the N atom and the center of partial positive charge is in the center of the triangle formed by the three H atoms.
4. Water: a bent molecule.
    a. Water acts like a dipole (a molecule with positive and negative portions). The water molecule must be bent; otherwise the charge on the bonds would cancel each other.
    b. According to VSEPR, the two bonds and the two nonbonding pairs of electrons on water's oxygen should have a shape based on a tetrahedron. The nonbonding pairs occupy a larger volume and push the two bonds closer together, forming an angle of 104.5° instead of the tetrahedral angle of 109.5°.

Answers to Self-Assessment Questions

1. d NH_3 has polar bonds. Its pyramidal shape results in a polar molecule.
2. c CCl_4 has polar bonds. Its tetrahedral shape results in a symmetry that causes the bond polarities to cancel one another, resulting in a nonpolar molecule.
3. c HCN (H-C≡N:) is a linear molecule. The C≡N bond is polar, so the molecule is polar.
4. a SiF_4 has polar bonds. Its tetrahedral shape gives a net nonpolar molecule.

Green Chemistry and Chemical Bonds

A. Molecular shape and composition of compounds control reactivity and interactions with other substances. Manipulating the geometry and composition of molecules alters their properties.
B. Interaction between molecules caused by geometrical orientation of their atoms is called molecular recognition.
C. New production methods that take advantage of molecule recognition typically use lower energy, solvent-free processes that do not require harsh conditions, reducing the environmental impact of manufacturing processes.
D. By understanding molecular shape, the nature of chemical bonds, and the way molecules interact with other molecules through noncovalent forces, chemists are able to design new medicine, molecules, and materials that benefit society and, at the same time, have minimal impact on the environment and human health.

DISCUSSION

Electrons are the glue that holds atoms together. Atoms achieve a stable octet of electrons in their outermost shell by transferring electrons (forming ions) from one to another or by sharing electrons (forming covalent bonds). Because only valence electrons (electrons outside the filled electron shells) are involved in either of these processes, an abbreviated notation called Lewis symbols and Lewis formulas (or structures) provide a convenient way of visualizing the process and its results. The group number of the main-group elements gives the number of valence electrons each has. For example, NaCl is formed from the reactive elements Na (a metal) and Cl_2 (a nonmetal). Electron configurations or electron-dot symbols illustrate the properties of Na, Cl_2, and NaCl. Covalent compounds are formed when the atoms share electrons. Reactive free radicals are exceptions to the octet rule. The VSEPR theory is used to predict shapes of molecules according to the number of electron sets around the atoms that are central to the structure. The shape (and polarity) of molecules can be used to help explain the states and properties of matter. The shapes of molecules become the basis for molecular recognition and the manner in which molecules interact with one another in many biological situations. Understanding molecular shape, the nature of chemical bonds, and the way that molecules interact with other molecules is the basis of the design of medicines and other substances that benefit society.

Many of the problems at the end of the chapter are drills. These include practice in drawing ions, putting together molecules, or naming compounds. This practice should help firmly establish the rules and guidelines governing chemical structure in your mind. You must recognize, however, that there are many exceptions to the "rules," so you should be open to those as well. Here are answers to some often-asked questions.

First, how do you know whether a compound is ionic or covalent? You can usually follow these general rules:

1. When hydrogen bonds with other nonmetals, except for group 7A, the compounds are covalent.
   Examples: H_2O, NH_3, and CH_4.
2. When hydrogen bonds with group 1A or 2A metals, the compounds are ionic.
   Examples: NaH, CaH_2, and LiH.
3. When Group 1A or 2A metals bond with group 5A, 6A, or 7A nonmetals, the compounds are ionic.
   Examples: NaCl, CaF_2, K_2S, and Na_3N.
4. When nonmetals bond with other nonmetals, the compounds are covalent.
   Examples: CO_2, CCl_4, and PCl_3.

Second, if the compound is covalent, how do you know whether it is polar or nonpolar?

1. If both (or all) of the atoms in the molecule are the same, the compound is nonpolar because all of the bonds are nonpolar.
   Examples: O_2, N_2, Br_2, and S_8.
2. If the atoms in the molecule are not the same, consider each **bond** separately. Determine the electronegativities of the two atoms sharing the electron pair. Follow the guidelines to determine whether the electronegativity difference falls in the range of nonpolar or polar covalent bonding.
   Examples: NO, PCl_3, SO_2, H_2, and HCl.
3. To decide if the **molecule** is polar or nonpolar, look first at the bonds.
   a. If the **bonds** in the molecule are nonpolar, then the **molecule** is nonpolar.
      Example: C_2H_6.
   b. If the **bonds** in the molecule are polar, then you must determine the **molecule**'s shape.
   c. If the shape is symmetrical, the **molecule** is nonpolar.
      Example: CO_2.
   d. If the shape is not symmetrical, the **molecule** is polar.
      Example: HCN.

Notice the emphasis on bonds and molecules. It is important to understand the connection between the polarity (or nonpolarity) of individual bonds molecules and the overall effect of those bonds in determining the polarity (or nonpolarity) of the molecules themselves.

LEARNING OBJECTIVES

You should be able to . . .

1. Determine the number of electrons in an ion. (4.1)
2. Write the Lewis symbol for an atom or ion. (4.2)
3. Distinguish between an ion and an atom. (4.3)
4. Describe the nature of the attraction that leads to formation of an ionic bond. (4.3)
5. Write symbols for common ions, and determine their charges. (4.4)
6. Describe the relationship between the octet rule and the charge on an ion. (4.4)
7. Name and write formulas for binary ionic compounds. (4.5)
8. Explain the difference between a covalent bond and an ionic bond. (4.6)
9. Name and write formulas for covalent compounds. (4.6)
10. Classify a covalent bond as polar or nonpolar. (4.7)
11. Use electronegativities of elements to determine bond polarity. (4.7)
12. Predict the number of bonds formed by common nonmetals (the HONC rules). (4.8)
13. Recognize common polyatomic ions and be able to use them in naming and writing formulas for compounds. (4.9)
14. Write Lewis formulas for simple molecules and polyatomic ions. (4.10)
15. Identify free radicals. (4.10)
16. Predict the shapes of simple molecules from their Lewis formulas. (4.11)
17. Classify a simple molecule as polar or nonpolar from its shape and the polarity of its bonds. (4.12)
18. Describe the concept of molecular recognition.
19. Explain the green chemistry advantages of using production methods based on molecular recognition.

ADDITIONAL PROBLEMS

1. Compounds are formed from the following pairs of elements. Indicate whether the compound would be ionic or covalent. (Note that you're not being asked to draw the compounds, but rather just to evaluate their tendency to form ionic or covalent bonds.)

a. Mg and O	b. F and Ca	c. Li and S	d. Br and Cl
e. Na and H	f. S and H	g. N and Cl	h. C and O

2. You should be able to draw electron-dot symbols for any element in the A groups (1A, 2A, etc.). To do this, write the symbol for the element and surround it with dots representing the valence (outermost) electrons. The number of valence electrons is given by the group number. For practice, draw electron-dot structures for the atoms of these elements.

a. barium	b. carbon	c. xenon	d. nitrogen
e. silicon	f. hydrogen	g. potassium	h. chlorine

3. Electron-dot structures for ions are drawn either by adding electron dots to complete the octet or by removing electron dots to empty the outermost level. Electrons are added to elements with five or more valence electrons; they are subtracted from elements with three or fewer valence electrons. For the following elements, how many electron dots would you add to or remove from the electron-dot symbols of the atoms to form the ions?

a. barium	b. chlorine	c. nitrogen	d. iodine
e. potassium	f. magnesium	g. sulfur	h. aluminum

4. The charge is written to the upper right of the electron-dot symbol for an ion. The charge is equal to the number of electrons added to or removed from the neutral atom to form the ion. The charge is positive if electrons are removed and negative if electrons are added. Write electron-dot symbols for the ions that are formed from the eight elements listed in Problem 3.

ANSWERS TO ADDITIONAL PROBLEMS

1. a. Whether a bond is nonpolar, polar, or ionic depends on the electronegativity difference between the two bonding atoms. If the electronegativity difference (ΔEN) is less than 0.5, the bond is nonpolar covalent. If ΔEN is between 0.5 and 2.0, the bond is polar covalent, and if ΔEN is greater than 2.0, the bond is ionic. According to Figure 4.5, the electronegativity of Mg is 1.2 and that of O is 3.5. $\Delta EN = 4.5 - 1.2 = 3.3$, so the Mg–O bond is ionic.

   b. EN of F = 4.0; EN of Ca = 1.0; ΔEN = 3.0, so a Ca–F bond is ionic.

   c. EN of Li = 1.0; EN of S = 2.5; ΔEN = 1.5, so a Li–S bond is polar covalent.

   d. EN of Br = 2.8; EN of Cl = 3.0; ΔEN = 0.2, so a Br–Cl bond is nonpolar covalent.

   e. EN of Na = 0.9; EN of H = 2.1; ΔEN = 1.2, so a Na–H bond is polar covalent.

   f. EN of S = 2.5; EN of H = 2.1; ΔEN = 0.4, so a S–H bond is nonpolar covalent.

   g. EN of N = 3.0; EN of Cl = 3.0; ΔEN = 0.0, so an N–Cl bond is nonpolar covalent.

   h. EN of C = 2.5; EN of O = 3.5; ΔEN = 1.0, so a C–O bond is polar covalent.

2. a. Barium atoms have 2 valence electrons. The Lewis symbol for a Ba atom is Ba surrounded by 2 dots.

   b. Carbon atoms have 4 valence electrons. The Lewis symbol for a C atom is C surrounded by 4 dots.

c. Xenon atoms have 8 valence electrons. The Lewis symbol for an Xe atom is Xe surrounded by 8 dots.

d. Nitrogen atoms have 5 valence electrons. The Lewis symbol for an N atom is N surrounded by 5 dots.

e. Silicon atoms have 4 valence electrons. The Lewis symbol for a Si atom is Si surrounded by 4 dots.

f. Hydrogen atoms have 1 valence electron. The Lewis symbol for an H atom is H with 1 dot.

g. Potassium atoms have 1 valence electron. The Lewis symbol for a K atom is K with 1 dot.

h. Chlorine atoms have 7 valence electrons. The Lewis symbol for a Cl atom is Cl surrounded by 7 dots.

3. a. Achieving an octet of outer electrons is generally easier for metal atoms if they lose their valence electrons to expose an octet of electrons at a lower level than to gain enough electrons to reach an octet. It is easier for nonmetal atoms to gain enough electrons so that they have a total of 8 valence electrons. Ba is a metal, so it would lose 2 electrons to form an ion.

b. Cl is a nonmetal with 7 valence electrons. Cl forms an ion by gaining 1 electron.

c. N is a nonmetal with 5 valence electrons. N forms an ion by gaining 3 electrons.

d. I is a nonmetal with 7 valence electrons. I forms an ion by gaining 1 electron.

e. K is a metal with 1 valence electron. K forms an ion by losing 1 electron.

f. Mg is a metal with 2 valence electrons. Mg forms an ion by losing 2 electrons.

g. S is a nonmetal with 6 valence electrons. S forms an ion by gaining 2 electrons.

h. Al is a metal with 3 valence electrons. Al forms an ion by losing 3 electrons.

4. a. Ba^{2+} b. $:\overset{\cdot\cdot}{\underset{\cdot\cdot}{\text{Cl}}}:^{-}$ c. $:\overset{\cdot\cdot}{\underset{\cdot\cdot}{\text{N}}}:^{3-}$ d. $:\overset{\cdot\cdot}{\underset{\cdot\cdot}{\text{I}}}:^{-}$ e. K^{+} f. Mg^{2+} g. $:\overset{\cdot\cdot}{\underset{\cdot\cdot}{\text{S}}}:^{2-}$ h. Al^{3+}

ANSWERS TO ODD-NUMBERED REVIEW QUESTIONS AND SOLUTIONS FOR ODD-NUMBERED END-OF-CHAPTER PROBLEMS

Review Questions

1. Na metal is a reactive, soft, silver metal. Na^+ cations are nonreactive and independently stable because, by losing one electron, they have achieved an octet of electrons.

3. Chlorine atoms have 7 valence electrons, one of which is unpaired. As such, they are free radicals and highly reactive. They do not naturally exist under normal circumstances (temperature, pressure, etc.). Chlorine molecules, Cl_2, form when two chlorine atoms share their unpaired electrons to form a nonpolar covalent bond giving both Cl atoms an octet of electrons. This nonpolar molecule is the elemental form of chlorine, which is a gas at room temperature. Chloride ions, Cl^-, form when Cl atoms each accept an electron from a metal atom to complete their octets. Unlike the situation in a Cl_2 molecule where each Cl atom is dependent on the shared electron pair for a complete octet, a Cl^- ion is independently stable because the added electron serves to complete its octet. Therefore, Cl^- ions are not reactive; they are strongly attracted to positive ions such as Na^+ with which they form ionic bonds.

5. a. Group 1A ions have a 1+ charge.
   b. Group 6A ions have a 2− charge.
   c. Group 5A ions have a 3− charge.
   d. Group 2A ions have a 2+ charge.

7. a. The elements in group 6A are nonmetals with 6 valence electrons and form 2− ions.
   b. The elements in group 3A are metals with 3 valence electrons and form 3+.
   c. The elements in group 1A are metals with 1 valence electron and form 1+ ions.
   d. The elements in group 2 A are metals with 2 valence electrons and form 2+ ions.

9. Carbon and nitrogen atoms can easily form triple bonds, either with other identical atoms or with atoms of other elements. Examples are H—C≡C—H, :N≡N:, and H—C≡N:. Oxygen very rarely forms triple bonds although one common example is :C≡O:.

Problems

11. a. Phosphorus: $\cdot\dot{\mathbf{P}}\cdot$ (with a lone pair below)

    b. Fluorine: $:\dot{\mathbf{F}}:$ (with a lone pair below)

    c. Boron: $\cdot\dot{B}\cdot$

13. a. Na^+ + $[:\ddot{I}:]^-$

    b. $2K^+$ + $[:\ddot{S}:]^{2-}$

    c. Ca^{2+} + $2\,[:\ddot{Cl}:]^-$

    d. Al^{3+} + $3\,[:\ddot{F}:]^-$

15. b. A neutral Mg atom (group 2A) has two valence electrons. The species shown in answer b is not Mg, but Mg^{2+}, indicating that the Mg atom has lost its valence electrons and has become isoelectronic with Ne. The other answers are correct.

17. a. S^{2-}

b. Potassium ion

c. Bromide ion

d. F^-

e. Calcium ion

f. Iron(III)

19. Mo^{4+} is the molybdenum(IV) ion.

Mo^{6+} is the molybdenum(VI) ion.

21 . a. Mn^{2+}

b. Mn^{3+}

c. Mn^{7+}

23. a matches with d: $FeCl_2$ is iron(II) chloride; c matches with h: AgF is silver fluoride; e matches with b: $FeBr_3$ is iron(III) bromide; and g matches with f: Na_2O is sodium oxide.

25. Cr_2O_3 is called chromium(III) oxide and CrO_3 is called chromium(VI) oxide.

27. a matches with d: NH_4NO_3 is ammonium nitrate; c matches with h: $CaCO_3$ is calcium carbonate; e matches with b: $FePO_4$ is iron(III) phosphate; and g matches with f: $Ca(CH_3CO_2)_2$ is calcium acetate.

29. a. $AgNO_2$ is silver nitrite.

b. The formula of lithium chromate is Li_2CrO_4.

c. $(NH_4)_2SO_3$ is ammonium sulfite.

d. The formula of magnesium hydrogen carbonate is $Mg(HCO_3)_2$.

e. The formula of copper(I) phosphate is Cu_3PO_4.

f. $Al(MnO_4)_3$ is aluminum permanganate.

31. The Lewis dot structure for the compound formed with H and F (HF) is

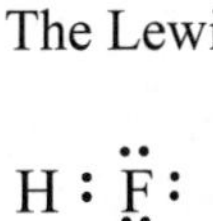

33. The Lewis dot structure for the compound formed from P and H (PH_3) is

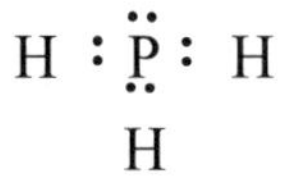

35. The Lewis dot structure for the compound formed from C and Cl (CCl_4) is

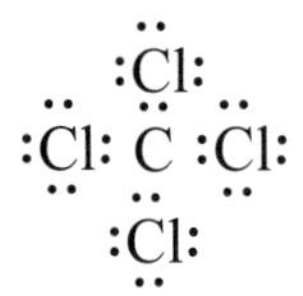

37 A hydrogen atom has only one valence electron. In a covalent bond involving H, the H atom pairs this single valence electron with a valence electron contributed by the atom with which it is forming the bond.

39. a. N_2O_4

b. $BrCl_3$

c. Oxygen difluoride

d. NI_3

e. Carbon tetrabromide

f. Dinitrogen tetrasulfide

41. a. The Lewis dot structure for SiH_4 is:

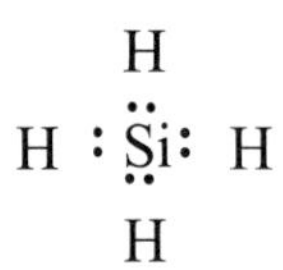

b. The Lewis dot structure for N_2F_4 is:

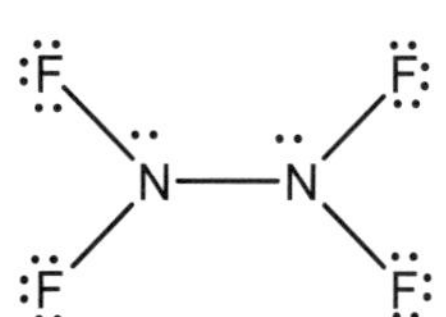

c. The Lewis dot structure for CH_5N (CH_3NH_2) is:

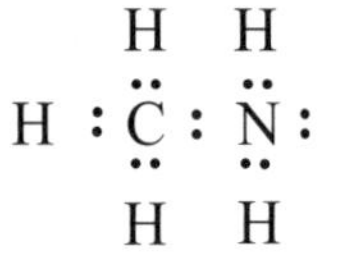

d. The Lewis dot structure for H_2CO is: 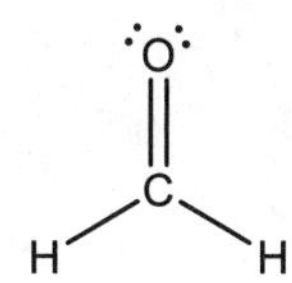

e. The Lewis structure for NOH_3 is 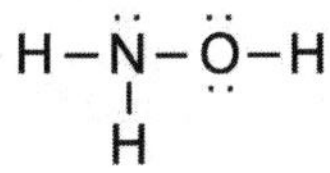

f. The Lewis structure for H_3PO_3 is

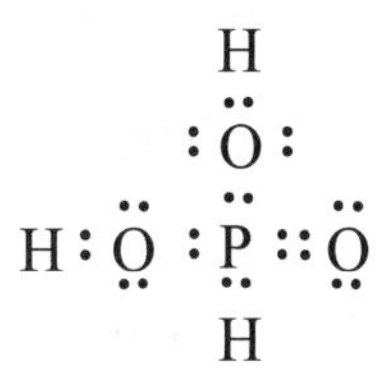

43. a. The Lewis dot structure for ClO^- is:

b. The Lewis dot structure for HPO_4^{2-} is:

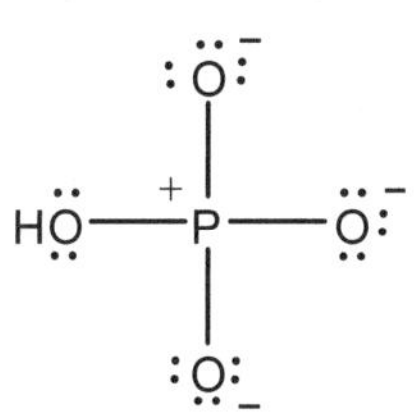

c. The Lewis dot structure for BrO_3^- is

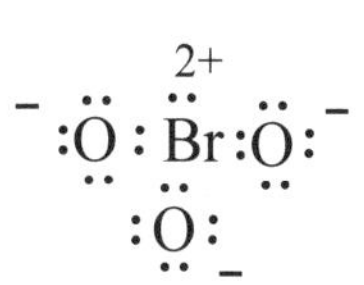

45. In order for a covalent bond between elements A and B to be polar, the electronegativity difference between A and B must be between 0.5 and 2.0. If the electronegativity difference between A and B is less than 0.5, the bond will be considered nonpolar; if the electronegativity difference is greater than 2.0, then the bond will likely be ionic.

47. a. The bond between H and O is polar. The electronegativities of H and O are 2.1 and 3.5 respectively. The electronegativity difference between H and O is 1.4.

b. The bond N–F is polar. The electronegativities of N and F are 3.0 and 4.0, respectively. The electronegativity difference between N and F is 1.0.

c. The bond Cl–B is polar. The electronegativites of Cl and B are 3.0 and 2.0, respectively. The electronegativity difference between Cl and B is 1.0.

49. a. Dipole for H—O

b. Dipole for N—F (arrow $\longmapsto$ pointing toward F)

c. Dipole for Cl—B (arrow pointing toward Cl, + at B)

51. a. $Si^{\delta+}$— $O^{\delta-}$ Si has an electronegativity of 1.8, and O has an electronegativity of 3.5.

b. F—F The difference in electronegativity is zero, so there are no partial charges.

c. $F^{\delta-}$—$N^{\delta+}$ F has an electronegativity of 4.0, and N has an electronegativity of 3.0.

53. a. The bond in K_2O is ionic. The electronegativity difference between K (0.8) and O (3.5) is 2.7, which is greater than 2.0.

b. The bond in BrCl is nonpolar covalent. The electronegativity difference between Br (2.8) and Cl (3.0) is 0.2, which is less than 0.5, the minimum electronegativity difference for a bond to be considered polar.

c. The bond in MgF_2 is ionic. The electronegativity difference between Mg (1.2) and F (4.0) is 2.8, which is greater than 2.0.

d. The bond between the two iodine atoms in I_2 is nonpolar covalent. The electronegativities of the two atoms (2.5) are identical, so the electron pair is shared equally by the bonding atoms.

55. We can judge the polarity of each of the bonds by the difference in electronegativity of the two bonding atoms. The bond with the smallest EN difference is the least polar; the bond with the greatest EN difference is the most polar. In order of increasing polarity: P-Cl (EN difference = 0.9) < N-F (EN difference = 1.0) < Si-Cl (EN difference = 1.2) < H-F (EN difference = 1.9) < B-F (EN difference = 2.0) < Si-F (EN difference + 2.2).

57. a. Silicon in SiH_4 has four bonded atoms, four bonded electron sets, and no nonbonded electrons. SiH_4 is tetrahedral.

b. Selenium in H_2Se has two bonded atoms and four electron sets counting the two shared pairs and the two lone pairs. H_2Se is bent.

c. Phosphorus in PH_3 has three bonded atoms and four sets of electrons, including one lone pair. PH_3 is pyramidal.

d. Silicon in SiF_4 has four bonded electron sets of electrons and no nonbonded electrons. SiF_4 is tetrahedral.

e. Oxygen in OF_2 has two bonded atoms and four electron sets counting the two shared pairs and the two lone pairs. OF_2 is bent.

f. C in H_2CO has three bonded atoms and three electron sets, one of which is a double bond between the C and O atoms. H_2CO is triangular.

59. BeF_2 is linear and therefore nonpolar because the polarity cancels out. F^{-}–Be^{+}—F^{-}. The center of partial positive charge is on the Be atom; the center of partial negative charge is also on the Be atom. Note: This is an interesting and a somewhat surprising case of a compound formed from atoms with very different electronegativities, fully expected to be ionic and yet, molecular in nature.

61. Dichlorofluoroborane, BCl_2F, has a triangular structure with the B atom in the center and the two Cl atoms and F atom at the three corners. The electronegativities of B, Cl, and F are 2.0, 3.0, and 4.0, respectively, so all three bonds to boron are polar. IF the electronegativities of the three atoms attached to the central B were all identical, then the molecule would be symmetrical and, therefore, nonpolar. However, there is a significant difference between the polarities of the B–Cl and B–F bonds, which means that the overall molecule does have separated centers of partial positive and partial negative charge. Therefore, the molecule is polar.

a. The electron configuration of Ca is $1s^22s^22p^63s^23p^64s^2$. The Ca^{2+} ion forms when Ca loses its two valence electrons. The electron configuration of Ca^{2+} is $1s^22s^22p^63s^23p^6$.

b. The most stable ion of Rb forms when a Rb atom loses its one valence electron. The electron configuration of Rb^+ is: $1s^22s^22p^63s^23p^63d^{10}4s^24p^6$.

c. The most stable ion of S is S^{2-}: $1s^22s^22p^63s^23p^6$.

d. The most stable ion of I is I^-: $1s^22s^22p^63s^23p^63d^{10}4s^24p^64d^{10}5s^25p^6$.

e. The most stable ion of N is N^{3-}: $1s^22s^22p^6$.

f. The most stable ion of Se is Se^{2-}: $1s^22s^22p^63s^23p^63d^{10}4s^24p^6$.

63. a. Chloroform, $CHCl_3$: The C-H bond is not polar because the electronegativity difference, 0.4 unit, is less than 0.5. The C—Cl bonds are right at the cutoff between polar and nonpolar designations, so they are very slightly polar. Because the central C atom is surrounded by four electron sets, the bond angles are approximately 109°. Overall, the molecule is essentially nonpolar because none of the bonds is more than very slightly polar.

b. Boron trichloride, BCl_3: The electronegativity difference between B and Cl is 1.0, so each of the three B—Cl bonds is polar. B is an exception to the octet rule in the sense that it often forms stable compounds in which the central B atom is surrounded by six electron sets. BCl_3 is an example of this exception. Because the central B atom is surrounded by three electron sets, the bond angles in the compound are 120°. Because the molecule is symmetrical, the molecule is nonpolar in spite of the fact that each of its three covalent bonds is polar.

c. Carbon tetrachloride, CF_4: The electronegativity difference between C and F is 1.5, so C–F bonds are polar. The central C atom is surrounded by four electron sets, so the bond angles in the compound are approximately 109°. Because the molecule is symmetrical, the molecule is nonpolar in spite of the fact that each of its four covalent bonds is very polar.

d. Sulfur difluoride, SF_2: The electronegativity difference between S and O is 1.0, so the S–F bonds are polar. The central S atom in SF2 is surrounded by four electron sets, two single covalent bonds to F, and two nonbonded electron pairs. Therefore the F–S–F bond angles in the compound are approximately 109°. The molecule is not symmetrical (its structure is much like the structure of H_2O), so SF_2 is a polar molecule.

e. Nitrogen triiodide, NI_3:The electronegativity difference between N and I is 0.5, so the N–I bonds are barely polar. The central N atom in the structure is surrounded by two electron sets, three single covalent bonds and one nonbonded electron pair. Therefore, the bond angles are approximately 109°. The molecule is not symmetrical, but the bonds are so slightly polar that the molecule itself is essentially nonpolar.

f. Dichlorodifluoromethane, CCl_2F_2: The electronegativity difference between C and Cl is 0.5 and that between C and F is 1.5, so the two C–Cl bonds are essentially nonpolar and the two C–F bonds are polar. The central C atom is surrounded by four electron sets so the bond angles are all approximately 109°. Because only two of the four covalent bonds are polar, the molecule is not symmetrical and, therefore, is polar.

65. The electronegativity difference between S and O is 1.0, so the three S-O bonds are polar. The central S atom is surrounded by three electron sets, two S–O single bonds, and one S–O double bond (S=O), resulting in bond angles of approximately 120°. Because the SO_3 molecule is symmetrical, it is nonpolar in spite of the fact that it contains three polar bonds.

67. The central atom in both molecules is surrounded by four electron sets, two of which are covalent bonds and the other two of which are nonbonded electron pairs, so the approximate bond angles in both compounds are 109°. Both compounds have bent structures and are nonsymmetrical. The electronegativity difference between S and Cl is 0.5, and that between O and H is 1.4, so the S—Cl bonds are essentially nonpolar while the O–H bonds are polar. Therefore, H_2O is a polar molecule and SCl_2 is essentially nonpolar.

69. A free radical is a species with an unpaired electron that results when the species has an uneven number of valence electrons. Sulfur atoms have 6 valence electrons, NO_2 molecules have 17 valence electrons, and N_2O_4 molecules have 34 valence electrons. Therefore NO_2 molecules are free radicals.

71. a. XeF_4:

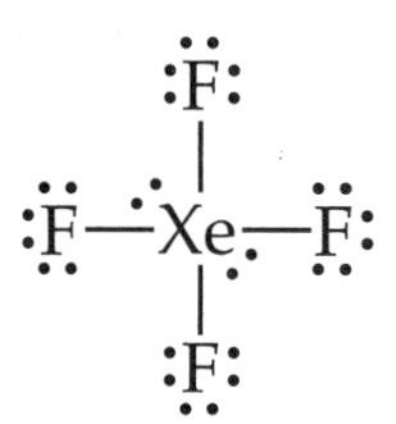

b. I_3^-:

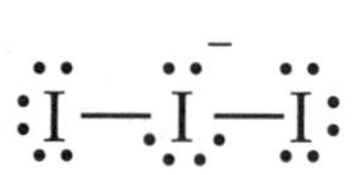

c. SF_4:

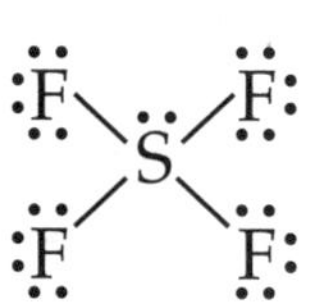

d. KrF_2:

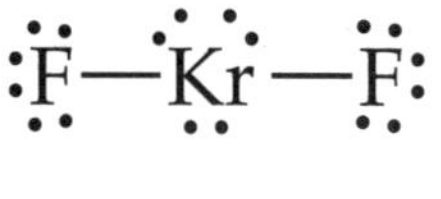

73.

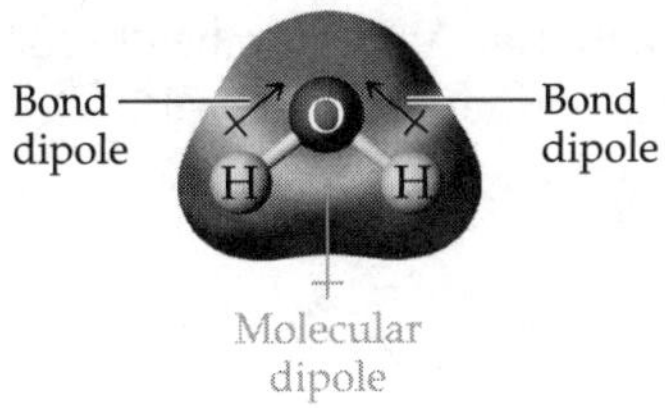

75. The central S atom in SCl_2 is surrounded by four electron sets giving, it bond angles of approximately 109°. Two of the electron sets are nonbonded electron pairs, so their electron density is located closer to the S atom than is the case with the other two electron pairs, the density of which is drawn toward the Cl atoms. The "extra" electron repulsion caused by the nonbonded electron pairs forces the Cl–S–Cl bond angle to contract slightly, making it a bit less than 109°.

77. The bond between the I and Cl atoms in ICl is essentially nonpolar covalent because the electronegativity difference between I and Cl is 0.5. Therefore, the ICl molecule is essentially nonpolar.

79. Taking K, metallic potassium, would not be a wise idea as it would react violently with the water in your mouth (saliva is an aqueous solution). Instead, the potassium your doctor prescribes is a salt containing K^+ ions, such as KCl.

81. a. Ca atoms have 20 electrons. The subshell notation for the electron arrangement around a Ca nucleus is: $1s^2 2s^2 2p^6 3s^2 3p^6 4s^2$. Ca is a metal. When a Ca atom loses its two valence electrons, it becomes isoelectronic with the noble gas Ar. The subshell notation for the electron arrangement for Ca^{2+} is: $1s^2 2s^2 2p^6 3s^2 3p^6$.

b. Rb atoms have 37 electrons. The subshell notation for the electron arrangement around a Rb nucleus is: $1s^2 2s^2 2p^6 3s^2 3p^6 4s^2 3d^{10} 4p^6 5s^1$. Rb is a metal. When an Rb atom loses its single valence electron, it becomes isoelectronic with the noble gas Kr. The subshell notation for the electron arrangement in Rb^+ is $1s^2 2s^2 2p^6 3s^2 3p^6 4s^2 3d^{10} 4p^6$.

c. S atoms have 16 electrons. The subshell notation for the electron arrangement around an S nucleus is: $1s^2 2s^2 2p^6 3s^2 3p^4$. S is a nonmetal, and nonmetals gain electrons to become isoelectronic with noble gas atoms. The subshell notation for the electron arrangement in S^{2-} is $1s^2 2s^2 2p^6 3s^2 3p^6$.

d. I atoms have 53 electrons. The subshell notation for the electron arrangement around an I nucleus is: $1s^2 2s^2 2p^6 3s^2 3p^6 4s^2 3d^{10} 4p^6 5s^2 4d^{10} 5p^5$. I is a nonmetal that needs to gain one electron to become isoelectronic with the noble gas Xe. The subshell notation for the election arrangement in I^- is $1s^2 2s^2 2p^6 3s^2 3p^6 4s^2 3d^{10} 4p^6 5s^2 4d^{10} 5p^6$.

e. N atoms have 7 electrons. The subshell notation for the electron arrangement around an N nucleus is $1s^2 2s^2 2p^3$. N is a nonmetal that must gain 3 electrons to become isoelectronic with the noble gas Ne. The subshell notation for the electron arrangement in N^{3-} is $1s^2 2s^2 2p^6$.

f. Se atoms have 34 electrons. The subshell notation for the electron arrangement around a Se nucleus is $1s^2 2s^2 2p^6 3s^2 3p^6 4s^2 3d^{10} 4p^4$. Se is a nonmetal that must gain two electrons to become isoelectronic with the noble gas Kr. The subshell notation for the electron arrangement in Se^{2-} is $1s^2 2s^2 2p^6 3s^2 3p^6 4s^2 3d^{10} 4p^6$.

83. Halogen atoms (F, Cl, Br, I) each have 7 valence electrons. They need to share their unpaired electron with one from another atom to form a single covalent bond to compete the octet.

85. Two Na^+ ions are required to balance the negative charge of the tungstate ion, WO_4, so the charge on this ion must be 2− and its formula must be WO_4^{2-}. The charge on an aluminum ion is 3+ (Al^{3+}). Using the crossover method, the formula for aluminum tungstate must be $Al_2(WO_4)_3$.

87. Use the cross-over method to get $Q_3(ZX_4)_2$.

89. A central atom adopting a tetrahedral geometry is one surrounded by four electron sets. While those four electron sets complete the octet for the central atom, they are mutually repulsive because electrons are electrically negative and like charges repel one another. The most efficient way of separating the four negative areas around a central atom is to space them so that the central atom occupies the center and the four electron sets occupy the vertices of a tetrahedron.

91. The term used to describe the interaction between molecules caused by the geometric orientations of their atoms is *molecular recognition.*

93. Medicinal chemists design new medicines and drug molecules, so they will resemble biological molecules and bind to enzymes or receptors in our body. Enzymes often recognize the molecules with which they interact by their shapes and partial charges.

CHAPTER

5

Chemical Accounting

Mass and Volume Relationships

CHAPTER SUMMARY

5.1 Chemical Sentences: Equations

Learning Objective: • Identify balanced and unbalanced chemical equations, and balance equations by inspection.

A. A chemical equation is a shorthand notation for describing the chemical change as the reactants (shown on the left of the arrow) are converted to products (shown on the right of the arrow).
   1. Reactants → Products
   2. We can indicate the physical states of reactants and products by writing the initial letter of the state immediately following the formula: (g) indicates the substance is a gas, (l) a liquid, and (s) a solid. The label (aq) means the substance is dissolved in water (producing an aqueous solution).

B. When balancing chemical equations, make sure that the same number of each kind of atom appears on both sides of the equation.

$$2\,H_2 + O_2 \rightarrow 2\,H_2O$$

   1. When counting atoms, multiply the coefficients preceding formulas by the subscripts for a given atom in the formula.
   2. Polyatomic ions are treated as units whenever possible in the process of balancing chemical equations.

Answers to Self-Assessment Questions

1. The answer is b. Looking at the various options:
   I. All elements are balanced.
   II. Al and Cl are not balanced.
   III. All elements are balanced.
   IV. O is not balanced.

2. c The balanced equation is: $4\ PH_3 + 8\ O_2 \rightarrow P_4O_{10} + 6\ H_2O$, so 6 molecules of H_2O are produced every time one P_4O_{10} molecule forms.
3. b The balanced reaction is $3\ Mg(OH)_2 + 2\ H_3PO_4 \rightarrow 6\ H_2O + Mg_3(PO_4)_2$. 2 molecules of H_3PO_4 produce 1 formula unit of $Mg_3(PO_4)_2$.
4. c The balanced reaction is $3\ Mg(OH)_2 + 2\ H_3PO_4 \rightarrow 6\ H_2O + Mg_3(PO_4)_2$. Six H_2O molecules are formed for every 2 H_3PO_4 molecules that react, so 3 H_2O molecules form for each H_3PO_4 molecule that reacts.
5. d The balanced reaction is $3\ Mg(OH)_2 + 2\ H_3PO_4 \rightarrow 6\ H_2O + Mg_3(PO_4)_2$. According to this equation, one formula unit of $Mg(OH)_2$ reacts with 2 H_3PO_4 molecules, so 9 formula units of $Mg(OH)_2$ would be required to react with 6 H_3PO_4 molecules.
   Number of formula units of $Mg(OH)_2$ = (6 molecules H_3PO_4)
   (3 formula units $Mg(OH)_2$/2 molecules H_3PO_4)
   = 9 formula units $Mg(OH)_2$.

5.2 Volume Relationships in Chemical Equations

Learning Objective: • Determine volumes of gases that react, using a balanced equation for a reaction.

A. Gay-Lussac's law of combining volumes: When all measurements are made at the same temperature and pressure, the volumes of gaseous reactants and products are in small whole-number ratios.
B. Avogadro's hypothesis: Equal volumes of all gases (at the same temperature and pressure) contain the same number of molecules.

Answers to Self-Assessment Questions

1. d Use the coefficients of the balanced equation, $2\ H_2 + O_2 \rightarrow 2\ H_2O$, for a 2 to 1 to 2 ratio of H_2 to O_2 to H_2O.
2. c Use the coefficients in the balanced equation $2\ C_8H_{18} + 25\ O_2 \rightarrow 16\ CO_2 + 18\ H_2O$.
   number of CO_2 molecules = (75 O_2 molecules)(16 CO_2 molecules/25 O_2 molecules)
   = 48 CO_2 molecules.
3 b Use the coefficients in the balanced equation, $N_2 + 3\ H_2 \rightarrow 2\ NH_3$.
   volume of NH_3 = (4.50 L N_2)(2 L NH_3/1 L N_2) = 9.00 L NH_3.

5.3 Avogadro's Number and the Mole

Learning Objectives: • Calculate the formula mass, molecular mass, or molar mass of a substance.
• Use Avogadro's number to determine the number of particles of different types in a mass of a substance.

A. The number of C-12 atoms in a 12.0-g sample of C-12 is called Avogadro's number and is equal to 6.02×10^{23}.
B. One mole (mol) of a substance is the amount of that substance that contains the same number of elementary units as there are atoms in exactly 12.0 g of C-12.

C. Each element has a characteristic atomic mass; the mass of a compound is the sum of the masses of its constituent atoms, called a formula mass. If the substance is a molecule, its mass is called its molecular mass.
   1. For example, S has an atomic mass of 32.0 u. A 32.0-g sample of S contains 6.02×10^{23} atoms of S.
   2. The formula, or molecular mass, of NO is 14.0 u + 16.0 u = 30.0 u. A 30.0-g sample of NO contains 6.02×10^{23} NO molecules.

D. We can calculate the percent of a formula (molecular) mass that is attributed to each element in the compound by comparing the contribution of that element to the overall formula mass and then multiplying by 100.
   1. This is called the mass percent composition.
   2. For example, the mass percent of nitrogen in NO is (14.0 u N/30.0 u NO) × 100% = 46.7%; the mass percent of O in NO is (16.0 u O/30.0 u NO) × 100% = 53.3%. Note that because NO contains only N and O, the sum of the two mass percentages is 100% (46.7% + 53.3% = 100%).

Answers to Self-Assessment Questions

1. d 16 g of O–16 is one mole of O–16 and contains Avogadro's number (6.02×10^{23} atoms/mole) of O–16 atoms..
2. a By definition, the mass of 1 mole of C–12 atoms is exactly 12 g.
3. a The molecular masses are CH_4 = 16.0 u, HCl = 36.5 u, H_2O = 18.0 u, and PH_3 = 34.0 u. The molecular mass of CH_4 is the smallest.
4. b According to the chemical formula, there are 3 mol S atoms/1 mol $Cr(SO_4)_3$.
5. b number of H atoms = (2 mol H_2O)(2 mol H atoms/1 mol H_2O) = 4 mol H atoms.
6. a The formula mass of NaCl is 58.5 u (23.0 u + 35.5 u = 58.5 u).
   The mass percent of Na in NaCl is (23.0 u/58.5 u)(100%) = 39.3%.
   The mass percent of Cl in NaCl is (35.5 u/58.5 u)(100%) = 60.7%.

5.4 Molar Mass: Mole-to-Mass and Mass-to-Mole Conversions

Learning Objectives: • Convert from mass to moles and from moles to mass of a substance. • Calculate the mass or number of moles of a reactant or product from the mass or number of moles of another reactant or product.

A. Mole (abbreviation: mol) is a term that refers to 6.02×10^{23} (to three significant figures; 6.022×10^{23} to four significant figures) things. It is the amount of substance containing as many elementary units as there are atoms in exactly 12 g of the carbon-12 isotope.
   1. One mole of Mg atoms = 6.02×10^{23} atoms.
   2. One mole of H_2O molecules = 6.02×10^{23} molecules.

B. The molar mass is equal to the mass of 1 mol of a substance, in grams.
   1. The molecular mass of NO is 30.0 u; the molar mass of NO is 30.0 g.

C. Calculations: grams to moles and moles to grams.
   1. The key to solving these problems is to determine a conversion ratio between the numbers of moles and grams of a substance.
   2. Use the periodic table to calculate the molar mass of a substance, which is the number of grams of the substance in 1 mole of the substance.
      a. This ratio can then be used in either type of conversion problem (moles to grams or grams to moles).

D. Mole and Mass Relationships in Chemical Equations.
   1. Molar masses of substances give the mole-to-gram ratio of a substance.
   2. Coefficients in balanced equations provide mole ratios of the substances in the equation.
   3. Stoichiometry: the quantitative or mass relationships in chemical equations.
      a. Typically, the amount of a substance in a problem is given in grams rather than moles, but when you are asked for grams:
         i. Write a balanced equation.
         ii. Use the periodic table to determine the molar masses of substances involved in the problem.
         iii. Convert the quantity given to the equivalent number of moles by using its molar mass.
         iv. Use the mole ratio from the coefficients in the balanced equation to convert from the number of moles of the substance given to the number of moles of the substance asked for in the problem.

Answers to Self-Assessment Questions

1. b mass of H_2SO_4 = (4.65 mol H_2SO_4)(98.1 g H_2SO_4/1 mol H_2SO_4) = 456 g H_2SO_4.
2. b For each element, divide 10.0 g by its molar mass. F has the lowest molar mass, resulting in the largest number of moles.
   number of moles of F = (10.0 g F)(1 mol F/19.0 g F) = 0.526 mol F.
3. c number of moles of CO_2 = (453.6 g CO_2)(1 mol CO_2/44.0 g CO_2) = 10.3 mol CO_2.
4. a number of moles of C_6H_6 = (7.81 g C_6H_6)(1 mol C_6H_6/78.1 g C_6H_6) = 0.100 mol C_6H_6.
5. c The balanced equation is $2\ H_2 + O_2 \rightarrow 2\ H_2O$. Use the coefficients in the equation to interpret the equation to read 2 mol of H_2 reacts with 1 mol O_2 to form 2 mol of H_2O.
6. a The balanced equation is $2\ H_2 + O_2 \rightarrow 2\ H_2O$.
   number of moles of H_2O = (0.500 mol H_2)(2 mol H_2O/2 mol H_2) = 0.500 mol H_2O.
7. a The balanced equation is $2\ H_2 + O_2 \rightarrow 2\ H_2O$.
   number of moles of O_2 = (0.222 mol H_2O)(1 mol O_2/2 mol H_2O) = 0.111 mol O_2.

5.5 Solutions

Learning Objectives: Calculate the concentration (molarity, percent by volume, or percent by mass) of a solution in a solution. Calculate the amount of solute or solution given the concentration and the other amount.

A. Solutions are homogeneous mixtures of two or more substances. The substance being dissolved is the solute, and the substance doing the dissolving is the solvent. When the solvent is water, the solutions are called aqueous.
B. Solution concentrations.
   1. Molarity is defined as the number of solute moles divided by the volume of the solution measured in liters. Molarity is abbreviated as M and has the units of moles/liter (mol/L). For example:
      a. The concentration of a solution in which 15 mol of salt are dissolved in a volume of 5 L is 3 M. (M = 15 mol/5 L = 3 M).
      b. To find the molarity of 54.5 g of KBr dissolved in 3.00 L of solution, first convert the grams of KBr to mol of KBr, then divide by the volume, in liters.
         1 mol of KBr = 119.0 g KBr
         number of mol of KBr = (54.5 g KBr)(1 mol KBr/119.0 g KBr) = 0.458 mol KBr
         0.458 mol of KBr/3.00 L = 0.153 mol/L or 0.153 M

c. To find the number of grams in a given volume of solution, calculate the number of moles from molarity: M = mol/L so mol = (M)(volume, L) and convert to grams. For example, to calculate the mass of LiCl dissolved in 300 mL (0.300 L) of an 0.400 M solution, start by calculating the number of mol of LiCl:

number of mol of LiCl = (0.400 M LiCl)(0.300 L) = 0.120 mol LiCl
molar mass of LiCl = 6.9 + 34.5 = 42.4 g/mol
(0.120 mol LiCl)(42.4 g LiCl/1 mol LiCl) = 5.09 g LiCl

2. Percent concentrations.
   a. Percent is a fraction (or part/whole) times 100%.
   b. Percent by volume is the ratio of the volume of solute to the volume of solution, times 100. Note that the volume of the solution is the sum of the volumes of all its components.
   c. Percent by mass is the mass of a component divided by the mass of the solution, times 100. Note that the mass of the solution is the sum of the mass of the solute(s) and the mass of the solvent.
   d. What is the mass percent $MgCl_2$ in a solution that contains 15 g of $MgCl_2$ in 250 g of solution?

   % $MgCl_2$ = (15 g $MgCl_2$/250 g solution)(100%) = 6.0%

Answers to Self-Assessment Questions

1. b Sucrose, the solute, is dissolved in water, the solvent.
2. d molarity = 0.500 mol NaOH/0.0625 L solution = 8.00 M
3. a number of moles of sugar = (2.00 L)(0.600 mol sugar/L) = 1.20 mol sugar
4. d 1 mol of NaCl = 58.5 g
   number of moles of NaCl = (2.500 L)(0.800 mol NaCl/L) = 2.00 mol NaCl
   mass of NaCl = (2.00 mol NaCl)(58.5g NaCl/mol NaCl) = 117 g NaCl
5. b M = mol/L so L = mol/M
   volume of HCl solution = 1.80 mol HCl/(6.00 mol HCl/L solution)
   volume = 0.300 L or 300 mL
6. b % = (part/whole) × 100%
   volume percent ethanol = (50.0 mL ethanol/250 mL solution)(100%) = 20%
7. b % = (part/whole) × 100%
   mass percent sucrose = (15.0 g sucrose)/(15.0 g sucrose + 60.0 g water)(100%) = 20.0%
8. d The molar mass of NaCl is 58.6 g/mol. First calculate the number of moles of NaCl represented by 23.4 g NaCl:
   number of mol = (23.4 g NaCl)(1 mol NaCl/58.6 g NaCl) = 0.400 mol
   The volume of a 4.00 M NaCl solution containing 0.400 mol NaCl is:
   volume, L = (0.400 mol/4.00 mol/L) = 0.100 L or 100 mL.
   The final solution needs to have a volume of 100 mL.

Green Chemistry

A. Atom Economy (AE)
1. Atom economy is a calculation of the number of atoms that are conserved in the desired product rather than in waste; a measure of reaction efficiency.
2. The traditional way of describing the efficiency of a chemical reaction is to determine the ratio of the amount of product actually collected from a reaction to the amount of product that could, theoretically, have been produced, times 100. In the somewhat rare instances when reaction yields are extremely high, it is still possible that, along with the desired product, there are many by-products that end up being discarded.

3 The green chemical approach is to minimize waste so that the emphasis is to design reaction schemes in which the largest number of reactant atoms is incorporated into the desired reaction product.
4. Therefore, rather than calculate percent yield, a better way to measure reaction efficiency is to calculate percent atom economy, % AE.

% AE = [(molar mass of desired product/sum of the molar masses of all reactants)](100%)

LEARNING OBJECTIVES

You should be able to . . .

1. Identify balanced and unbalanced chemical equations, and balance equations by inspection. (5.1)

2. Determine volumes of gases that react, using a balanced equation for a reaction. (5.2)

3. Calculate the formula mass, molecular mass, or molar mass of a substance. (5.3)

4. Determine the percentage composition of a substance from its formula. (5.3)

5. Use Avogadro's number to determine the number of particles of different types in a mass of a substance. (5.3)

6. Convert from mass to moles and from moles to mass of a substance. (5.4)

7. Calculate the mass or number of moles of a reactant or product from the mass or number of moles of another reactant or product. (5.4)

8. Calculate the concentration (molarity, percent by volume, or percent by mass) of a solute in a solution. (5.5)

9. Calculate the amount of solute or solution given the concentration and the other amount. (5.5)

10. Explain how the concept of atom economy can be applied to pollution prevention and environmental protection.

11. Calculate the atom economy for chemical reactions.

DISCUSSION

Chapter 5 introduces the language of chemistry. If, instead of chemistry, English literature were our area of study, we would just be at the point of having learned to read. In earlier chapters, we learned to use chemical symbols to represent elements. The symbols are the alphabet of chemistry. In this chapter, we learn to use the "words" of chemistry—the formulas for compounds, to compose chemical "sentences"—that is, to write chemical equations. This chapter also introduces the math of chemistry.

You may never have realized how much information is contained in a chemical equation. Much of the chapter is devoted to a discussion of the concepts and terminology needed to extract every last bit of information from an equation. Therefore, one of the first things you should do is learn the key terms.

Molar mass and mole are interrelated terms. The molar mass of a compound (calculated the same way as formula mass but with different units) is the mass of 1 mol of the compound expressed in grams.

$$1 \text{ formula mass} = 1 \text{ mole}$$

You can treat this relationship as a conversion factor. If you calculate the formula mass of a compound and remember the relationship, you can interconvert units expressed in grams to moles and vice versa. This is enormously important because it allows us to convert back and forth from counting units (moles) and weighing units (grams). This is important because chemical equations are written in counting units (1 of these, 2 of those) but because atoms and molecules are so small, we can't measure them out by counting; the only way we can measure them out is by weighing.

Example: Conversions involving methane (CH_4).

Formula weight: atomic weight of C = 12.01 = 12.0. Atomic weight of H = 1.008, so 4 H atoms weigh (4)(1.008) = 4.03.The total is the molar mass of methane or 16.0 g. Now that we have the molar mass, we have our conversion factor: 16.0 g CH_4 = 1 mol CH_4.

How many moles of CH_4 are represented by 4.0 g of CH_4?

Number of mol of CH_4 = (4.0 g CH_4)(1 mol CH_4/16.0 g CH_4) = 0.25 mol CH_4

What is the mass of 4.0 moles of CH_4?

Mass = (4.0 mol CH_4)(16.0 g CH_4/mol CH_4) = 64 g CH_4

There are many ways of expressing concentrations, including molarity, mass percent composition, and volume percent composition. These various concentration units are interconvertible.

Realizing the environmental concerns associated with disposal of wastes, chemical and otherwise, the green chemistry approach to help solve this problem is to revise the standard synthesis routes to maximize the atom economy of the procedure. This means that synthesis routes are chosen in which the maximum number of reactant atoms is incorporated into the desired product, not into by-products that, ultimately, will be discarded. Reaction sequences are evaluated in terms of their percent atom economy rather than in terms of percent yield.

ADDITIONAL PROBLEMS

1. Calculate the formula masses of the following compounds. The formulas are relatively complicated just to make sure that you understand when a subscript applies to a particular atom in the formula and when it does not. You'll require a periodic table or a list of atomic masses.

a. $CaCO_3$ b. $(NH_4)_2CO_3$ c. $Be(NO_3)_2$ d. $(NH_4)_2C_2O_4$

e. $Al_2(C_2O_4)_3$ f. $Ca(C_2H_3O_2)_2$ g. $(CH_3)_2SO_4$

2. In this problem, all questions refer to the compound $C_5H_8O_2$.

   a. How many mol of $C_5H_8O_2$ are represented by 100 g of this compound? By 200 g of this compound? By 25.0 g of this compound? By 3.687 g of this compound?

   b. What is the mass of $C_5H_8O_2$, in grams, of 1 mol of this compound? Of 8 mol of this compound? Of 0.8 mol of this compound? Of 0.01 mol of this compound?

3. For additional practice in interconverting these units, answer the questions in Problem 2 for the compound H_2CO_3.

   Let's suppose you are now totally at ease with moles and formula weights and conversions. That brings us to equations. First, determine if the equation is balanced. (We'll supply the correct reactants and products.) If the equation isn't balanced, the quantitative information derived from it will be incorrect. As we indicated in the chapter, you won't be balancing extremely complex equations, but you should be able to handle those in Problem 4.

4. Balance the following chemical equations.

   a. $Zn + KOH \rightarrow K_2ZnO_2 + H_2$
   b. $HF + Si \rightarrow SiF_4 + H_2$
   c. $B_2O_3 + H_2O \rightarrow H_6B_4O_9$
   d. $SiCl_4 + H_2O \rightarrow SiO_2 + HCl$
   e. $SnO_2 + C \rightarrow Sn + CO$
   f. $Fe_2O_3 + CO \rightarrow FeO + CO_2$
   g. $Fe_3O_4 + C \rightarrow Fe + CO$
   h. $Fe(OH)_3 + H_2S \rightarrow Fe_2S_3 + H_2O$

Once you have a balanced equation, the coefficients in that equation give you the following information:

a. The combining ratio of molecules or other formula units.
b. The combining ratio of moles of molecules or other formula units.

The coefficients *do not* give you the combining mass ratios. Thus, from the following equation

$$CH_4 + 2\,O_2 \rightarrow CO_2 + 2\,H_2O$$

you know that:

a. 1 mol of methane (CH_4) reacts with 2 molecules of oxygen (O_2) to produce 1 molecule of carbon dioxide (CO_2) and 2 molecules of water (H_2O).
b. 1 mol of methane reacts with 2 mol of oxygen to give 1 mol of carbon dioxide and 2 mol of water.

The equation does *NOT* say that 1 g of methane reacts with 2 g of oxygen to produce 1 g of carbon dioxide and 2 g of water. If you want to find out how many grams of oxygen react with 1 g of methane, you must first convert mass to the equivalent number of moles and only then use the balanced chemical equation to determine the combining ratio in moles. After using the equation, you'll have the answer in moles and must then convert to grams.

The examples in the text demonstrate the use of equations to obtain information about combining ratios. Review those examples and then the problems at the end of the chapter. For more practice, try the following problems.

5. Refer to the equation

$$CS_2 + 2\ CaO \rightarrow CO_2 + 2\ CaS$$

a. How many moles of CO_2 are obtained from the reaction of 2 mol of CS_2? From the reaction of 2 mol of CaO?
b. How many moles of CaO are consumed if 0.3 mol of CS_2 react? If 0.3 mol of CaS are produced?
c. How many grams of CaS are obtained if 152 g of CS_2 are consumed in the reaction? If 7.6 g of CS_2 are consumed? If 22 g of CO_2 are produced? If 44 g of CO_2 are produced?
d. How many grams of CaO are required to react completely with 38 g of CS_2? With 152 g of CS_2? To produce 36 g of CaS?

ANSWERS TO ADDITIONAL PROBLEMS

1. a. $CaCO_3$ molar mass = 40.1 + 12.0 + (3)(16.0) = 100.1 g/mol

b. $(NH_4)_2CO_3$ molar mass = (2)(14.0) + (8)(1.0) + 12.0 + (3)(16.0) = 96.0 g/mol

c. $Be(NO_3)_2$ molar mass = 9.0 + (2)(14.0) + (6)(16.0) = 133.0 g/mol

d. $(NH_4)_2C_2O_4$ molar mass = (2)(14.0) + (8)(1.0) + (2)(12.0) + (4)(16.0) = 124.0 g/mol

e. $Al_2(C_2O_4)_3$ molar mass = (2)(27.0) + (6)(12.0) + (12)(16.0) = 318.0 g/mol

f. $Ca(C_2H_3O_2)_2$ molar mass = 40.1 + (4)(12.0) + (6)(1.0) + (4)(16.0) = 158.1 g/mol

g. $(CH_3)_2SO_4$ molar mass = (2)(12.0) + (6)(1.0) + 32.1 + (4)(16.0) = 126.1 g/mol

2. The molar mass of $C_5H_8O_2$ = (5)(12.0) + (8)(1.0) + (2)(16.0) = 100.0 g/mol

a. number of mol of $C_5H_8O_2$ = (100 g $C_5H_8O_2$)(1 mol $C_5H_8O_2$/100.0 g $C_5H_8O_2$) = 1.00 mol

number of mol of $C_5H_8O_2$ = (200 g $C_5H_8O_2$)(1 mol $C_5H_8O_2$/100.0 g $C_5H_8O_2$) = 2.00 mol

number of mol of $C_5H_8O_2$ = (25.0 g $C_5H_8O_2$)(1 mol $C_5H_8O_2$/100.0 g $C_5H_8O_2$) = 0.250 mol

number of mol of $C_5H_8O_2$ = (3.687 g $C_5H_8O_2$)(1 mol $C_5H_8O_2$/100.0 g $C_5H_8O_2$) = 0.03687 mol

b. mass of $C_5H_8O_2$ = (1 mol $C_5H_8O_2$)(100.0 g $C_5H_8O_2$/mol $C_5H_8O_2$) = 100 g $C_5H_8O_2$

mass of $C_5H_8O_2$ = (8 mol $C_5H_8O_2$)(100.0 g $C_5H_8O_2$/mol $C_5H_8O_2$) = 800 g $C_5H_8O_2$

mass of $C_5H_8O_2$ = (0.8 mol $C_5H_8O_2$)(100.0 g $C_5H_8O_2$/mol $C_5H_8O_2$) = 80 g $C_5H_8O_2$

mass of $C_5H_8O_2$ = (0.01 mol $C_5H_8O_2$)(100.0 g $C_5H_8O_2$)(mol $C_5H_8O_2$) = 1.0 g $C_5H_8O_2$

3. The molar mass of H_2CO_3 = (2)(1.0) + 12.0 + (3)(16.0) = 62.0 g/mol

a. number of mol of H_2CO_3 = (100 g H_2CO_3)(1 mol H_2CO_3/62.0 g H_2CO_3) = 1.61 mol

number of mol of H_2CO_3 = (200 g H_2CO_3)(1 mol H_2CO_3/62.0 g H_2CO_3) = 3.23 mol

number of mol of H_2CO_3 = (25.0 g H_2CO_3)(1 mol H_2CO_3/62.0 g H_2CO_3) = 0.403 mol

number of mol of H_2CO_3 = (3.687 g H_2CO_3)(1 mol H_2CO_3/62.0 g H_2CO_3) = 0.0595 mol

b. mass of H_2CO_3 = (1 mol H_2CO_3)(62.0 g H_2CO_3/mol H_2CO_3) = 62.0 g H_2CO_3

mass of H_2CO_3 = (8 mol H_2CO_3)(62.0 g H_2CO_3/mol H_2CO_3) = 500 g H_2CO_3

mass of H_2CO_3 = (0.8 mol H_2CO_3)(62.0 g H_2CO_3/mol H_2CO_3) = 50 g H_2CO_3

mass of H_2CO_3 = (0.01 mol H_2CO_3)(62.0 g H_2CO_3)(mol H_2CO_3) = 6.2 g H_2CO_3

4. a. $Zn + 2\ KOH \rightarrow K_2ZnO_2 + H_2$
   b. $4\ HF + Si \rightarrow SiF_4 + 2\ H_2$
   c. $2\ B_2O_3 + 3\ H_2O \rightarrow H_6B_4O_9$
   d. $SiCl_4 + 2\ H_2O \rightarrow SiO_2 + 4\ HCl$
   e. $SnO_2 + 2\ C \rightarrow Sn + 2\ CO$
   f. $Fe_2O_3 + CO \rightarrow 2\ FeO + CO_2$
   g. $Fe_3O_4 + 4\ C \rightarrow 3\ Fe + 4\ CO$
   h. $2\ Fe(OH)_3 + 3\ H_2S \rightarrow Fe_2S_3 + 6\ H_2O$

5. $CS_2 + 2\ CaO \rightarrow CO_2 + 2\ CaS$; molar masses: ($CS_2$: 76.2 g/mol; CaO: 56.1 g/mol; CO_2: 44.0 g/mol; CaS: 72.2 g/mol

   a. number of mol of CO_2 = (2 mol CS_2)(1 mol CO_2/1 mol CS_2) = 2 mol CO_2

      number of mol of CO_2 = (2 mol CaO)(1 mol CO_2/2 mol CaO) = 0.5 mol CO_2

   b. number of mol of CaO = (0.3 mol CS_2)(2 mol CaO/1 mol CS_2) = 0.6 mol CaO

      number of mol of CaO = (0.3 mol CaS)(2 mol CaO/2 mol CaS) = 0.3 mol CaO

   c. mass of CaS = (152 g CS_2)(1 mol CS_2/76.2 g CS_2)(2 mol CaS/1 mol CS_2)
      (72.2 g CaS/1 mol CaS) = 288 g CaS

      mass of CaS = (7.6 g CS_2)(1 mol CS_2/76.2 g CS_2)(2 mol CaS/1 mol CS_2)
      (72.2 g CaS/1 mol CaS) = 14 g CaS

      mass of CaS = (22 g CO_2)(1 mol CO_2/44.0 g CO_2)(2 mol CaS/1 mol CO_2)
      (72.2 g CaS/1 mol CaS) = 72 g CaS

      mass of CaS = (44 g CO_2)(1 mol CO_2/44.0 g CO_2)(2 mol CaS/1 mol CO_2)
      (72.2 g CaS/1 mol CaS) = 140 g CaS

   d. mass of CaO = (38 g CS_2)(1 mol CS_2/76.2 g CS_2)(2 mol CaO/1 mol CS_2)
      (56.1 g CaO/1 mol CaO) = 56 g CaO

      mass of CaO = (152 g CS_2)(1 mol CS_2/76.2 g CS_2)(2 mol CaO/1 mol CS_2)
      (56.1 g CaO/1 mol CaO) = 224 g CaO

      mass of CaO = (36 g CaS)(1 mol CaS/72.2 g CaS)(2 mol CaO/2 mol CaS)
      (56.1 g CaO/1 mol CaO) = 28 g CaO

ANSWERS TO ODD-NUMBERED QUESTIONS AND SOLUTIONS FOR ODD-NUMBERED END-OF-CHAPTER PROBLEMS

Review Questions

1. a. A formula unit is the smallest unit of an ionic compound.
   b. Formula mass is the sum of the masses of the atoms represented in a formula unit.
   c. One mole is 6.02×10^{23} particles.
   d. Avogadro's number is the name we give to the number of ^{12}C atoms in exactly 12 g of ^{12}C, or 6.02×10^{23}. It is also the number of particles in one mole of those particles.

e. Molar mass is the mass of one mole of a substance, expressed in grams.
f. The molar volume of a gas is the volume occupied by Avogadro's number of atoms or molecules of a gas at a specified temperature and pressure.

3. According to Avogadro's hypothesis, equal volumes of all gases, measured at the same pressure and temperature, contain the same number of molecules. Gay-Lussac's law of combining volumes says that when all measurements are made at the same temperature and pressure, the volumes of gaseous reactants and products are in small, whole-number ratios. Because there is a relationship between the volume of a gas and the number of particles it contains, it follows that the volumes of gases would be in the same, whole-number ratios as the coefficients in the balanced chemical equation.

5. a. A solution is a homogeneous mixture of two or more substances.
   b. The solvent is the substance present in the greatest quantity in a solution.
   c. A solute is a substance dissolved in a solvent to form a solution.
   d. An aqueous solution is a solution in which water is the solvent.

Problems

7. a. There are 12 oxygen atoms in $Al(H_2PO_4)_3$.
   b. There are 3 oxygen atoms in $CH_3C(O)CH_2COOCH_3$.
   c. There are 6 oxygen atoms in $(BiO)_2SO_4$.

9. One formula unit of $(NH_4)_2HPO_4$ contains 2 N atoms, 1 P atom, 9 H atoms, and 4 O atoms, so 3 formula units would contain three times as many of each kind of atom or has 6 N, 3 P, 27 H, and 12 O atoms.

11. For the reaction: $2\ H_2O_2 \rightarrow 2\ H_2O + O_2$.
    a. At the molecular level: 2 molecules of H_2O_2 produce 2 molecules of H_2O and one molecule of O_2.
    b. In terms of molar relationships, 2 mol of H_2O_2 produce 2 mol of H_2O and one mol of O_2.
    c. In terms of mass relationships, 68 g of H_2O_2 produce 36 g of H_2O and 32 g of O_2.

13. a. $2\ Mg + O_2 \rightarrow 2\ MgO$
    b. $C_3H_8 + 5\ O_2 \rightarrow 3\ CO_2 + 4\ H_2O$
    c. $3\ H_2 + Ta_2O_3 \rightarrow 2\ Ta + 3\ H_2O$

15. a. $N_2 + O_2 \rightarrow 2\ NO$
    b. $2\ O_3 \rightarrow 3\ O_2$
    c. $UO_2 + 4\ HF \rightarrow UF_4 + 2\ H_2O$

17. $2\ H_2(g) + O_2(g) \rightarrow 2\ H_2O(g)$ Sketch should show two volumes of hydrogen, one volume of oxygen reacting to produce two volumes of water.

19. Use the law of combining volumes and the balanced equation:
    $C_7H_{16}(g) + 11\ O_2(g) \rightarrow 7\ CO_2(g) + 8\ H_2O(g)$
    a. volume of CO_2, L = (20.6 L C_7H_{16})(7 L CO_2/1 L C_7H_{16}) = 144 L CO_2
    b. volume C_7H_{16}, mL = (29.0 mL O_2)(1 mL C_6H_{14}/11 mL O_2) = 2.64 mL

21. Use the law of combining volumes and the balanced equation:

$$C_7H_{16} + 11\ O_2 \rightarrow 7\ CO_2 + 8\ H_2O$$

7 mol of CO_2 are produced for every mole of C_7H_{16} burned. Therefore, the ratio of the volume of CO_2 produced to the volume of C_7H_{16} reacted is 7:1.

23. a. number of S_8 molecules = (1.00 mol S_8)(6.02×10^{23} molecules/mol S_8)
= 6.02×10^{23} molecules of S_8

b. number of S atoms =
(1.00 mol S_8)(6.02×10^{23} molecules S_8/mol S_8)(8 S atoms/molecule S_8) = 4.82×10^{24} S atoms

25. The answer is c. There are two Br atoms in one Br_2 molecule and 6.02×10^{23} Br_2 molecules in one mole of Br_2. Using the conversion factors:
number of Br atoms = (1 mol Br_2)(6.02×10^{23} Br_2 molecules/mol Br_2)(2 Br atoms/Br_2 molecule)
= 1.20×10^{24} Br atoms

27 a. The molar mass of $AgNO_3$ is 169.9 g/mol.
Ag (107.9 g/mol), N (14.0 g/mol), O (16.0 g/mol): 107.9 + 14.0 + (3)(16.0) = 169.9

b. The molar mass of $Zn(IO)_2$ is 351.2 g/mol.
Zn (65.4 g/mol), I (126.9 g/mol), O (16.0 g/mol): 65.4 + (2)(126.9) + (2)(16.0) = 351.2

c. The molar mass of $Sn(ClO_4)_2$ is 317.7 g/mol.
Sn (118.7 g/mol), Cl (35.5 g/mol), O (16.0 g/mol): 118.7 + (2)(35.5) + (8)(16.0) = 317.7

d. The molar mass of $CH_3(CH_2)_3COF$ is 104.0 g/mol.
C (12.0 g/mol), H (1.0 g/mol), O (16.0 g/mol), F (19.0 g/mol): (5)(12.0) + (9)(1.0) + 16.0 + 19.0 = 104.0 g/mol

29. a. The molar mass of $BaSO_4$ is 233.4 g/mol.
mass of $BaSO_4$ = (7.57 mol $BaSO_4$)(233.4 g $BaSO_4$/mol $BaSO_4$) = 1770 g

b. The molar mass of $CuCl_2$ is 134.5 g/mol.
mass of $CuCl_2$ = (4.72 mol $CuCl_2$)(134.5 g $CuCl_2$/mol $CuCl_2$) = 635 g.

c. The molar mass of $C_{12}H_{22}O_{11}$ is 342.0 g/mol
mass of $C_{12}H_{22}O_{11}$ = (0.250 mol $C_{12}H_{22}O_{11}$)(342.0 g $C_{12}H_{22}O_{11}$/mol $C_{12}H_{22}O_{11}$) = 85.5 g

31. a. The molar mass of Sb_2S_3 is 339.9 g/mol.
number of moles = (6.63 g Sb_2S_3)(1 mol Sb_2S_3/339.72 g Sb_2S_3) = 0.0195 mol

b. The molar mass of MoO_3 is 143.9 g/mol.
number of moles = (19.1 g MoO_3)(1 mol MoO_3/143.9 g MoO_3) = 0.133 mol.

c. The molar mass of $AlPO_4$ is 122.0 g/mol
number of moles = (43.4 g $AlPO_4$)(1 mol $AlPO_4$ /122.0 g $AlPO_4$) = 0.356 mol

33. a. The molar mass of $Ca(NO_3)_2$ is 164.1 g/mol, of which (2)(14.0) = 28.0 is contributed by the N atoms. The mass percent of N in $Ca(NO_3)_2$ is (28.0 g/164.1 g)(100%) = 17.1%.

b. The molar mass of NH_4Cl is 53.5 g/mol, of which 14.0 is contributed by the N atom. The mass percent of N in NH_4Cl is (14.0 g/53.5 g)(100%) = 26.2%.

35. Use the coefficients of the balanced equation: $2\ C_4H_{10} + 13\ O_2 \rightarrow 8\ CO_2 + 10\ H_2O$.

a. number of moles of CO_2 = (0.845 mol C_4H_{10})(8 mol CO_2/2 mol C_4H_{10}) = 3.38 mol CO_2
b. number of moles of O_2 = (2.54 mol C_4H_{10})(13 mol O_2/2 mol C_4H_{10}) = 16.5 mol O_2

37. The balanced chemical equation is $N_2 + 3\ H_2 \rightarrow 2\ NH_3$.
   a. mass of NH_3 = (250 g H_2)(1 mol H_2/2.0 g H_2)(2 mol NH_3/3 mol H_2)(17.0 g NH_3/mol NH_3)
      = 1400 g NH_3
   b. mass of H_2 = (923 g N_2)(1 mol N_2/28.0 g N_2)(3 mol H_2/1 mol N_2)(2.0 g H_2/1 mol H_2)
      = 200 g H_2

39. M = mol of solute/L of solution
   a. molarity = 23.4 mol/12.0 L = 1.95 M
   b. molarity = 0.875 mol/1.032 L = 0.848 M

41. M = mol of solute/L of solution; mol = (M)(L)
   a. number of mol of NaOH = (0.500 mol NaOH/L)(5.50 L) = 2.75 mol NaOH
      mass of NaOH = (2.75 mol NaOH)(40.0 g NaOH/mol NaOH) = 110 g NaOH
   b. number of mol of $C_6H_{12}O_6$ = (1.55 mol $C_6H_{12}O_6$/L)(0.065 L) = 0.101 mol $C_6H_{12}O_6$
      mass of $C_6H_{12}O_6$ = (0.101 mol $C_6H_{12}O_6$)(180.0 g $C_6H_{12}O_6$ /mol $C_6H_{12}O_6$)
      = 18.2 g $C_6H_{12}O_6$

43. M = mol of solute /L of solution; L = mol/M
   a. volume of solution, L = (2.00 mol NaOH)/1 L solution/6.00 mol NaOH) = 0.333 L
   b. The molar mass of KH_2AsO_4 is 180.0 g/mol.
      number of moles of KH_2AsO_4 = (8.10 g KH_2AsO_4)(1 mol KH_2AsO_4/180 g KH_2AsO_4)
      = 0.0450 mol KH_2AsO_4
      volume of solution = 0.0450 mol KH_2AsO_4/0.070 M = 0.64 L.

45. V% = $(V_{solute}/V_{solution}) \times 100$
   a. V% = [(68.0 mL water/650 mL solution)](100%) = 10.5%
   b. V% = [(35.5 mL methanol/755 mL solution)](100%) = 4.70%

47. 8.2% NaCl = 8.2 g NaCl/100 g solution
   mass of NaCl = (3750 g solution)(8.2 g NaCl/100 g solution) = 310 g NaCl
   To prepare the solution, dissolve 310 g of NaCl in 3440 g of water.

49. 2.00% acetic acid, by volume = 2.00 L acetic acid/100 L of solution
   volume of acetic acid = (2.50 L solution)(2.00 L acetic acid/100 L solution)
   = 0.0500 L acetic acid
   Transfer 1 L of water to a 2-L volumetric flask. Add 0.0400 L (40.0 mL) of acetic acid to the water. Then add enough water to bring the total volume of the solution to 2.00 L.

51. The first equation is correct. The second equation is not correct because it is not balanced. One of the criteria for a balanced equation in which some/all of the species are charged is that the sum of

the charges on each side of the equation must be the same. In the case of the second equation, the charge on the reactant side (left side) of the equation is +2, while the charge on the product side (right side) of the equation is 3+.

53. a. $Hg(NO_3)_2(s) \rightarrow Hg(l) + 2\ NO_2(g) + O_2(g)$
b. $Na_2CO_3(aq) + 2\ HCl(aq) \rightarrow H_2O(l) + CO_2(g) + 2\ NaCl(aq)$

55. a. Each mol of H_2 contains 2 mol of H atoms, so 1.00 mol of H_2 contains 2.00 mol of atoms. He is only composed of one atom, so 2.00 mol He is composed of 2.00 mol of He atoms. There are 4 mol of atoms/mol of C_2H_2, so 0.50 mol of C_2H_2 contains (0.50 mol)(4 mol atoms/mol C_2H_2) = 2.0 mol of atoms. Therefore, all three samples contain the same number of atoms.
b. The molar mass of H_2 is 2.0 g/mol, so 1.00 mol of H_2 has a mass of 2.0 g. The molar mass of He is 4.00 g/mol, so 2.00 mol of He has a mass of 8.00 g. The molar mass of C_2H_2 is 26.0 g/mol so 0.50 mol of C_2H_2 has a mass of 13.0 g. The C_2H_2 sample has the largest mass.

57. The balanced equation is $CaCO_3 \rightarrow CaO + CO_2$.
number of moles of $CaCO_3$ = $(2.5 \times 10^5$ g $CaCO_3)(1$ mol $CaCO_3/100.1$ g $CaCO_3)$
= 2.5×10^3 mol
number of moles of CaO = $(2.5 \times 10^3$ mol $CaCO_3)(1$ mol CaO/1 mol $CaCO_3)$
= 2.5×10^3 mol
mass of CaO = $(2.5 \times 10^3$ mol CaO)(56.1 g CaO/mol CaO) = 1.4×10^5 g CaO

59. mass of H_2O_2 solution in 1 bottle =
(16.0 oz solution)(29.6 mL/1.00 oz)(1.00 g solution/mL solution)
= 474 g solution
mass of H_2O_2 in 1 bottle of solution = (474 g solution)(3 g H_2O_2/100 g solution) = 14.2 g
number of mol of H_2O_2 in 1 bottle of solution = (14.2 g H_2O_2)(1 mol H_2O_2/34.0 g H_2O_2)
= 0.418 mol H_2O_2

61. V% = $(V_{solute}/V_{solution})(100\%)$
a. V% = (355 mL – 18.0 mL/355 mL solution)(100%) = 94.9% ethanol
b. V% = (0.00400 L acetone/1.55 L solution)(100%) = 0.258% acetone

63. a. number of moles of U = (1.50 ymol U)$(1 \times 10^{-24}$ mol/1 ymol) = 1.50×10^{-24} mol
mass of U = $(1.50 \times 10^{-24}$ mol U)(238.0 g U/mol U) = 3.57×10^{-22} g U
= $(3.57 \times 10^{-22}$ g U)(1 yg/1×10^{-24} g) = 357 yg

b. number of U atoms = (1.20 zmol U)$(1 \times 10^{-21}$ mol/zmol)$(6.02 \times 10^{23}$ U atoms/mol U)
= 722 U atoms

65. Volume of ethanol = (5.0 L blood)(1000 mL/L)(0.165 mL alcohol/100 mL blood) = 8.25 mL

67. According to Problem 25 in Chapter 1, the volume of water in all of Earth's oceans is 1.44×10^9 km^3. To evaluate the statement we need to convert the units of this volume to cm^3.
volume, cm^3 = $(1.44 \times 10^9$ km^3)(1000 m/km)^3(100 cm/m)^3 = 1.44×10^{24} cm^3

We can now calculate the number of cups of water present in all of Earth's oceans:
number of cups $= (1.44 \times 10^{24}\ cm^3)(1\ cup/236\ cm^3) = 6.10 \times 10^{21}$ cups
To calculate the number of molecules of water in 1 cup of water, we use the density of water, 1.0 g/1.0 mL, to convert from volume to mass, and then we can convert the mass of water to the number of moles of water, using the molar mass of water, 18.0 g/mol. Finally, we can use Avogadro's number to calculate the number of water molecules in one cup of water.
mass of water in 1 cup $= (236\ mL)(1.0\ g\ H_2O/mL\ H_2O) = 236\ g\ H_2O$
number of moles of water $= (236\ g\ H_2O)(1\ mol\ H_2O/18.0\ g\ H_2O) = 13.1\ mol\ H_2O$
number of H_2O molecules $= (13.1\ mol\ H_2O)(6.02 \times 10^{23}$ molecules/mol)
$= 7.89 \times 10^{24}$ molecules
The number of cups of water in all the oceans is 6.10×10^{21} and the number of H_2O molecules in a cup of water is $7{,}89 \times 10^{24}$, so the statement is valid.

69. The balanced chemical equation for the neutralization reaction is:
$$H_2SO_4 + 2\ NaHCO_3 \rightarrow Na_2SO_4 + 2\ H_2O + 2\ CO_2$$
mass of $H_2SO_4 = (50{,}000\ kg)(1000\ g/1\ kg) = 5.0 \times 10^7\ g\ H_2SO_4$
number of moles of H_2SO_4 spilled $= (5.0 \times 10^7\ g\ H_2SO_4)(1\ mol\ H_2SO_4/98.1\ g\ H_2SO_4)$
$= 5.10 \times 10^5\ mol\ H_2SO_4$
number of moles of $NaHCO_3 = (5.10 \times 10^5\ mol\ H_2SO_4)(2\ mol\ NaHCO_3/1\ mol\ H_2SO_4)$
$= 1.02 \times 10^6$ mol
mass of $NaHCO_3$ required $= (1.02 \times 10^6\ mol\ NaHCO_3)(84.0\ g\ NaHCO_3/mol\ NaHCO_3)$
$= 8.57 \times 10^7$ g or 8.57×10^4 kg

71. mass of water $= (2\ L)(1000\ mL/L)(1.00\ g/L) = 2000$ g
number of moles of $H_2O = (2000\ g\ H_2O)(1\ mol\ H_2O/\ 18.0\ g\ H_2O) = 111\ mol\ H_2O$

73. a. The balanced equations for each of the reactions are:
$$C_6H_{12}O_6 \rightarrow 2\ C_2H_6O + 2\ CO_2$$
$$C_2H_4 + H_2O \rightarrow C_2H_6O$$
b. For the first reaction, the molar masses of ethanol (C_2H_6O) and glucose ($C_6H_{12}O_6$) are 46.0 g/mol and 180.0 g/mol, respectively.
% AE $= [(2\ mol\ C_2H_6O)(46.0\ g\ C_2H_6O)/$
$(1\ mol\ C_6H_{12}O_6)(180.0\ g\ C_6H_{12}O_6/mol\ C_6H_{12}O_6)](100\%) = 51.1\%$
For the second reaction, the molar masses of ethylene (C_2H_4) and water are 28.0 g/mol and 18.0 g/mol, respectively.
% AE $= (1\ mol\ C_2H_6O)(46.0\ g\ C_2H_6O/mol\ C_2H_6O)/$
$[(1\ mol\ C_2H_4)(28.0\ g\ C_2H_4/mol\ C_2H_4) + (1\ mol\ H_2O)(18.0\ g\ H_2O/mol\ H_2O)](100\%)$
$= 100\%$
c. In the case of the first reaction, for every mole of ethanol produced, two moles of CO_2 are also produced and these are, in some respects, "waste." Therefore, the %AE for this reaction is, by definition, less than 100%. In the case of the second reaction, one mole of water is added to one mole of ethylene to form ethanol with no by-products. Every reactant atom is incorporated into the reaction product, making the %AE = 100%.
d. The first preparation process is sustainable because the reactant, glucose, is obtainable from renewable crops (corn, switch grass, etc.). The second reaction is not sustainable because the reactant, ethylene, is a compound derived from a nonrenewable resource, petroleum.

e. For the long term, in spite of the fact that it produces CO_2, a compound that contributes to climate change, the first reaction is the reaction of choice. Finding ways of disposing of or utilizing CO_2 will be a challenge if large-scale ethanol production continues, but the process has the advantage of not depleting our nonrenewable resource that is also used extensively for polymer and pharmaceutical production.

75. a. The balanced chemical equation is: $CuCl_2(aq) + Na_2CO_3(aq) \rightarrow \underline{CuCO_3}(s) + 2\ NaCl(aq)$.
molar mass $CuCO_3$ = 123.6 g/mol; molar mass $CuCl_2$ = 134.8 g/mol; molar mass Na_2CO_3 = 106.0 g/mol

percent atom economy = (molar mass of desired product/molar masses of all reactants)(100%)
= [123.6/(134.8 + 106.0)](100%) = 51.3%

b. The balanced chemical equation is: $4\ NH_3(g) + 5\ O_2(g) \rightarrow 4\ \underline{NO}(g) + 6\ H_2O(g)$.
molar mass NO = 30.0 g/mol; molar mass NH_3 = 17.0 g/mol; molar mass H_2O = 18.0 g/mol
percent atom economy = (molar mass of desired product/molar masses of all reactants)(100%)
Note that the coefficients in the balanced equation are taken into account as well. For this reaction, based on the NO produced,
mass of NO = (4 mol NO)(30.0 g NO/mol NO) = 120.0 g
The masses of all the reactants = (4 mol NH_3)(17.0 g NH_3/mol NH_3) +
(5 mol O_2)(32.0 g O_2/mol O_2) = 228.0 g
Percent atom economy for this reaction = (120 g/228 g)(100%) = 52.6%.

c. The balanced chemical equation is:
$6\ KMnO_4(aq) + 6\ KOH(aq) + KI(aq) \rightarrow 6\ K_2MnO_4(aq) + KIO_3(aq) + 3\ H_2O(l)$
molar mass K_2MnO_4 = 197.1 g/mol; molar mass $KMnO_4$ = 158.0 g/mol;
molar mass KOH = 56.1 g/mol; molar mass KI = 166.0 g/mol
percent atom economy = (molar mass of desired product/molar masses of all reactants)(100%)
Note that the coefficients in the balanced equation are taken into account as well. For this reaction, based on the K_2MnO_4 produced,
mass of K_2MnO_4 = (6 mol K_2MnO_4)(197.1 g K_2MnO_4/mol K_2MnO_4) = 1182.6 g K_2MnO_4
The masses of all the reactants = (6 mol $KMnO_4$)(158.0 g $KMnO_4$/mol $KMnO_4$) +
(6 mol KOH)(56.1 g KOH/mol KOH) + (1 mol KI)(166.0 g KI/mol KI) = 1450.6 g.
percent atom economy for this reaction = (1182.6 g/1450.6 g)(100%) = 81.5%

CHAPTER

6

Gases, Liquids, Solids . . . and Intermolecular Forces

CHAPTER SUMMARY

6.1 Solids, Liquids, and Gases

Learning Objective: • Explain how the different properties of solids, liquids, and gases are related to the motion and spacing of atoms, molecules, or ions.

A. In solids, particles are:
   1. Highly ordered.
   2. Close together, making them difficult to compress.
   3. Held together by ionic or intermolecular forces.

B. In liquids, particles are:
   1. More randomly arranged.
   2. Very close together, making them difficult to compress.
   3. Less tightly held by ionic or intermolecular forces so they can flow and conform to the container shape.

C. In gases, particles:
   1. Are moving rapidly in random directions.
   2. Are separated by great distances.
   3. Experience little or no attraction to one another.

D. The three physical states can be changed from one to the other.
   1. Melting: the process of a solid changing into a liquid.
   2 Melting point: temperature at which a solid becomes a liquid.
   3. Vaporization: the process of a liquid changing to a gas.
   4. Boiling point: temperature at which a liquid becomes a gas.
   5. Condensation: the process of a gas changing to a liquid.
   6. Sublimation: the process of a solid changing directly to a gas, bypassing the liquid state.
   7. Deposition: the reverse of sublimation; the process of a gas changing directly to a solid.

Answers to Self-Assessment Questions

1. d Molecules are farther from each other in the gas phase than in any other phase.
2. a Molecules in the liquid phase are difficult to compress because they are very close together but have no definite shape because their organization is random. They do have a definite volume, however.
3. d Vaporization occurs when the particles in the liquid phase gain enough energy to overcome attractions and enter the gas phase.
4. b Freezing occurs when the particles in a liquid lose energy and become more influenced by ionic or intermolecular forces, causing them to slow down so the molecules can establish fixed positions.
5. c Sublimation occurs when a solid changes directly into a gas without proceeding through a liquid phase.

6.2 Comparing Ionic and Molecular Substances

Learning Objective: • Identify some differences between ionic and molecular substances, and explain why these differences exist.

A. Compounds that experience strong intermolecular forces are either solids or liquids at room temperature and pressure. Under the same conditions, compounds without significant intermolecular forces are usually in the gas phase at room temperature.

B. The differences between ionic and molecular substances:
   1. Almost all ionic compounds are solids at room temperature. While some molecular substances are solid at room temperature (for example, sulfur, S_8, glucose, $C_6H_{12}O_6$, and iodine, I_2), many are liquids (for example ethanol, C_2H_5OH, and bromine, Br_2) and some are gases (for example, ethane, C_2H_6, chlorine, Cl_2, and hydrogen sulfide, H_2S).
   2. Ionic compounds generally have much higher melting points and boiling points than molecular compounds.
      a. The amount of energy required to overcome ionic interactions is much greater than the amount of energy required to overcome intermolecular interactions
   3. It generally takes between 10 and 100 times as much energy to melt one mole of an ionic solid than it takes to melt one mole of a molecular substance.
   4. Many ionic compounds dissolve in water and form solutions that conduct electricity because they dissociate into ions. Molecular compounds usually form solutions that do not conduct electricity.
   5. As solids, most ionic compounds are crystalline, hard, and often brittle, while solid molecular compounds are usually much softer.

Answers to Self-Assessment Questions

1. c NaBr is an ionic compound, not a molecule, so it does not exhibit inter*molecular* forces.
2. c KBr is an ionic compound and has the strongest forces (ionic interactions) holding its ions together. Therefore more energy is required to separate these ions from one another than is required to separate molecules from one another.

6.3 Forces between Molecules

Learning Objectives: • Classify forces between molecules as dipole-dipole forces, dispersion forces, or hydrogen bonds. • Explain the effects that intermolecular forces have on melting and boiling points.

A. Intramolecular versus Intermolecular Forces
   1. Intramolecular forces (forces *within* molecules) are the covalent bonds that hold the atoms in a molecule together. Intermolecular forces (forces *between* molecules) are the attractions one molecule feels for another.
   2. There are three fundamental types of intermolecular forces: dipole-dipole forces, dispersion forces, and hydrogen bonds.

B. Dipole-Dipole Forces.
   1. Unsymmetrical molecules containing polar bonds are dipoles with centers of partially negative and partially positive charges.
   2. Polar molecules attract one another as the partially positive end of one molecule interacts with the partially negative end of another molecule.
   3. Dipole-dipole forces occur between any two polar molecules.
   4. Dipole-dipole forces are weaker than ionic bonds but stronger than forces between nonpolar molecules of comparable size.

C. Dispersion Forces.
   1. Nonpolar compounds experience attractive intermolecular forces owing to momentary induced dipoles arising from the motions of electrons around the nuclei of atoms in the compound.
   2. These transient attractive forces, called dispersion forces, are fairly weak but are present in all molecules and increase as the size and number of electrons in the molecule increases.
   3. Dispersion forces can be substantial between large molecules, such as those in polymers.
   4. Although, individually, they are much weaker than dipole-dipole or ionic forces, dispersion forces exist between any two particles, whether polar, nonpolar, or ionic.

D. Hydrogen Bonds.
   1. Compounds containing H attached to the small electronegative elements N, O, or F exhibit stronger intermolecular attractive forces than would be expected on the basis of dipole-dipole forces alone. These forces are called hydrogen bonds.
   2. A hydrogen bond is much weaker than a covalent bond. The hydrogen bond is an interaction of the partially positive hydrogen of the donor molecule with the lone pair of nonbonding electrons on the F, O, or N of the acceptor molecule.
      a. Hydrogen bonds are usually represented by dotted lines.
      b. Hydrogen bonds are extremely important in biological molecules, including in establishing the three-dimensional structure of proteins, and the arrangement of the double strands of DNA.
      c. The relative strength of hydrogen bonds accounts for the unexpectedly high boiling points of HF, NH_3, and H_2O.

Answers to Self-Assessment Questions

1. c Hydrogen bonding causes water to be held together much more strongly than is the case for other molecules of similar molar mass.
2. a Covalent bonds, which are involved in joining the atoms in a molecular structure, are much stronger than the forces that attract one molecule to another.
3. d Ethanol has an –OH group which can be involved in hydrogen bonding, as well as carbon-hydrogen bonds which can be involved in dispersion forces. The C–O bond is polar, which allows for some dipole-dipole interaction between CH_3CH_2OH molecules.

4. c Water molecules are attracted to other water molecules by intermolecular forces called hydrogen bonds.
5. c Hydrogen bonding occurs between molecules that include a hydrogen atom bonded to O, N, or F.
6. b Chloroform is a polar molecule, but its structure does not include a hydrogen atom bonded to O, N, or F.

6.4 Forces in Solutions

Learning Objective: • Explain why nonpolar solutes tend to dissolve in nonpolar solvents and polar and ionic solutes tend to dissolve in polar solvents.

A. Solutions.
   1. A solution is a homogeneous mixture of two or more substances.
   2. The solute is the substance being dissolved (the minor component).
   3. The solvent is the substance doing the dissolving (the major component).

B. "Like dissolves like."
   1. Nonpolar substances dissolve best in nonpolar solvents; polar substances dissolve best in polar solvents.
   2. Salts dissolve in water because the ion-dipole forces overcome the ion-ion attractions.

Answers to Self-Assessment Questions

1. b Solutions are uniform throughout or are homogeneous.
2. a The positively and negatively charged ions in ionic compounds are strongly attracted to each other. In order to dissolve in a solvent, the solvent-solute interactions must be as strong as the interactions between solute particles. The only interactions that exist in nonpolar solvents are weak dispersion forces.
3. b Both I_2 and C_6H_{14} are nonpolar covalent substances that are attracted to one another by dispersion forces.
4. b $CaCl_2$ is an ionic compound and water is polar covalent, giving rise to ion-dipole interactions.
5. c Both acetic acid and water can form hydrogen bonds.
6. a Octane is a nonpolar covalent substance and will dissolve nonpolar covalent substances such as $CH_3(CH_2)_4CH_3$.

6.5 Gases: The Kinetic-Molecular Theory

Learning Objective: • List the five basic concepts of the kinetic-molecular theory of gases.

A. Kinetic-Molecular Theory.
   1. Gas particles are in rapid constant motion and move in straight lines.
   2. Particles of a gas are small compared with the distances between them.
   3. Because the particles of a gas are so far apart, there is very little attraction between them.
   4. Particles of a gas collide with one another. Energy is conserved in these collisions; energy lost by one particle is gained by the other.
   5. Temperature is a measure of the average kinetic energy (energy of motion) of the gas particles.

Answers to Self-Assessment Questions

1. b The kinetic molecular theory does not address the pressure and volume relationships of a gas.
2. c The gas particles are very much smaller than the volume occupied by the gas.
3. d Energy is conserved when gas particles collide—the energy lost by one molecule is gained by the other.

6.6 The Simple Gas Laws

Learning Objectives: • State the four simple gas laws, by name and mathematically. • Use any gas law to find the value of one variable if the other values are given.

A. The condition of a gas sample can be described by four variables: pressure (P), temperature (T), volume (V), and the number of moles of sample present (n). The simple gas laws describe the way one of these variables changes with a change in a second variable, assuming that the third and fourth variables are kept constant. Note: In all cases, when temperature is used in a gas law, it must be expressed as absolute temperature or in Kelvins (T (K) = T (°C) + 273).
B. Boyle's law: For a given amount of gas at constant temperature (n and T are constant), the volume of the gas (V) varies inversely with its pressure (P): ($P_1V_1 = P_2V_2$).
C. Charles's law: The volume (V) of a fixed amount of a gas at constant pressure (n and P are constant) is directly proportional to its absolute (Kelvin) temperature (T): ($V_1/T_1 = V_2/T_2$).
   1. When the volume of gas reaches zero, the temperature is at absolute zero
D. Gay-Lussac's law: The pressure (P) of a fixed amount of gas at a constant volume (n and V are constant) is directly proportional to its absolute temperature (T): ($P_1/T_1 = P_2/T_2$).
E. Avogadro's law: At a fixed temperature and pressure (T and P are constant) the volume of a gas (V) is directly proportional to the number of moles of gas in the sample (n). At what is known as standard temperature and pressure (STP, 0 °C and 1 atm) the molar volume of a gas is equal to 22.4 L.
   1. We can use the molar volume of a gas to calculate its density at STP if we know its molecular formula from which we can calculate its molar mass. The molar volume is the volume occupied by 1 mole of any gas if measured at STP.

Answers to Self-Assessment Questions

1. b Gas particles striking vessel walls cause pressure.
2. d Boyle's law states that the pressure of a gas is inversely proportional to its volume (the larger the sample volume, the lower will be its pressure).
3. b $P_1V_1 = P_2V_2$, $P_2 = (P_1V_1)/V_2 = (2.00 \text{ atm})(6.00 \text{ L})/(1.50 \text{ L}) = 8.00 \text{ atm}$.
4. d The Kelvin temperature scale must be used. To halve the volume, the Kelvin temperature must be halved. Note: One way to think this through is to substitute arbitrary numbers into Charles's law: $V_1/T_1 = V_2/T_2$ Assume, for example, that the volume of the first sample is 10 L at a temperature of 20 K and that of the second sample is half that, or 5 L. Now solve for T_2: $T_2 = (V_2)(T_1)/V_1 = (5 \text{ L})(20 \text{ K})/10 \text{ L} = 10 \text{ K}$. The temperature was halved.
5. b Both flasks have the same number of particles because they have the same pressure and temperature. NO has a smaller molar mass than NO_2, so flask A weighs less than flask B.
6. a 1 mol of a gas at STP occupies 22.4 L.
   number of mol of gas = (5.60 L)(1 mol gas/22.4 L) = 0.250 mol

6.7 The Ideal Gas Law

Learning Objective: • State the ideal gas law, and use it to calculate one of the quantities if the others are given.

When Boyle's, Charles's, Gay-Lussac's, and Avogadro's laws are combined, the equation PV = nRT is obtained in which *n* is the number of moles of gas in the sample and *R* is the universal gas constant 9a proportionality constant) with a value of 0.0821 L atm/mol·K. Note that when using this value of R, pressure must be expressed in atmospheres, volume in liters, and temperature in Kelvins. If any of these variables is given in different units, we must convert those values to L, atm, and K before using them in the expression.

Answers to Self-Assessment Questions

1. c *n* stands for the number of moles of gas in the sample.
2. c $(P_1V_1)/T_1 = (P_2V_2)/T_2$
   $V_2 = (P_1V_1T_2)/(T_1P_2)$
   $V_2 = [(1.00\ atm)(22.4\ L)(293\ K)]/[(273\ K)(1.00\ atm)] = 24.0\ L$
   Note: You could also solve this problem using the ideal gas law:
   PV = nRT so V = (nRT)/P.
   $V = [(1.00\ mol)(0.0821\ L{\cdot}atm/mol{\cdot}K)(273 + 20\ K)]/1\ atm = 24.1\ L$
3. b $P_1V_1/T_1 = P_2V_2/T_2$
   $V_2 = P_1V_1T_2/T_1P_2$
   $V_2 = (1.00\ atm)(22.4\ L)(546\ K)/(273\ K)(2.00\ atm) = 22.4\ L$
   Note: You could also solve this problem using the ideal gas law:
   PV = nRT so V = (nRT)/P.
   $V = [(1.00\ mol)(0.0821\ L{\cdot}atm/mol{\cdot}K)(546\ K)]/2.00\ atm = 22.4\ L$

Green Chemistry: Supercritical Fluids

A. Supercritical fluids
   1. The critical point for a substance is the highest temperature and pressure at which a substance can exist in the liquid and gas phases.
   2. Above the critical point, matter exists as neither a gas nor a liquid but as a hybrid called a supercritical fluid.
   3. Supercritical fluids have properties of both gases and liquids.
   4. Supercritical carbon dioxide ($scCO_2$) is generally regarded as environmentally safe (GRAS) and is particularly useful for many industrial processes, including decaffeinating coffee, isolating flavors and fragrances, and as the active agent in dry cleaning.

LEARNING OBJECTIVES

You should be able to . . .

1. Explain how the different properties of solids, liquids, and gases are related to the motion and spacing of atoms, molecules, or ions. (6.1)
2. Identify some differences between ionic and molecular substances, and explain why these differences exist. (6.2)
3. Classify forces between molecules as dipole-dipole forces, dispersion forces, or hydrogen bonds. (6.3)
4. Explain the effects that intermolecular forces have on melting and boiling points (6.3)
5. Explain why nonpolar solutes tend to dissolve in nonpolar solvents and polar and ionic solutes tend to dissolve in polar solvents. (6.4)
6. List the five basic concepts of the kinetic-molecular theory of gases. (6.5)
7. State the four simple gas laws, by name and mathematically. (6.6)
8. Use any gas law to find the value of one variable if the other values are given. (6.6)
9. State the ideal gas law, and use it to calculate one of the quantities if the others are given. (6.7)
10. Describe how green chemistry principle 3 must be considered in designing chemical reactions and processes.
11. Identify the properties that make supercritical fluids applicable in greener chemical processes.

DISCUSSION

Chapter 6 discusses the properties of the physical states: solids, liquids, and gases. These properties are explained by the type of ionic or intermolecular attractions between particles. These forces are determined by the type of bonds within the particle: Polar molecules give rise to dipole-dipole interactions; molecules, including one or more hydrogen atoms bonded to oxygen, nitrogen, or fluorine atoms, gives rise to hydrogen bonding; and nonpolar molecules give rise to dispersion interactions. The gas laws summarize data concerning volume, temperature, pressure, and the number of moles of gases.

Utilization of supercritical fluids to replace otherwise hazardous solvents illustrates the green chemical approach to using less hazardous materials and recycling the materials that are used.

Sample Gas Law Problems

Boyle's law, Charles's law, and the ideal gas law quantitatively describe the behavior of gases. Here are some worked examples:

1. The pressure of a 4.0 L sample of He gas is increased from 5.0 atm to 7.0 atm. What is the new volume?

   $P_1V_1 = P_2V_2$ so $V_2 = (P_1V_1)/P_2$
   $V_2 = (5.0 \text{ atm})(4.0 \text{ L})/7.0 \text{ atm} = 2.9 \text{ L}$

2. The temperature of a 4.0 L sample of He gas is increased from 10 °C to 100 °C. What is the new volume, assuming the pressure is held constant?

   First, convert the temperature to Kelvin: 10 °C = 283 K, 100 °C = 373 K
   Then list your variables: $T_1 = 283$ K; $T_2 = 373$; $V_1 = 4.0$ L; and V_2 is the unknown
   $V_1/T_1 = V_2/T_2$ so $V_2 = T_2V_1/T_1$
   $V_2 = (373 \text{ K})(4.0 \text{ L})/283 \text{ K} = 5.3 \text{ L}$

3. Use the ideal gas law to calculate the volume of 3.0 mol of He gas at 473 K and 2.0 atm.

   Always list your variables. This helps you to know which formula to use.
   V= unknown; P = 2 atm; n = 3 mol; T = 473 K, R = 0.0821 L·atm/mol·K
   PV = nRT so V = nRT/P
   $V = (3.0 \text{ mol})(0.0821 \text{ L} \cdot \text{atm/mol} \cdot \text{K})(473\text{K})/(2.0 \text{ atm}) = 58 \text{ L}$

ANSWERS TO ODD-NUMBERED REVIEW QUESTIONS AND SOLUTIONS FOR ODD-NUMBERED END-OF-CHAPTER PROBLEMS

Review Questions

1. Solids and liquids are both composed of particles (atoms, molecules, and/or ions). In both cases, the particles are tightly held together, compact, difficult to compress, and have a definite volume. Most solids are highly ordered with their particles in fixed positions, giving the solid a definite shape. In the liquid state, particles have more energy than is the case in the solid state. They are randomly arranged and freer to move, leading to indefinite shapes which are determined by the shape of the space they fill in their container.

3. The conversion from a solid to a liquid is called melting, and the conversion from a liquid to a gas is called vaporization. The conversion of a gas to a liquid is called condensation, and the conversion from a liquid to a solid is called freezing.

5. a. Ionic interactions are the forces that attract cations (positively charged ions) to anions (negatively charged ions) in ionic compounds. Sodium chloride, NaCl, is an example of an ionic compound, in which Na^+ is the cation and Cl^- is the anion.
   b. Dipole-dipole interactions are the attractions between nonsymmetrical polar covalent compounds. The attractions between molecules of formaldehyde, $H_2C{=}O$, are an example of dipole-dipole interactions.
   c. Hydrogen bonding is a special kind of dipole-dipole interaction that exists between compounds, with one or more H atoms bonded to N, O, or F atoms. Water (H_2O), ethyl alcohol (CH_3CH_2OH), and ammonia (NH_3) are substances that can form hydrogen bonds.

d. Dispersion forces are the strongest interactions that exist between nonpolar molecules. Although a dispersion force is weaker than a hydrogen bond, a dipole-dipole interaction, or an ionic interaction, when nonpolar molecules with high molar masses experience dispersion forces along their entire lengths, the additive effect of the forces can cause the substances to be solids at room temperature. Carbon tetrachloride, CCl_4, is an example of a compound that experiences dispersion forces.

7. In the combined gas law, the volume is inversely proportional to pressure. In the combined gas law, the volume is directly proportional to the absolute temperature.

Problems

9. NBr_3 has polar covalent bonds leading to dipole-dipole interactions. N_2 is a nonpolar covalent substance, and NaBr is an ionic compound.

11. N_2 is a nonpolar covalent substance, so the strongest intermolecular interactions it can experience are dispersion forces. Both NH_3 and NCl_3 are nonsymmetrical polar covalent compounds. The strongest interactions between NCl_3 molecules are dipole-dipole interactions because its structure does not include any H—F, H—O, or H—N bonds. In the case of NH_3, the strongest interactions are hydrogen bonds because the compound includes one or more (three, in this case) N—H bonds.

13. Based on the strength of their intermolecular interactions, KCl, an ionic compound, would have the highest melting point. Ammonia (NH_3) molecules form hydrogen bonds with one another, while the strongest interactions between C_2H_6 molecules are dispersion forces. Therefore, more energy is required to disrupt the molecules in solid NH_3 than to disrupt those in solid C_2H_6 so the melting point of NH_3 is higher than that of C_2H_6. Helium (He) is a monoatomic substance and the extent of dispersion forces between atoms is minimum, so the melting point He would be the lowest of the four substances.

15. Water is a polar solvent, while carbon tetrachloride, CCl_4, is a nonpolar solvent. We expect ionic or polar solutes to dissolve in polar solvents and nonpolar solutes to dissolve in nonpolar solvents. Both NaCl and Li_2CO_3 are ionic compounds, so we would expect each of them to dissolve in water and not in CCl_4. C_5H_{12} and C_8H_{18} are both nonpolar covalent compounds, so we would expect each of them to dissolve in CCl_4 and not in water.

17. Boyle's law: At constant temperature and sample size, $P_1V_1 = P_2V_2$ so $V_2 = P_1V_1/P_2$
$V_2 = (10.1\ m^3)(4.25\ atm)/(1.00\ atm) = 42.9\ m^3$

19. Boyle's law: At constant temperature and sample size, $P_1V_1 = P_2V_2$ so $P_2 = P_1V_1/V_2$
$P_2 = (1.10\ atm)(4.50\ L)/9.95\ L = 0.497\ atm$

21. Boyle's law: At constant temperature and sample size, $P_1V_1 = P_2V_2$ so $V_2 = P_1V_1/P_2$
$V_2 = (0.994\ atm)(1.88\ L)/0.497\ atm = 3.76\ L$

23. Charles's law: At constant pressure and sample size, $V_1/T_1 = V_2/T_2$ assuming temperature is expressed in Kelvins. Therefore, $V_2 = V_1T_2/T_1$ and 100 °C = 100 + 273 = 373 K and 10 °C = 283 K.
$V_2 = (154\ mL)(283\ K)/373\ K = 117\ mL$.

25. Charles's law: At constant pressure and sample size, $V_1/T_1 = V_2/T_2$ assuming temperature is expressed in Kelvins. Therefore, $V_2 = V_1T_2/T_1$ and 18 °C = 18 + 273 = 291 K and 37 °C = 310 K
$V_2 = (1.00\ L)(310\ K)/291\ K = 1.07\ L$.

27. Charles's law: At constant pressure and sample size, $V_1/T_1 = V_2/T_2$ assuming temperature is expressed in Kelvins. Therefore, $T_2 = V_2T_1/V_1$ and 305 °C = 305 + 273 = 578 K
T_2 = (425 mL)(578 K)/576 mL = 433 K or 433 – 273 = 160 °C.

29. a. 22.4 L Note: The molar volume of any gas is 22.4 L/mol at STP.
b. V = (1.75 mol)(22.4 L/mol) = 39.2 L
c. V = (0.225 mol)(22.4 L/mol) = 5.04 L

31. Density is mass/volume (d = m/V). 1 mol of Rn has a mass of 222 g and a volume of 22.4 L at STP. The density is (222 g/mol)/(22.4 L/mol) = 9.91 g/L.

33. Density = mass/volume or d = m/V so m = dV The molar volume of a gas at STP is 22.4 L/mol.
a. molar mass = (2.12 g/L)(22.4 L/mol) = 47.4 g/mol
b. molar mass = (2.97 g/L)(22.4 L/mol) = 66.5 g/mol

35. The ideal gas law is PV = nRT. If n is constant:
a. The volume of the gas will decrease if the pressure increases and the temperature remains constant. Volume and pressure are inversely proportional at constant temperature. When the pressure is increased, the volume must decrease.
b. The volume of the gas will decrease if the temperature is decreased and the pressure is kept constant. Volume and temperature are directly proportional to one another at constant pressure; when one decreases the other must also decrease.
c. The volume of the gas will increase. The pressure decrease will cause an increase in volume, and the temperature increase will also cause an increase in volume. One way to test this would be to make up some values, for example, for one mole of a gas (n = 1), V = 22.4 L at T = 273 K and P = 1 atm. Then calculate a new volume for the same 1-mol sample when T = 373 and P = 0.5 atm.

37. a. If the molecules of a gas move more slowly, then the temperature will decrease.
b. If the molecules of a gas hit the walls of a container less often, then the pressure will decrease.

39. PV = nRT
a. V = nRT/P
V = (0.00600 mol)(0.0821·L atm/mol·K)(304 K)/(0.870 atm) = 0.172 L
b. P = nRT/V
P = (0.0108 mol)(0.0821·L atm/mol·K)(310 K)/(0.265 L) = 1.04 atm

41. PV = nRT so n = PV/(RT) and 25 oC = 25 + 273 = 298 K
n = (0.918 atm)(0.555 L)/(0.0821 L atm/mol K)(298 K) = 0.0208 mol

43. PV = nRT so n = PV/(RT). Temperature is given in Kelvins, but volume must be converted from mL to L. V, L = (660 mL)(1 L/1000 mL 0 0.660 L
n = (0.154 atm)(0.0.660 L)/(0.0821 L atm/mol K)(298 K) = 0.00415 mol

45. PV = nRT so V = nRT/P
V = (4.55 mol)(0.0821 L atm/mol K)(285 K)/7.32 atm = 14.5 L

47. The five basic postulates of the kinetic-molecular theory of gases are:
   1. Gas particles are in rapid constant motion and move in straight lines.
   2. Particles of a gas are small compared with the distances between them.
   3. Because the particles of a gas are so far apart, there is very little attraction between them.

4. Particles of a gas collide with one another. Energy is conserved in these collisions; energy lost by one particle is gained by the other.
5. Temperature is a measure of the average kinetic energy (energy of motion) of the gas particles.

The second of these postulates indicates that there is a great deal of empty space in a gas sample, which explains why gases can be compressed.

49. a. Boyle's law describes the relationship between the pressure and volume of a sample, assuming that the sample size and pressure are held constant.
b. Charles's law describes the relationship between the volume and Kelvin temperature of a sample, assuming that the sample size and pressure are held constant.
c. Avogadro's law describes the relationship between the number of moles of gas present and the volume of the sample, assuming that the temperature and pressure are kept constant.

51. Equal molar amounts of gases occupy the same volume, so 0.75 mol of He would occupy the same volume as is occupied by 0.75 mol of H_2. The mass of He in that volume would be:
mass = (0.75 mol He)(4.0 g He/mol He) = 3.0 g He.

53. $V_1/T_1 = V_2/T_2$
$V_2 = T_2V_1/T_1$
$V_2 = (77\text{ K})(1.50\text{ L})/(293\text{ K}) = 0.394\text{ L}$

55. We can use the ideal gas law to determine the number of moles of gas present in the sample. Because the units of the proportionality constant, R, are L atm/mol K, we must convert the pressure from mmHg to atm and temperature to Kelvins.
P, atm = (760 mmHg) (1 atm/760 mmHg) = 1.00 atm and T, K = 25 °C + 273 = 298 K.
n = PV/RT = (1.00 atm)(8.00 L)/(0.0821 L atm/mol K)(298 K) = 0.327 mol
To determine the molar mass of the gas, we can set up a proportion:
14.4 g/0.327 mol = x g/ 1.00 mol. x = 44.0 g so the molar mass of the gas is 44.0 g/mol.

57. a. The volume of the room, in m^3, is the length times the width times the height:
V, m^3 = (4.6 m)(4.9 m)(3.0 m) = 68 m^3
b. To convert this volume to liters, we use the conversion factor 1 L = 0.001 m^3.
V, L = (68 m^3)(1 L/0.001 m^3) = 68,000 L
c. We can use the ideal gas law to calculate the number of moles of CO_2 present in this volume at STP
n = PR/RT n = (1 atm)(68,000 L)/(0.0821 L atm/mol K)(273 K) = 3000 mol
Finally, we can calculate the mass of CO_2 present, in kg, using the molar mass of CO_2 (44.0 g/mol) and the conversion of grams to kilograms.
Mass of CO_2, kg = (3000 mol CO_2)(44.0 g CO_2/mol CO_2)(1 kg/1000 g) = 132 kg CO_2

59. We can estimate these various pressures by comparing the mass of the air above each of the levels, with the mass of air above a point at ground level where the pressure is 760 mmHg.
a. The overall mass of air above an altitude of 5.5 km is 50% of that at ground level, so the pressure would be 50% of that at ground level or (760 mmHg)(0.50) = 380 mmHg.
b. The overall mass of air above an altitude of 25 km is only 5% of that at ground level, so thc pressure would be only 5% of that a ground level or (760 mmHg)(0.050) = 38 mmHg.
c. The overall mass of air above an altitude of 30 kg is only 1% of that at ground level, so the pressure would be only 1% of that at ground level or (760 mmHg)(0.010) = 7.6 mmHg.

61. Answers b, c, and d all result in a decrease in volume of a gas. Option a results in an increase in the volume of a gas.

63. We can calculate the mass of C_4H_{10} in the lighter using the volume and density.
d = m/V so m = dV.
mass of C_4H_{10}, g = (0.601 g/mL)(5.00 mL) = 3.01 g
We can convert this mass to the number of moles of C_4H_{10} in the lighter, using the molar mass of C_4H_{10}, which is (4)(12.0) + (10)(1.0) = 58.0 g/mol.
Number of moles = (3.01 g C_4H_{10})(1 mol C_4H_{10}/58.0 g C_4H_{10}) = 0.0519 mol
Finally, we can use the ideal gas law to determine the volume of this sample at STP.
PV = nRT so V = nRT/P
V = (0.0519 mol)(0.0821 L atm/mol K)(283 K)/1 atm = 1.16 L

65. Supercritical carbon dioxide is inexpensive, convenient to use, is flammable, and nontoxic—all characteristics that support its use in greener chemical processing.

67. Perchloroethylene, previously used in the dry cleaning industry, has been replaced with $scCO_2$, and methylene chloride has been replaced by $scCO_2$ for the decaffeination of coffee.

CHAPTER

7

Acids and Bases

Please Pass the Protons

CHAPTER SUMMARY

7.1 Acids and Bases: Experimental Definitions

Learning Objectives: • Distinguish between acids and bases using their chemical and physical properties. • Explain how an acid–base indicator works.

A. Characteristic Properties of Acids.

Acids:

1. Taste sour.
2. Turn blue litmus indicator dye red.
3. Dissolve active metals to produce H_2 gas.
4. React with bases to produce water and ionic compounds called salts.

B. Characteristic Properties of Bases.

Bases:

1. Taste bitter.
2. Turn red litmus indicator dye blue.
3. Feel slippery.
4. React with acid to form water and ionic compounds called salts.

Answers to Self-Assessment Questions

1. a Bases, not acids, feel slippery on the skin.
2. b Bases do not react with salts to form acids. Bases react with acids to form salts and water.
3. d Acids and bases neutralize each other to form a salt (an ionic compound) and water.
4. d Yogurt is a milk product containing lactic acid.
5. d Citric acid is a component of grapefruit.

7.2 Acids, Bases, and Salts

Learning Objectives: • Identify Arrhenius and Brønsted–Lowry acids and bases. • Write a balanced equation for a neutralization or an ionization.

A. The Arrhenius Theory of Acids and Bases.
   1. An acid is a molecular substance that breaks up (ionizes) in aqueous solution into hydrogen ions (H^+, also called protons) and anions.
      a. Chemists often indicate an acid by writing the formula with the H atom(s) that ionize first: $HC_2H_3O_2$, HNO_3.
   2. In water, the properties of acids are those of the H^+ ions.
   3. Bases are substances that release hydroxide ions (OH^-) in aqueous solution.
      a. Ionic hydroxides (for example, NaOH, KOH) dissolve in water rather than reacting with it.
   4. In water, the properties of bases are those of the OH^- ions.
   5. Neutralization is the reaction of an acid and a base: the cation that was part of the base combines with the anion that was part of the acid to form a salt; the H^+ ion(s) from the acid combines with the OH^- ion(s) from the base to form water.
B. Limitations of the Arrhenius Theory.
   1. Free protons (H^+) do not exist in water. H^+ ions react with H_2O molecules to form H_3O^+ ions, called hydronium ions.
   2. The Arrhenius theory does not explain why ammonia (NH_3) is basic.
   3. The Arrhenius theory applies only to aqueous solutions and not to solutions with other kinds of solvents.
C. Brønsted-Lowry Acid–Base Theory.
   1. An acid is a proton donor.
   2. A base is a proton acceptor.
      a. This definition explains why NH_3 is basic: $NH_3 + H_2O \leftrightarrow NH_4^+ + OH^-$.
D. Salts.
   1. Salts are ionic compounds formed from neutralization reactions of acids and bases.
   2. Salts that dissolve in water form solutions that conduct electricity.

Answers to Self-Assessment Questions

1. e CH_3COOH is acetic acid, which is found in vinegar.
2. c H_3BO_3 is a mild antiseptic.
3. d HCl reacts with $CaCO_3$, the main component of boiler scale.
4. a H_2SO_4 is found in car batteries.
5. b NaOH is used in soap-making to convert fats into soaps and glycerol.
6. d When dissolved in water, HBr forms H_3O^+ and Br^- ions.
7. c An Arrhenius acid produces H^+ in water.
8. d An Arrhenius base produces OH^- in water.
9. b A Brønsted acid is a proton (H^+) donor.
10. a A Brønsted base is a proton (H^+) acceptor.

7.3 Acidic and Basic Anhydrides

Learning Objective: • Identify acidic and basic anhydrides, and write equations showing their reactions with water.

A. Nonmetal oxides are called acid anhydrides. They form acids when added to water—for example, $SO_2 + H_2O \rightarrow H_2SO_3$.
B. Metal oxides are called basic anhydrides. They form bases when added to water—for example, $BaO + H_2O \rightarrow Ba(OH)_2$.

Answers to Self-Assessment Questions

1. c Subtract 2 H atoms and 1 O atom from H_2SeO_4 to get SeO_3.
2. a Subtract 2 H atoms and 1 O atom from $Zn(OH)_2$ to get ZnO.

7.4 Strong and Weak Acids and Bases

Learning Objective: • Define and identify strong and weak acids and bases.

A. Strong acids are those that ionize completely in aqueous solutions. One mole of HCl in 1 L of water is virtually 100% ionized so the reaction below goes to completion.

$$HCl + H_2O \rightarrow H_3O^+ + Cl^-$$

1. Strong acids (a list worth remembering): HCl, H_2SO_4, HNO_3, HBr, HI, $HClO_4$.
2. Note: The term *strong* relates only to the extent to which an acid ionizes in water and NOT to the concentration of the acid solution. The terms we use to denote relative concentrations are *concentrated* and *dilute*.

B. Weak acids ionize only slightly in aqueous solutions. One mole of HCN in 1 L of water solution is only 1.0% ionized, so the reaction below does not go to completion.

$$HCN + H_2O \leftrightarrow H_3O^+ + CN^-$$

C. Strong bases are those that ionize completely, or nearly so, in water.

1. Strong bases (information worth remembering): All group 1A metal hydroxides, all group 2A metal hydroxides except $Be(OH)_2$, remembering that both $Ca(OH)_2$ and $Mg(OH)_2$ are only slightly soluble in water so the OH^- concentrations in those solutions is not high. Thus, when 1 mol of NaOH, a strong base, dissolves in water it produces 1 mol of OH^-.
2. Note: The term *strong* relates only to the extent to which the dissolved base provides OH^- ions in solution, NOT to the concentration of the base solution. The terms we use to denote relative concentrations are *concentrated* and *dilute*.

D. Weak bases are those that yield relatively few hydroxide ions when dissolved in water.

Answers to Self-Assessment Questions

1. b $Ca(OH)_2$ is a strong base (see the list of strong bases: a group 2A metal hydroxide).
2. c HCN is a weak acid (HCN is not on the list of strong acids).
3. c HF is a weak acid (HF is not on the list of strong acids).
4. a HNO_3 is a strong acid (see the list of strong acids).
5. d NH_3 is a weak base (NH_3 is not on the list of strong bases).
6. b KOH is a strong base (see the list of strong bases: a group 1A metal hydroxide).
7. d CH_3NH_2 is a weak base (CH_3NH_2 is not on the list of strong bases).
8. c The reaction is $CH_3COOH + H_2O \leftrightarrow CH_3COO^- + H_3O^+$.
9. b The reaction is $NH_3 + H_2O \leftrightarrow NH_4^+ + OH^-$.

7.5 Neutralization

Learning Objective: • Identify the reactants and predict the products in a neutralization reaction.

A. In water, an acid will release H_3O^+. When combined with an equivalent amount of base that releases OH^-, a neutralization reaction combining H_3O^+ (acid) and OH^- (base) will produce water and a salt.

$$HCl + NaOH \rightarrow H_2O + NaCl$$
$$\text{Acid} + \text{Base} \rightarrow \text{Water} + \text{Salt}$$
$$H_3O^+ + OH^- \rightarrow H_2O$$

Answers to Self-Assessment Questions

1. d Neutralization occurs when an acid and base are mixed.
2. b The balanced chemical equation for the reaction of HCl with NaOH is
   $HCl + NaOH \rightarrow NaCl + H_2O$.
   We can calculate the number of moles of HCl required using the stoichiometric ratio of HCl to NaOH in this equation.
   number of mol of HCl = (1.5 mol NaOH)(1 mol HCl/1 mol NaOH) = 1.5 mol HCl
3. d The balanced chemical equation for the neutralization of $Ca(OH)_2$ with HCl is:
   $$2\ HCl + Ca(OH)_2 \rightarrow CaCl_2 + 2\ H_2O.$$
   We can use the stoichiometric relationship to calculate the number of moles of HCl required to neutralize 2.4 mol of $Ca(OH)_2$:
   number of mol of HCl = (2.4 mol $Ca(OH)_2$)(2 mol HCl/mol $Ca(OH)_2$) = 4.8 mol HCl
4. c The balanced chemical equation for the neutralization of H_3PO_4 with $Sr(OH)_2$ is:
   $$2\ H_3PO_4 + 3\ Sr(OH)_2 \rightarrow Sr_3(PO_4)_2 + 6\ H_2O.$$
   We can use the stoichiometric relationship to calculate the number of moles of $Sr(OH)_2$ required to neutralize 2 mol of H_3PO_4:
   number of moles of $Sr(OH)_2$ = (2 mol H_3PO_4)(3 mol $Sr(OH)_2$/2 mol H_3PO_4) = 3 mol $Sr(OH)_2$

7.6 The pH Scale

Learning Objectives: • Describe the relationship between the pH of a solution and its acidity or basicity. • Find the molar concentration of hydrogen ion, $[H^+]$, from a pH value or the pH value from $[H^+]$.

A. pH is a measure of the molar concentration (M = mol/L) of H_3O^+. We express the molar concentration of a solute by putting its chemical formula in square brackets: $[H_3O^+]$.
   1. pH is defined as the negative logarithm of the molar concentration of hydrogen ion:
      $$pH = -\log [H^+].$$
   2. On the pH scale, a H^+ concentration of 1×10^{-7} mol/L of hydrogen ions becomes a pH of 7; a H^+ concentration of 1×10^{-10} mol/L becomes a pH of 10; and so on.
      a. Note that the pH scale is logarithmic. A decrease of one pH unit represents a ten-fold increase in the H^+ ion concentration.
      b. It is useful to remember that in aqueous solutions $[H^+][OH^-] = 1.0 \times 10^{-14}$
   3. Interpreting pH.
      a. Solutions with pH < 7.0 are acidic.
      b. Solutions with pH = 7.0 are neutral.
      c. Solutions with pH > 7.0 are basic.
   4. Acidic solutions contain a higher concentration of H^+ ions than of OH^- ions.
      Basic solutions contain higher concentration of OH^- ions than of H^+ ions.
      Neutral solutions contain equal concentrations of H^+ and OH^- ions.

Answers to Self-Assessment Questions

1. b HNO_3 is a strong acid. Change 0.0010 M HNO_3 to scientific notation to get 1.0×10^{-3} M.
2. d The negative logarithm of 1×10^{-11} is 11.0.

3. c To calculate pH, we need to find $[H^+]$, but we are given $[OH^-]$. We can calculate $[H^+]$ from the relationship $[H^+][OH^-] = 1 \times 10^{-14}$ where $[H^+] = (1 \times 10^{-14})/[OH^-]$.
For this example, $[H^+] = (1 \times 10^{-14})/(1 \times 10^{-5}) = 1 \times 10^{-9}$.
$pH = -\log[H^+] = -\log 1 \times 10^{-9} = 9.0$.
4. c Find the antilog of –8 to get 1.0×10^{-8}, or the molar concentration of H^+ in the pool water.
5. c The concentrations of H^+ and OH^- ions in pure water are the same: 1.0×10^{-7} M. The pH of pure water is 7.0.
6. b HCl is a strong acid, so the H^+ ion concentration in 0.015 M HCl is 0.015 M. The pH of this solution is somewhat lower than the pH of a solution with $[H^+] = 0.010$ M, which would be 2.0 but higher than the pH of a solution, with $[H^+] = 0.1$ M, which would be 1 so pH = 0.182 makes sense.
7. d An 0.015 M solution of NaOH should be very basic. pH = 12.18 is the most reasonable answer.
8. e This is close to pH 7 but slightly basic. If the H^+ concentration were 1.0×10^{-8}, the pH would be 8.0, not 7.4. Because a solution with pH = 7.4 contains a slightly higher concentration of H^+ ions than one of pH = 8.0, $[H^+] = 4 \times 10^{-8}$ M is a reasonable choice.

7.7 Buffers and Conjugate Acid-Base Pairs

Learning Objectives: • Write the formula for the conjugate base of an acid or for the conjugate acid of a base. • Describe the action of a buffer.

A. A conjugate acid–base pair is a pair of compounds or ions, the formulas of which differ by one proton (H^+). For example, NO_3^- is the conjugate base of the acid HNO_3, and NH_4^+ is the conjugate acid of NH_3.
B. A buffer is a solution that resists change in pH upon the addition of small amounts of acid or base.
   1. A buffer solution contains a weak acid and the salt of that weak acid (for example, HF and NaF) or
   2. A buffer solution contains a weak base and the salt of that weak base (for example, NH_3 and NH_4Cl).

Answers to Self-Assessment Questions

1. b HCN and CN^- have a common anion and differ only by H^+. HCN is a weak acid, and CN^- is the conjugate base of that weak acid.
2. d Only NH_3 and H_3O^+ do not share a common ion.
3. a $HCOO^-$ is the base in this buffer and will react with added H^+ to form HCOOH.
4. a C_6H_5COOH and C_6H_5COONa share a common ion. C_6H_5COOH is a weak acid, and $C_6H_5COO^-$ is the conjugate base of that weak acid.

7.8 Acids and Bases in Industry and in Daily Life

Learning Objective: • Describe everyday uses of acids and bases and how they affect daily life.

A. Antacids: A Basic Remedy.
   1. Antacids work to neutralize excess stomach acid.
   2. Sodium bicarbonate, $NaHCO_3$, was one of the earliest antacids. This compound works by combining with H^+ to form H_2CO_3 which breaks down to form CO_2 and H_2O.
   3. Other compounds commonly used in antacids are calcium carbonate, $CaCO_3$, aluminum hydroxide, $Al(OH)_3$, magnesium carbonate, $MgCO_3$, and magnesium hydroxide, $Mg(OH)_2$.

4. Although antacids are generally safe for occasional use, they can interact with certain medications.

B. Why doesn't stomach acid dissolve the stomach?
   1. A mucus layer protects the cells lining the stomach from stomach acid.
   2. Bacterial infections can damage the mucus, exposing the cells to stomach acid and causing ulcer formation.

C. Acids and Bases in Industry and at Home.
   1. Sulfuric acid, H_2SO_4, is the leading chemical product in the United States and around the world. H_2SO_4 is used to make fertilizers and other industrial chemicals, and is also used in car batteries.
   2. Hydrochloric acid, HCl, also called muriatic acid, is used in construction.
   3. Lime, calcium oxide (CaO), is the most widely used commercial base. CaO is prepared by heating limestone ($CaCO_3$) to drive off CO_2.
   4. Sodium hydroxide, NaOH, is used in such products as oven cleaners and drain cleaners, as well as to make soaps.
   5. Ammonia, NH_3, is used for fertilizers.

D. Acid Rain.
   1. "Acid rain" is produced when acidic pollutants such as sulfur oxides and nitrogen oxides present in the atmosphere dissolve in the rain.
   1. All rain is slightly acidic because atmospheric CO_2 dissolves in the falling raindrops, forming carbonic acid, H_2CO_3, which ionizes to form H^+ and HCO_3^- ions.

E. Acids and Bases in Health and Disease.
   1. Strong acids and bases can break down, or denature, proteins.
   2. Living cells can function properly only at an optimal pH.

Green Chemistry: Acids and Bases, Greener Alternatives

1. While acids and bases are among the most commonly used substances, they can be hazardous, corrosive, and difficult to transport, and pose disposal problems.
2. There are greener alternatives
   a. Example: Conversion of brine (sodium chloride solution) to chlorine and sodium hydroxide, a process that produces vast amounts of alkaline waste that must be neutralized.
      i. Companies are now beginning to neutralize the waste with CO_2.
   b. Example: Dehydration of alcohols, a process that traditionally requires concentrated sulfuric and phosphoric acids and high temperatures.
      i. Recently, H_2SO_4 and H_3PO_4 are being replaced by a much safer substance – a type of clay called montmorillonite.

Answers to Self-Assessment Questions

1. a Calcium carbonate reacts with excess H^+ to form H_2CO_3 which then decomposes to produce CO_2 and H_2O.
2. a Acidic solutions have low pHs. When some of the acid is neutralized, the concentration of H^+ decreases slightly, causing the stomach pH to rise slightly.
3. d H_2SO_4 is the leading chemical product not only in the United States but also worldwide.
4. d One of the uses of H_2SO_4 is in automobile batteries.
5. c Aqueous solutions of lye (NaOH) are used to convert fats to soaps and glycerol.

LEARNING OBJECTIVES

You should be able to . . .

1. Distinguish between acids and bases using their chemical and physical properties. (7.1)
2. Explain how an acid–base indicator works. (7.1)
3. Identify Arrhenius and Brønsted-Lowry acids and bases. (7.2)
4. Write a balanced equation for a neutralization or an ionization. (7.2)
5. Identify acidic and basic anhydrides, and write equations showing their reactions with water. (7.3)
6. Define and identify strong and weak acids and bases. (7.4)
7. Identify the reactants and predict the products in a neutralization reaction. (7.5)
8. Describe the relationship between the pH of a solution and its acidity or basicity. (7.6)
9. Find the molar concentration of hydrogen ion, $[H^+]$, from a pH value or the pH value from $[H^+]$. (7.6)
10. Write the formula for the conjugate base of an acid or for the conjugate acid of a base. (7.7)
11. Describe the action of a buffer. (7.7)
12. Describe everyday uses of acids and bases and how they affect daily life. (7.8)
13. Recognize carbon dioxide in water as a useful, safe source of acid.
14. Describe alternate choices for greener reaction conditions when using acids and bases.

DISCUSSION

We will begin by reviewing the definitions of acids. Acids are defined in the chapter in two ways: as compounds that yield hydronium ions in aqueous solutions and as compounds that act as proton donors. It is now generally accepted that a hydrogen ion does not exist in solution as an independent unit. Thus, the second definition is an attempt to be a bit more accurate in describing the action of an acid.

Because H^+ ions are only protons (a nuclear particle), they do not exist as such in water but associate with H_2O molecules forming H_3O^+ ions, called hydronium ions. There are many practical aspects of acids and bases that you should learn: their properties, their sources (acid and base anhydrides), whether a particular acid or base is strong or weak (and what this means), the pH scale, and some of the more common applications of acids and bases.

ANSWERS TO ODD-NUMBERED REVIEW QUESTIONS AND SOLUTIONS FOR ODD-NUMBERED END-OF-CHAPTER PROBLEMS

Review Questions

1. a. Arrhenius said an acid is a substance that donates one or more H^+ ions to water. Brønsted-Lowry defined an acid as a substance that donates one or more H^+ ions to a solution. In essence, the two definitions are the same. H_2SO_4, sulfuric acid, is an example.
   b. An Arrhenius base is a compound that donates one or more OH^- ions to a solution. According to the Brønsted-Lowry definition, a base is a proton acceptor. KOH, potassium hydroxide, is an example of an Arrhenius base. KOH is also an example of a Brønsted-Lowry base because OH^- ions can accept H^+ ions to form H_2O. NH_3 is an example of a Brønsted-Lowry base but not of an Arrhenius base.
   c. A salt is an ionic compound formed from the neutralization of an acid with a base. K_2SO_4, potassium sulfate, is the salt formed by the reaction of H_2SO_4 and KOH.

3. A neutralized solution would have a pH = 7, and this solution would have no effect on either iron or zinc. Although both of these metals would react with an acidic solution, neither reacts in a solution with pH = 7.

5. We refer to H^+ ions as protons when describing acid–base chemistry. Protons are the positively charged nuclear particles that are counted when determining the atomic number of an element. The atomic number of hydrogen is 1, meaning that there is one proton in the nucleus of an H atom. Because the mass number (number of protons plus neutrons) of most H atoms is 1, most H nuclei contain only one proton. When the sole electron from an H atom is lost to form H^+, all that remains is a proton—hence the name.

7. Neutralization of an acid occurs when an equivalent amount of base is added so that $[H^+] = [OH^-]$. Bases are neutralized when an equivalent amount of acid is added so that $[H^+] = [OH^-]$.

9. Strong bases ionize in water, producing OH^- to the extent that they are soluble. Weak bases also react with water to produce OH^- ions, but only 5% or less of the weak base species in solution undergo the reaction.

11. Strong acids and bases are corrosive and can cause burns.

Problems

13. a. $$HClO_4 \xrightarrow{H_2O} H^+ + ClO_4^-.$$

    b. $$RbOH \xrightarrow{H_2O} Rb^+ + OH^-$$

15. $$H_2PO_4^- \xrightarrow{H_2O} HPO_4^{2-} + H^+$$

17. a. The N atom in $(CH_3)_2NH$ shares its nonbonded electron pair on the N atom with an H^+ ion released from H_2O. Therefore, $(CH_3)_2NH$ is a base.
    b. The N atom in $C_6H_5NH_2$ shares its nonbonded electron pair on the N atom with an H^+ ion released from H_2O. Therefore, $C_6H_5NH_2$ is a base.
    c. $C_6H_5CH_2NH_2$ shares its nonbonded electron pair on the N atom with an H^+ ion released from H_2O. Therefore, $C_6H_5CH_2NH_2$ is a base.

19. a. $HCOOH(aq) + H_2O(l) \rightarrow H_3O^+(aq) + COO^-(aq)$.
    b. $C_5H_5N(aq) + H_2O(l) \rightarrow C_5H_5NH^+(aq) + OH^-(aq)$

21. $NH_3(aq) + H_2O \longrightarrow NH_4{}^+(aq) + OH^-(aq)$

23. a. HCl: HCl(g) is hydrogen chloride; HCl(aq) is hydrochloric acid.
    b. $Sr(OH)_2$
    c. KOH is potassium hydroxide.
    d. H_3BO_3

25. a. H_3PO_4 is phosphoric acid, an acid.
    b. CsOH is cesium hydroxide, a base.
    c. H_2CO_3 is carbonic acid, an acid.

27. a. HNO_3 is nitric acid, so the formula for nitrous acid is HNO_2.
    b. H_3PO_4 is phosphoric acid, so the formula for phosphorus acid is H_3PO_3.

29. a. $SO_3 + H_2O \rightarrow H_2SO_4$. An acid is produced when a nonmetal oxide reacts with water. H_2SO_4 is sulfuric acid, an acid.
    b. $MgO + H_2O \rightarrow Mg(OH)_2$. A base is produced when a metal oxide reacts with water. $Mg(OH)_2$ is magnesium hydroxide, a base.

31. HI is a strong acid because all the HI molecules ionize and form H^+ and I^- ions when the substance is added to water.

33. CH_3NH_2 is a weak base because only a few of the dissolved molecules react with water to form $CH_3NH_3{}^+$ and OH^- ions.

35. a. LiOH is a strong base (it's a Group 1A hydroxide).
    b. HBr is a strong acid (it's on the list of strong acids).
    c. HNO_2 is a weak acid. It is not on the list of strong acids, although you have to look carefully because nitric acid, HNO_3, is on the list.
    d. $CuSO_4$ is copper(II) sulfate, a salt.

37. a. 0.10 M HNO_3: Nitric acid is a strong acid and ionizes completely in water. This solution has the highest concentration of H^+ ions.
    b. 0.10 M NH_3: Ammonia is a weak base. When dissolved in water, NH_3 produces OH^- ions so, of the three solutions, the H^+ ion concentration in this solution is the lowest.
    c. 0.10M CH_3COOH: Acetic acid is a weak acid, so its ionization is not complete, but it does increase the H^+ ion concentration in aqueous solution.

39. a. $KOH(aq) + HCl(aq) \rightarrow KCl(aq) + H_2O(l)$
    b. $LiOH(aq) + HNO_3(aq) \rightarrow LiNO_3(aq) + H_2O(l)$

41. a. $H_2SO_3(aq)$ + $Mg(OH)_2(aq)$ $\rightarrow$ 2 $H_2O(aq)$ + $MgSO_3(aq)$

43. a. Solutions with pH < 7.0 are acidic; those with pH > 7.0 are basic. A solution with pH = 4 is acidic.
b. Solutions with pH = 7 are said to be neutral because $[H^+] = [OH^-]$.
c. A solution with pH = 3.5 is acidic. In fact, it's more acidic than the solution in (a).
d. A solution with pH = 9 is basic because solutions with pH > 7 are basic.

45. $pH = -\log [H^+] = -\log 1.0 \times 10^{-5} = 5.0$

47. $pH = -\log [H^+]$. We are given $[OH^-]$, so we have to calculate $[H^+]$ from the relationship $[H^+][OH^-] = 1 \times 10^{-14}$ where $[H^+] = (1 \times 10^{-14})/[OH^-] = (1 \times 10^{-14})/1 \times 10^{-5} = 1 \times 10^{-9}$.
$pH -\log 1 \times 10^{-9} = 9$.

a. A solution with pH 4 is acidic (pH < 7.0).
b. A solution with pH 7 is neutral (pH = 7.0).
c. A solution with pH 3.5 is acidic (pH < 7.0).
d. A solution with pH 9.0 is basic (pH > 7.0).

49. The pH of the solution is 5.0. The pH is the negative logarithm of H^+ concentration, or 5.0.

51. $[H^+] = 1 \times 10^{-3}$ M. The concentration is the negative antilog of the pH.

53. The pH of milk of magnesia is between 10 and 11. The pH is the negative logarithm of H^+ concentration.

55. $HNO_3(aq)$ + $NH_3(aq)$ $\rightarrow$ $NO_3^-(aq)$ + $NH_4^+(aq)$
a. HNO_3, nitric acid, is the acid in this reaction because it releases an H^+ ion. Ammonia, NH_3, is the base in this reaction because it accepts an H^+ ion.
b. The conjugate base of $HNO_3(aq)$ is $NO_3^-(aq)$.
c. The conjugate acid of $NH_3(aq)$ is $NH_4^+(aq)$.

57. $Al(OH)_3(s)$ + 3 $HCl(aq)$ $\rightarrow$ $AlCl_3(aq)$ + 3 H_2O
$Mg(OH)_2(s)$ + 2 $HCl(aq)$ $\rightarrow$ $MgCl_2(aq)$ + 2 H_2O

59. No. To be an Arrhenius base, the OH^- group must be paired with a metal. For example, methyl alcohol, CH_3OH, has an OH^- group but is not a base.

61. a. The Cl^- ion is the conjugate base of a strong acid (HCl) and is, therefore, a weak base.
b. The CN^- ion is the conjugate base of a weak acid (HCN) and is, therefore, a strong base.

63. a. pOH = 3 The pOH is the negative logarithm of OH^- concentration.
b. pOH = 2 The pOH is the negative logarithm of OH^- concentration.

65. To solve this, first write the equation for the ionization of $Ba(OH)_2$:
$$Ba(OH)_2(aq) \rightarrow Ba^{2+} + 2\ OH^-(aq).$$
Therefore, one mole $Ba(OH)_2$ produces 2 mol OH^- ions, so $[OH^-] = (2)(0.5\ M) = 1\ M$
$pH = -\log [H_3O^+] = -\log [1.0 \times 10^{-14}/1] = -\log 1.4 \times 10^{-14} = 14$.

67. A buffer solution is made from a mixture of a weak acid and its salt or a weak base and its salt. Ammonia is a weak base, while sodium hydroxide is a strong base. Therefore, the combination of ammonia (NH_3) and ammonium acetate ($NH_4C_2H_3O_2$) would meet the requirement of a weak base and its salt, while the combination of sodium hydroxide (NaOH) and sodium acetate ($NaC_2H_3O_2$) would not.

69. $Al_2(CO_3)_3(s) + 6\ HCl(aq) \rightarrow 2\ AlCl_3(aq) + 3\ H_2O(l) + 3\ CO_2(g)$

71. $S(s) + O_2(g) \rightarrow SO_2(g)$

$2\ SO_2(g) + O_2 \xrightarrow{V_2O_5} 2\ SO_3(g)$

$SO_3(g) + H_2O(l) \rightarrow H_2SO_4(l)$

CHAPTER

8

Oxidation and Reduction

Burn and Unburn

CHAPTER SUMMARY

8.1 Oxidation and Reduction: Four Views

Learning Objectives: • Classify a particular change within a redox reaction as either oxidation or reduction. • Identify an oxidation–reduction reaction.

A. Combustion reactions are one of the most important examples of the redox process. During combustion, the substance (or element) that gains oxygen atoms is oxidized. Substances that lose oxygen atoms are reduced.
   1. $4\ Fe + 3\ O_2 \rightarrow 2\ Fe_2O_3$
      Fe is oxidized.
      O_2 is reduced.
   2. $2\ H_2 + O_2 \rightarrow 2\ H_2O$
      H_2 is oxidized.
      O_2 is reduced.

B. A second view of the redox process involves H atoms. Oxidation is a loss of hydrogen atoms. Reduction is a gain of hydrogen atoms.
   1. $CH_4O + \frac{1}{2}\ O_2 \rightarrow CH_2O + H_2O$
      CH_4O is oxidized; O_2 is reduced.
   2. $CO + 2\ H_2 \rightarrow CH_4O$
      CO is reduced.

C. A third view of the redox process involves a transfer of electrons from one species to another. The species that loses electrons is oxidized. The species that gains electrons is reduced. Think of LEO says GER (loss of electrons is oxidation; gain of electrons is reduction) or, if you prefer, OIL RIG (Oxidation Is Loss of electrons; Reduction Is Gain of electrons.)
   1. $Mg + Cl_2 \rightarrow Mg^{2+} + 2\ Cl^-$
      Cl_2 is reduced.

D. The fourth view involves oxidation numbers of a substance before and after the reaction. The substance whose oxidation number increases is oxidized; the substance whose oxidation number decreases is reduced.
   $Mg + \frac{1}{2}\ O_2 \rightarrow Mg^{+2}O^{-2}$
   Mg is oxidized (its oxidation number goes from zero to +2).
   O is reduced (its oxidation number goes from zero to –2).

1. When an oxidation number increases, the perceived charge of an atom (or ion) becomes more positive or less negative as a result of the loss of one or more electrons. The charge of an atom or ion becomes less positive or more negative as a result of the gain of one or more electrons.
2. The following rules usually allow us to calculate meaningful oxidation numbers (ON):
    a. An atom in its elemental form has an oxidation number of zero (0).
    b. The oxidation number of a monoatomic ion is equal to its charge.
    c. Hydrogen and oxygen in compounds almost always have oxidation numbers of +1 and –2, respectively.
    d. The sum of the oxidation numbers of the atoms in a polyatomic ion (or molecule) is equal to the charge on that ion (or molecule).

Answers to Self-Assessment Questions

1. a "Electrons gained" is reduction, not oxidation.
2. d N loses one electron to go from +4 to +5. Alternatively, in the reactant, there are 2 O atoms for each N atom while in the product there are 5 O atoms for each 2 N atoms, representing a gain in O atoms.
3. b The Cr atom in CrO_4^- loses oxygen atoms to become Cr^{3+} which represents a reduction of Cr.
4. b The Mo atom loses oxygen atoms and so is reduced. Looked at another way, Mo gains four electrons to go from an oxidation state of +4 to an oxidation state of 0.
5. d Cl goes from an oxidation state of zero in Cl_2 to an oxidation state of –1 in Cl^-, which is a reduction.
6. c Mn^0 loses two electrons in the oxidation to Mn^{2+}.
7. d An example is #6 above.
8. c This is an ionic compound composed of Na^+ and O^{2-} ions.
9. a K forms a 1+ ion, K^+, and in most compounds the oxidation number of O is – 2. $KMnO_4$ has no charge, so the sum of the oxidation numbers of K, O, and Mn must equal zero. Therefore, the oxidation number of Mn in this compound must be +7: $(+1) + (4)(-2) + (\text{ON of Mn}) = 0$. ON of $\text{Mn} = 0 - 1 + 8 = +7$.
10. d When they form ions, alkaline earth elements lose their two valence electrons forming 2+ ions.
11. c If the oxidation number of P is +5 and the oxidation numbers of the four O atoms are each –2, then the charge on the phosphate ion must be $+5 + (4)(-2) = -3$.

8.2 Oxidizing and Reducing Agents

Learning Objective: • Identify the oxidizing agent and the reducing agent in a redox reaction.

A. Oxidation and reduction must occur together.
   1. Substance being oxidized is the reducing agent.
   2. Substance being reduced is the oxidizing agent.

Answers to Self-Assessment Questions

1. c A reducing agent supplies the electrons for the substance that is reduced.
2. a A compound that gains one or more electrons takes them from the substance that is oxidized, so the compound is an oxidizing agent.
3. a A substance loses electrons when its oxidation number increases, so it is a reducing agent.
4. b Ag^+ takes the electrons released when Cu is oxidized.
5. a Al(s) supplies electrons to Cr^{3+}, so Al is the reducing agent.
6. b Tl^+ takes an electron released by the oxidization of Zn.
7. a Fe^{2+} provides the electrons required to reduce I in IO_3^-.

8.3 Electrochemistry: Cells and Batteries

Learning Objectives: • Balance redox equations. • Identify and write the half-reactions in an electrochemical cell.

A. An electric current in a wire is a flow of electrons. Oxidation–reduction reactions can be used to produce electricity, as is done in dry cell and storage batteries.

B. When a reactive metal (zinc) is placed in contact with the ions of a less reactive metal (copper), the more active metal will give up its electrons (oxidation) to the ions of the less active metal (which are reduced).

C. If the two metals (zinc and copper) are placed in solutions of their metal ions in separate containers, the electrons traveling between the two containers must flow through an external circuit and can be harnessed to do work.
   1. The metal electrode where oxidation takes place (more active metal) is the anode.
   2. The metal electrode where reduction takes place (less active metal) is the cathode.
   3. This arrangement is an electrochemical cell.

D. We balance redox reactions by separating them into two half-reactions, the half-reaction of oxidation and the half-reaction of reduction. For example, consider the reaction of Fe^{2+} with IO_3^- in acid solution to form Fe^{3+} and I_2.
   1. Write the oxidation half reaction: $Fe^{2+} \rightarrow Fe^{3+} + 1\ e^-$. Note that this half reaction is balanced: There is one Fe on each side of the equation, and the net charge on both sides of the equation is the same, +2.
   2. Write the reduction half reaction, noting that the oxidation number of I in IO_3^- is +5 while that in I_2 is zero. Note also that both I atoms in I_2 undergo this reduction: $2\ IO_3^- + 10\ e^- \rightarrow I_2$. The oxygen in this equation is not balanced so, if the reaction is carried out in an acidic solution, we add H^+ and H_2O as necessary to balance the equation. If the reaction is carried out in a basic solution, we add OH^- and H_2O as necessary to balance the equation. Thus, the oxidation and reduction half reactions are:

      Oxidation: $Fe^{2+} \rightarrow Fe^{3+} + 1\ e^-$

      Reduction: $2\ IO_3^- + 10\ e^- + 12\ H^+ \rightarrow I_2 + 6\ H_2O$
   3. If the numbers of electrons exchanged in the oxidation and reduction half reactions are equal, then we add the half reactions together, canceling the electrons, to produce the completed balanced equation. If the numbers of electrons exchanged in the two half reactions are not equal, then we multiply each reaction by a coefficient that will make the number of electrons equal. In this case we have to multiply the oxidation half reaction by 10 so that there are 10 electrons on both sides of the equation and then we add the two half reactions.

$$10[\ Fe^{2+} \rightarrow Fe^{3+} + 1\ e^-\ ]$$
$$2\ IO_3^- + 10\ e^- + 12\ H^+ \rightarrow I_2 + 6\ H_2O$$
$$10\ Fe^{2+} + 2\ IO_3^- + 10\ e^- + 12\ H^+ \rightarrow 10\ Fe^{3+} + I_2 + 6\ H_2O + 10\ e^-$$

   Canceling the electrons, we end up with the balanced equation:

$$10\ Fe^{2+} + 2\ IO_3^- + 12\ H^+ \rightarrow 10\ Fe^{3+} + I_2 + 6\ H_2O.$$

E. A battery is a series of electrochemical cells. (In everyday life, however, we refer to a single electrochemical cell, such as that used in flashlights, as a "battery.")

F. Dry cells are the common batteries used in flashlights.

G. Lead storage batteries are the rechargeable batteries found in cars. These batteries can be recharged but are heavy and contain sulfuric acid.

H. Other common batteries are lithium-SO_2 cells used in submarines and rockets, lithium iodine cells used in pacemakers, lithium-FeS_2 batteries used in cameras, radios, and compact disc players, rechargeable Ni-Cad batteries for portable radios and cordless tools, and small "button" batteries used in watches, calculators, and cameras.

I. In fuel cells, fuel is oxidized at the anode, and oxygen is reduced at the cathode.
   1. Fuel cells are a much more efficient way of using fuel.

J. Whereas the chemical reactions in batteries occur spontaneously, electrolysis reactions occur when electricity is supplied to cause a chemical reaction to occur.
   1. Aluminum is produced from Al_2O_3, chromium (for chrome plating) from Cr^{6+} solutions, and copper from Cu(II) salts by electrolysis.

Answers to Self-Assessment Questions

1. a Electrons flow from the substance that is oxidized (at the anode) to the substance that is reduced (at the cathode).
2. c Zinc is oxidized at the anode, which is negative. In a dry cell battery, zinc acts as the anode and a carbon rod is the cathode.
3. b In a zinc-carbon cell Zn is the anode and C is the cathode. Electrons flow from the Zn to the carbon cathode where they reduce manganese.
4. c The porous membrane allows ions to pass through in order to have a complete circuit.
5. d Mg loses two electrons to become Mg^{2+}. Loss of electrons is oxidation.
6. d. $PbSO_4$ is produced at both the cathode and anode of a lead storage battery. Sulfuric acid is used in lead storage batteries.
7. d Lithium storage batteries are rechargeable.
8. c Reduction occurs at the cathode where O_2 is reduced to O^{2-}.

8.4 Corrosion and Explosion

Learning Objectives: • Describe the reactions that occur when iron rusts. • Explain why an explosive reaction is so energetic.

A. The rusting of iron is an electrochemical process that requires water, oxygen, and an electrolyte.
   1. Oxidation and reduction often occur at different places on the metal's surface.

B. Aluminum is more reactive than iron but is protected by an aluminum oxide film on its surface that forms from the reaction of Al with oxygen in the air.

C. Silver tarnish is largely silver sulfide (Ag_2S).
   1. It is formed by the reaction of silver with hydrogen sulfide (H_2S) in the air.
      a. Tarnish can be removed by reacting it with aluminum.

D. Except for nuclear reactions, explosive reactions are oxidation–reduction reactions that occur very rapidly (corrosion reactions occur slowly).
   1. Ammonium nitrate mixed with fuel oil (ANFO) is used for mining, earthmoving projects, and demolishing buildings.

Answers to Self-Assessment Questions

1. b Fe is oxidized when corroded, changing from Fe to Fe^{3+}.
2. a Al_2O_3 forms on the surface of Al and protects it from further oxidation.
3. a Hydrogen sulfide from the air or from foods reacts with silver to produce silver sulfide, which is the black substance we call tarnish. In the process, H^+ (from H_2S) is reduced to produce H_2.
4. c The formation of a large volume of gas from a small volume of solid or liquid reactant creates enormous pressure, which causes the damage of an explosion.
5. c The nitrate (NO_3^-) component of KNO_3 is the oxidizing agent in black powder just as NO_3^- is the oxidizing agent in ANFO.

8.5 Oxygen: An Abundant and Essential Oxidizing Agent

Learning Objectives: • Write equations for reactions in which oxygen is an oxidizing agent. • List some of the common oxidizing agents encountered in daily life.

A. Oxygen is one of the most important elements on Earth.
   1. Air: one-fifth (21%) elemental oxygen by volume.
   2. Water: 89% oxygen by mass.
   3. People: approximately two-thirds oxygen by mass.

B. Fuels such as natural gas, gasoline, coal, and the foods we eat all require oxygen to burn and release their stored chemical energy.

C. Pure oxygen is obtained by liquefying air, then allowing the nitrogen and argon to boil off.

D. Many metals and nonmetals react with oxygen.

E. Ozone (O_3) is a powerful oxidizing agent that is a pollutant at ground level and a beneficial chemical in the upper stratosphere.

F. Other Common Oxidizing Agents.
   1. Germicides (kill microorganisms), such as disinfectants (for nonliving tissue) and antiseptics (for living tissue), are oxidizing agents. Hypochlorous acid (HOCl) is used in swimming pools for this purpose.
   2. Hydrogen peroxide is used as a 3% or 30% aqueous solution. Hydrogen peroxide is reduced to water, and oxygen is produced.
   3. Iodine is used as an antiseptic.
   4. Potassium dichromate is a common laboratory oxidizing agent and is used in breathalyzer tests which make use of the color change associated with the reduction of dichromate ions.
   5. Acne ointments often contain benzoyl peroxide as the active ingredient.
   6. Bleaches, used to whiten paper or fabrics, are usually sodium hypochlorite (NaOCl) or calcium hypochlorite, $Ca(OCl)_2$. Nonchlorine bleaches contain sodium percarbonate (a combination of Na_2CO_3 and H_2O_2) or sodium perborate (a combination of $NaBO_2$ and H_2O_2).

Answers to Self-Assessment Questions

1. d Carbon dioxide and water are produced with efficient burning of CH_4.
2. d The reaction is $S + O_2 \rightarrow SO_2$.
3. b H_2O_2 is a common oxidizing agent used to clean wounds and as an ingredient in nonchlorinated bleach.
4. a Cl_2 is a strong oxidizing agent.
5. b. NaOCl is a common ingredient in chlorine bleach and is an oxidizing agent.
6. c Hydrogen peroxide is commonly used to bleach hair.
7. a Oxidation is an effective way to disinfect many surfaces.
8. d Manganese atoms gain 5 electrons when MnO_4^- (the oxidation number of Mn is +7) is converted to Mn^{2+} (the oxidation number of Mn is +2) in acidic solution. In basic solution, manganese atoms are converted from MnO_4^- (oxidation number of Mn is +7) to MnO_2 (oxidation number of Mn is +4) the Mn atoms gain 3 electrons.

8.6 Some Common Reducing Agents

Learning Objective: • Identify some common reducing agents.

A. Elemental carbon (coke) is used as a reducing agent for large-scale production of metals.

B. Antioxidants such as ascorbic acid (vitamin C) and vitamin E are reducing agents in food chemistry.

C. Hydrogen is an important reducing agent for metals, for reducing organic compounds, and for reducing nitrogen to ammonia (nitrogen fixing).
   1. Many reactions with H_2 require use of a catalyst, a substance that increases the rate of a reaction (by lowering the activation energy of the reaction) without being consumed by the reaction.
   2. Because of their affinity for H_2, metals such as iron, platinum, and palladium are often used as catalysts.

D. Catalysts are an important area of research in green chemistry.
   1. Catalysts increase the efficiency and efficacy of chemical and energy resources and reduce reaction time, leading to cost savings.
   2. Catalysts composed of using nontoxic materials are being designed to withstand thousands of catalytic cycles and can be effective at levels as low as 1 ppm.

E. A Closer Look at Hydrogen.
   1. Hydrogen represents only 0.9% of the Earth's crust (by weight), but is the most abundant element in the universe.
   2. Elemental hydrogen (uncombined) is rarely found on Earth.
   3. Combined hydrogen is found in water, natural gas, petroleum products, and all foodstuffs.
   4. Hydrogen can be ignited with a spark, as occurred in 1937 when the airship *Hindenburg* was destroyed in a fire and explosion over Lakehurst, New Jersey.

Answers to Self-Assessment Questions

1. c H_2 is often used to reduce metal ores to metals.
2. a C, often in the form of coke, is used to reduce metal ores to metals.
3. a Al does not interact with hydrogen.
4. d By lowering the activation energy, the catalyst speeds up the reaction.

8.7 Oxidation, Reduction, and Living Things

Learning Objective: • Write the overall equations for the metabolism of glucose and for photosynthesis

A. Reduced compounds represent a form of stored potential energy. The driving force to produce reduced compounds is ultimately derived from the sun in the processes of photosynthesis.
   1. Plants and the animals that feed on plants use the glucose to make other reduced compounds such as carbohydrates. The many steps of the metabolic process that releases the energy in carbohydrates, cellular respiration, are:

$$C_6H_{12}O_6 + 6\ O_2 \rightarrow 6\ CO_2 + 6\ H_2O + \text{energy}$$

   2. The overall photosynthesis reaction, the process by which plants utilize the energy from the sun to synthesize carbohydrates, is essentially the reverse of the cellular respiration reaction and is the only natural process that produces O_2.

$$6\ CO_2 + 6\ H_2O + \text{energy} \rightarrow C_6H_{12}O_6 + 6\ O_2$$

Answers to Self-Assessment Questions

1. b Cellular respiration, or the oxidation of glucose, is the main process used by animals to get energy.
2. b Photosynthesis by plants is the only natural process that produces oxygen.
3. b During photosynthesis, carbon is reduced from its highest oxidation form, CO_2. During respiration, the process is reversed and C is oxidized back to CO_2.

Green Chemistry: Green Oxidation Catalysts

Learning Objectives: • Distinguish between the objectives of chemists producing industrial chemicals and those of chemists producing specialty chemicals. • Explain how green chemistry can be applied to the design of new catalysts.

1. A catalyst provides a pathway for chemicals (reactants) to combine in a more effective manner than is possible without it.
   a. Catalysts are widely used in organic–oxidation reactions, especially those designed to produce industrial (produced on a large scale, often as starting materials for other substances) and specialty (complex compounds produced on a small scale) chemicals.
   b. Carefully chosen catalysts can be more effective in producing products with higher atom economy while, at the same time, be more environmentally friendly.

LEARNING OBJECTIVES

You should be able to . . .

1. Classify a particular change within a redox reaction as either oxidation or reduction. (8.1)
2. Identify an oxidation–reduction reaction. (8.1)
3. Identify the oxidizing agent and the reducing agent in a redox reaction. (8.2)
4. Balance redox equations. (8.3)
5. Identify and write the half-reactions in an electrochemical cell. (8.3)
6. Describe the reactions that occur when iron rusts. (8.4)
7. Explain why an explosive reaction is so energetic. (8.4)
8. Write equations for reactions in which oxygen is an oxidizing agent. (8.5)
9 List some of the common oxidizing agents encountered in daily life. (8.5)
10. Identify some common reducing agents. (8.6)
11. Write the overall equations for the metabolism of glucose and for photosynthesis. (8.7)
12. Distinguish between the objectives of chemists producing industrial chemicals and those of chemists producing specialty chemicals.
13. Explain how green chemistry can be applied to the design of new catalysts.

DISCUSSION

It is impossible to overemphasize the importance of oxidation–reduction processes. Think of it this way: You are powered by the energy of sunlight; only you can't simply unfold solar panels, as artificial satellites do, and convert sunlight to stored electrical energy to be tapped as necessary. You are a chemical factory and not an artificial satellite, solar-powered or otherwise. So somehow you have to tap that solar energy in a chemical way. This is precisely the role of oxidation–reduction reactions in life processes—they plug you into the sun.

You should practice writing the formulas and chemical equations for the oxidation–reduction reactions. Be sure to know the common oxidation and reduction agents and their applications. Remarkably, these applications range from large-scale industrial uses to batteries to household uses, and finally to the source of energy for living organisms, including us.

EXAMPLE PROBLEM

1. Identify the element being oxidized, the element being reduced, the oxidizing agent, and the reducing agent in the following reactions.
   a. $2\ Mg + CO_2 \rightarrow 2\ MgO + C$
   Mg gains oxygen; therefore, it is oxidized. CO_2 loses oxygen; therefore, it is reduced. If Mg is oxidized, CO_2 must be the oxidizing agent. If CO_2 is reduced, Mg must be the reducing agent.
   b. $C_2H_4 + N_2H_2 \rightarrow C_2H_6 + N_2$
   C_2H_4 gains hydrogen; therefore, it is reduced. N_2H_2 gives up hydrogen; therefore, it is oxidized. N_2H_2 is the reducing agent (it reduces C_2H_4), and C_2H_4 is the oxidizing agent (it oxidizes N_2H_2).
   c. $Cu + 2\ Ag^+ \rightarrow Cu^{2+} + 2\ Ag$
   Cu loses electrons to become Cu^{2+}; therefore, it is oxidized (and is the reducing agent). Ag^+ gains electrons to become Ag; therefore, it is reduced (and is the oxidizing agent).

ADDITIONAL PROBLEM

1. Identify the substance oxidized, the substance reduced, the oxidizing agent, and the reducing agent in each of the following equations.
   a. $CuO + H_2 \rightarrow Cu + H_2O$
   b. $C_2H_4 + 3\ O_2 \rightarrow 2\ CO_2 + 2\ H_2O$
   c. $Fe + Cu^{2+} \rightarrow Fe^{2+} + Cu$
   d. $5\ CO + I_2O_5 \rightarrow I_2 + 5\ CO_2$
   e. $CH_3CHO + H_2O_2 \rightarrow CH_3COOH + H_2O$
   f. $C_6H_{12}O_6 + 6\ O_2 \rightarrow 6\ CO_2 + 6\ H_2O$
   g. $16\ H^+ + 2\ Cr_2O_7^{2-} + 3\ C_2H_5OH \rightarrow 4Cr^{3+} + 3\ C_2H_4O + 11\ H_2O$

ANSWERS TO ADDITIONAL PROBLEM

a. $CuO + H_2 \rightarrow Cu + H_2O$
Cu is reduced from an ON of +2 to an ON of zero and is the oxidizing agent.
H_2 is oxidized from an ON of zero to an ON of +1 and is the reducing agent.

b. $C_2H_4 + 3\ O_2 \rightarrow 2\ CO_2 + 2\ H_2O$
Carbon is oxidized (gains O, loses H) and is the reducing agent.
O is reduced (gains H) and is the oxidizing agent.

c. $Fe + Cu^{2+} \rightarrow Fe^{2+} + Cu$
Fe is oxidized (loses electrons) and is the reducing agent.
Cu is reduced (gains electrons) and is the oxidizing agent.

d. $5\,CO + I_2O_5 \rightarrow I_2 + 5\,CO_2$
C is oxidized (gains O) and is the reducing agent.
I is reduced (loses O) and is the oxidizing agent.

e. $CH_3CHO + H_2O_2 \rightarrow CH_3COOH + H_2O$
One C atom in CH_3CHO is oxidized (is originally bonded to 1 O atom and, in the product, is bonded to 2 O atoms) and is the reducing agent. O in H_2O_2 is reduced. Its ON in H_2O_2 is –1 and, in H_2O is –2) and is the oxidizing agent.

f. $C_6H_{12}O_6 + 6\,O_2 \rightarrow 6\,CO_2 + 6\,H_2O$
C is oxidized (gains O) and is the reducing agent.
O in O_2 is reduced (gains H) and is the oxidizing agent.

g. $16\,H^+ + 2\,Cr_2O_7^{2-} + 3\,C_2H_5OH \rightarrow 4Cr^{3+} + 3\,C_2H_4O + 11\,H_2O$
Cr is reduced (goes from an ON of +6 to an ON of +3) and is the oxidizing agent.
C is oxidized (goes from an ON of –2 to an ON of –1) and is the reducing agent.

ANSWERS TO ODD-NUMBERED REVIEW QUESTIONS AND SOLUTIONS FOR ODD-NUMBERED END-OF-CHAPTER PROBLEMS

Review Questions

1. The oxidation number is increased during oxidation and decreased in reduction.

3. In an electrochemical cell, the anode is a conducting material in contact with a substance that will spontaneously release electrons to the substance in contact with the cathode. Electrons flow from the anode to the cathode because the substances that form as a result of the redox reaction are more energetically stable than the reactants.

5. The porous plate between two electrode compartments allows ions to pass through, thereby completing the electrical circuit and keeping the solutions electrically neutral.

7. The case (container) of a regular carbon–zinc dry cell battery is composed of Zn, which acts as the anode in the oxidation–reduction reaction. Zn is oxidized to Zn^{2+} in the reaction.

9. The reaction that occurs during the recharging of a lead storage battery is:
$2\,PbSO_4 + 2\,H_2O \rightarrow Pb + PbO_2 + 2\,H_2SO_4$. $PbSO_4$ is converted to Pb and PbO_2.

11. Lithium has a very low density and can provide higher voltages than other metals.

13. Ag is oxidized by H_2S to produce Ag_2S, which is black. Aluminum foil can be used to reduce Ag^+ back to Ag. In this way, the silver remains on the object rather than being abraded off with a rough cloth.

15. Through the process of photosynthesis, plants are able to store energy from the sun in the form of chemical bonds in reduced forms of carbon. We consume foods which contain these compounds and reverse the photosynthetic reaction by oxidizing the carbon atoms and releasing the stored energy. We use the released energy to fuel the physiological reactions that, among other things, enable our hearts to beat.

Problems

17. a. $C + 2\,Cl_2 \rightarrow CCl_4$ The oxidation number of carbon in C is zero; in CCl_4 the ON is +4. The ON increased so C is oxidized.
    b. $2\,C + O_2 \rightarrow 2\,CO$ The oxidation number of C increases from zero to +2. C is oxidized.
    c. $C + O_2 \rightarrow CO_2$ The oxidation number of C increases from zero to +4. C is oxidized.
    d. $C_2H_2 + I_2 \rightarrow C_2H_2I_2$ The oxidation of C increases from –1 to zero, so C is oxidized.
    e. $C + 2\,H_2 \rightarrow CH_4$ The oxidation number of C decreases from zero to –4, so C is reduced.

19. a. $Ca \rightarrow Ca^{2+} + 2\,e^-$
    b. $Al \rightarrow Al^{3+} + 3\,e^-$
    c. $Cu \rightarrow Cu^+ + e^-$
    $Cu \rightarrow Cu^{2+} + 2\,e^-$

21. a. $4\,Cu + O_2 \rightarrow 2\,Cu_2O$ Cu is oxidized; O_2 is reduced.
    b. $Mg + 2\,H^+ \rightarrow Mg^{2+} + H_2$ Mg is oxidized; H^+ is reduced.
    c. $Br_2 + H_2O_2 \rightarrow O_2 + 2\,H^+ + 2\,Br^-$ The O in H_2O_2 is oxidized; Br_2 is reduced.
    d. $4\,S_2O_3^{2-} + O_2 + 2H^+ \rightarrow 2\,S_4O_6^{2-} + 2\,H_2O$ S in $S_2O_3^{2-}$ is oxidized; O_2 is reduced.

23. a. $3\,C + Fe_2O_3 \rightarrow 3\,CO + 2\,Fe$ C is oxidized, so it is the reducing agent; Fe is reduced, so it is the oxidizing agent.
    b. $P_4 + 5\,O_2 \rightarrow P_4O_{10}$ P is oxidized. so it is the reducing agent; O_2 is reduced, so it is the oxidizing agent.
    c. $C + H_2O \rightarrow CO + H_2$ C is oxidized and is the reducing agent; H is reduced and is the oxidizing agent.
    d. $H_2SO_4 + Zn \rightarrow ZnSO_4 + H_2$ Zn is oxidized and is the reducing agent; H is reduced and is the oxidizing agent.

25. Copper reduces the silver from Ag^+ to Ag.

27. MnO_2 is reduced, so it is the oxidizing agent; Zn is oxidized, so it is the reducing agent.

29. The anode provides electrons, so it is the reducing agent; the cathode absorbs the electrons, so it is the oxidizing agent.

31. When glucose is oxidized, it provides electrons to reduce the enzyme, so the enzyme is the oxidizing agent.

33. a. $Fe \rightarrow Fe^{2+} + 2\,e^-$ half-reaction of oxidation.
    $2\,H^+ + 2\,e^- \rightarrow H_2$ half-reaction of reduction.
    b. $Cr^{2+}\ 2\,e^- \rightarrow Cr^0$ half-reaction of reduction.
    $Al \rightarrow Al^{3+} + 3\,e^-$ half-reaction of oxidation.

35. a. $Cu^{2+} + 2\,e^- \rightarrow Cu$ half reaction of reduction
    $Fe \rightarrow Fe^{2+} + 2\,e^-$ half reaction of oxidation
    Complete balanced reaction: $Cu^{2+} + Fe \rightarrow Cu + Fe^{2+}$
    b. $2\,H_2O_2 \rightarrow 2\,O_2 + 4\,H^+ + 4\,e^-$ half reaction of oxidation
    $Fe^{3+} + e^- \rightarrow Fe^{2+}$ half reaction of reduction (multiply by 4 to cancel electrons)
    Complete balanced reaction: $2\,H_2O_2 + 4\,Fe^{3+} \rightarrow 2\,O_2 + 4\,H^+ + 4\,Fe^{2+}$

c. $WO_3 + 6\,H^+ + 6\,e^- \rightarrow W + 3\,H_2O$ half reaction of reduction
$C_2H_6O \rightarrow C_2H_4O + 2\,H^+ + 2\,e^-$ half reaction of oxidation (multiply by 3, cancel H^+)
Complete balanced reaction: $WO_3 + 3\,C_2H_6O \rightarrow W + 3\,H_2O + 3\,C_2H_4O$

37. a. $Mn^{2+} \rightarrow Mn^{5+} + 3\,e^-$ Oxidation half reaction
$Pb^{4+} + 2\,e^- \rightarrow Pb^{2+}$ Reduction half reaction
Complete reaction: $2\,Mn^{2+} + 3\,Pb^{4+} \rightarrow 2\,Mn^{5+} + 3\,Pb^{2+}$
b. $2\,S^{2+} \rightarrow 2\,S^{2.5+} + e^-$ Oxidation half reaction
$O + 2\,e^- \rightarrow O^{2-}$ Reduction half reaction
Complete reaction: $4\,S^{2+} + O \rightarrow 4\,S^{2.5+} + O^{2-}$

39. a. $CH_3CHO + H_2O_2 \rightarrow CH_3COOH + H_2O$ One of the C atoms in CH_3CHO is oxidized; one of the O atoms in H_2O_2 is reduced.
b. $5\,C_2H_6O + 4\,MnO_4^- + 12\,H^+ \rightarrow 5\,C_2H_4O_2 + 4\,Mn^{2+} + 11\,H_2O$ C is oxidized; Mn is reduced.

41. a $2\,MnO_2 + NO_2^- + 2\,H^+ \rightarrow Mn_2O_3 + NO_3^-$ Mn is reduced from ON = +4 to ON = +3.
b $3\,MnO_2 + 2\,NH_4^+ + 4\,H^+ \rightarrow 3\,Mn^{2+} + N_2 + 6\,H_2O$ Mn is reduced from ON = +4 to ON = +2.

43. $C_{57}H_{104}O_6 + 3\,H_2 \rightarrow C_{57}H_{119}O_6$ The unsaturated oil is reduced because the number of H atoms per C atom has increased.

45. $2\,I^- + Cl_2 \rightarrow I_2 + 2\,Cl^-$ I^- is oxidized (loses electrons); Cl_2 is reduced (gains electrons)

47. The transformation is from $C_4H_6O_6 \rightarrow C_6H_{12}O_6$. The reactant, tartaric acid, is reduced to glucose in the ripening process. The C atoms in tartaric acid gain H and lose O atoms in the process.

49. In the transformation from $C_6H_8O_6 \rightarrow C_6H_6O_6$, H atoms were lost so the compound was oxidized.

51. $C_6H_{12}O_6 + 24\,Fe(OH)_3 + 48\,H^+ \rightarrow 6\,CO_2 + 24\,Fe^{2+} + 66\,H_2O$. In this reaction Fe is reduced as its ON changes from +3 to +2. Fe^{3+} is the oxidizing agent. C is reduced as its ON changes from zero in $C_6H_{12}O_6$ to +4 in CO_2.

53. a. $N_2 + 2\,O_2 \rightarrow 2\,NO_2$ (this is one possible answer)
b. $CS_2 + 3\,O_2 \rightarrow CO_2 + 2\,SO_2$
c. $C_5H_{12} + 8\,O_2 \rightarrow 5\,CO_2 + 6\,H_2O$

55. a. The oxidation number of C on CO is +2.
b. The oxidation number of C in CO_2 is +4.
c. The oxidation number of C in CH_4 is – 4.
d. The oxidation number of C in CCl_4 is +4.
e. The oxidation number of C in CH_2F_2 is +2.

57. The relative oxidizing strength of the halogens decreases down the group or family: F > Cl > Br > I.

59. The light grease or oil coating protects the metal tools from air oxidation.

61. $2\ CO_2 + e^- \rightarrow C_2O_4^{2-}$ The oxidation number of C goes from +4 in CO_2 to +3 in $C_2O_4^{2-}$ so CO_2 is reduced in this process.

63. volume of O_2/min = (15 breaths/min)(0.5 L air/breath)(21 L O_2/100 L air) = 1.6 L O_2/min
volume of O_2/day = (1.6 L O_2/min)(60 min/hr)(24 hr/day) = 2000 L O_2/day (one significant figure.)

65. The light reaction is given: $12\ H_2O \rightarrow 6\ O_2 + 24\ H^+ + 24\ e^-$. Electrons are released in this step of the reaction which means that this is an oxidation process. The oxidation number of O changes from –2 in H_2O to zero in O_2. The dark reaction must be a reduction half-reaction and include a combination of substances and electrons such that when the light and dark reactions are added together we end up with the reaction representing photosynthesis: $6\ CO_2 + 6\ H_2O \rightarrow C_6H_{12}O_6 + 6\ O_2$. This half reaction must have 24 H^+ and 24 e^- and 6 CO_2 as reactants and 6 H_2O (to cancel 6 of the H_2Os on the reactant side of the light reaction) plus $C_6H_{12}O_6$ on the product side. Thus, the dark reaction must be:
$6\ CO_2 + 24\ H^+ + 24\ e^- \rightarrow 6\ H_2O + C_6H_{12}O_6$. And the equation is balanced!

67. The overall photosynthetic reaction is $6\ CO_2 + 6\ H_2O \rightarrow C_6H_{12}O_6 + 6\ O_2$. The oxidation number of carbon in CO_2 is +4. The oxidation number of C in $C_6H_{12}O_6$ is zero.

69. Metals frequently act as reducing agents because they can often gain more stable valence electron configurations through the loss of electrons. Nonmetals can act as oxidizing or as reducing agents depending on the substance(s) they are paired with. For example, hydrogen peroxide, H_2O_2, can function as an oxidizing agent or as a reducing agent.

71. Burning H_2 reacts with Cl_2 to form HCl. Cl_2 is the oxidizing agent in this reaction.
$H_2 + Cl_2 \rightarrow 2\ HCl$

73. The balanced equation is: $2\ Al + 6\ H_2O \rightarrow 2\ Al(OH)_3 + 3\ H_2$.
Metallic Al is the reducing agent and H_2O is the oxidizing agent.

75.
a. NAD^+ is reduced to NADH and ethanol is oxidized to acetaldehyde.
The half-reaction of reduction is: $NAD^+ + H^+ + 2\ e^- \rightarrow NADH$.
The half-reaction of oxidation is: $CH_3CH_2OH \rightarrow CH_3CHO + 2\ H^+ + 2\ e^-$.
The complete reaction is: $NAD^+ + CH_3CH_2OH \rightarrow CH_3CHO + H^+ + NADH$.
b. The half-reaction of reduction is:
$CH_3COCOOH + 2\ H^+ + 2\ e^- \rightarrow CH_3CHOHCOOH$.
The half-reaction of oxidation is: $NADH \rightarrow NAD^+ + H^+ + 2\ e^-$.
The complete reaction is:
$CH_3COCOOH + NADH + H^+ \rightarrow CH_3CHOHCOOH + NAD^+$.

77.
a. Industrial chemicals are produced on a large scale.
b. Complex molecules are usually produced as specialty chemicals.
c. Specialty chemicals usually have a higher value in the marketplace than industrial chemicals.
d. Industrial chemicals are often the building blocks for more complex specialty chemicals.

79. Many oxidation catalysts have organic molecules built into their structures. These organic components may undergo chemical oxidation, causing the catalyst to lose effectiveness.

CHAPTER

9

Organic Chemistry

Organic Chemistry Versus Organic Foods

CHAPTER SUMMARY

Introduction

A. Organic chemistry is the chemistry of carbon-containing compounds.
   1. Carbon is unique because its atoms readily bond to one another, forming long, straight, and branched chains and ring structures (silica atoms can form only short chains).
   2. Carbon also frequently forms strong bonds to elements such as H, N, and O, and many times to the halogens (F, Cl, Br, and I).
   3. Nearly all common elements are found in at least a few organic compounds.

9.1 Aliphatic Hydrocarbons

Learning Objectives: • Define *hydrocarbon*, and recognize the structural features and properties of alkanes, alkenes, and alkynes. • Identify hydrocarbon molecules as alkanes, alkenes, or alkynes, and name them.

A. Hydrocarbons contain only hydrogen and carbon.
   1. Hydrocarbons are classified according to the type of bonding between carbon atoms.
   2. Hydrocarbons are divided into two subgroups: aliphatic compounds (with little multiple bonding between carbon atoms) and aromatic compounds (with extensive multiple bonding between C atoms. See Section 9.2).

B. Alkanes.
   1. Alkanes are hydrocarbons that contain only single bonds.
   2. Methane, CH_4, is the simplest example.
   3. The four attachments to the carbon atoms in alkanes are in a tetrahedral arrangement around the central carbon atom (see Chapter 4).
   4. Alkanes are considered to be saturated because each carbon atom is bonded to the maximum possible number of hydrogen atoms.
   5. Compounds are often represented with condensed structural formulas such as $CH_3CH_2CH_3$.
   6. Hydrocarbons are named according to the number of carbon atoms in the longest continuous chain of carbon atoms in the structure. The prefix to the name indicates the length of this chain, and the ending of the name, *ane*, indicates that the compound is a hydrocarbon. Any branching in the chain is both located with respect to its insertion into the longest chain and described in terms of its length.

7. It will be important to memorize the prefixes in Table 9.1 and to look carefully at the compound names in Table 9.2.

meth = 1	hex = 6
eth = 2	hept = 7
prop = 3	oct = 8
but = 4	non = 9
pent = 5	dec = 10

C. Homologous Series.
   1. A series of compounds whose properties vary in a regular and predictable manner.
   2. Example of a homologous series: CH_4, CH_3CH_3, $CH_3CH_2CH_3$, and $CH_3CH_2CH_2CH_3$, where each member differs from the one on either side of it by a CH_2 unit.
   3. Sometimes members of a homologous series are branched.

D. Isomerism.
   1. Because carbon chains can be straight or branched, often there are several different compounds with the same chemical formula. These compounds are referred to as isomers or constitutional isomers.
   2. As the number of carbon atoms in a series increases, the number of possible isomers increases dramatically.
   3. Examples of isomers:

butane 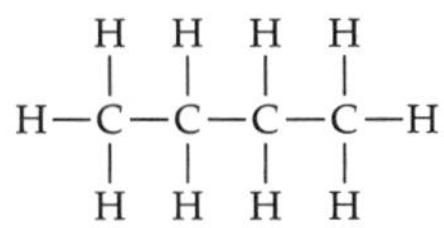

***iso*-butane** 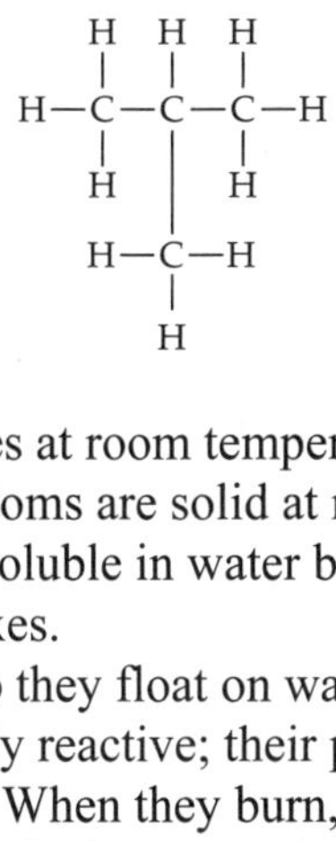

E. Properties of Alkanes.
   1. Alkanes with 1 to 4 C atoms are gases at room temperature; alkanes with 5 to 16 C atoms are liquids; alkanes with 17 or more C atoms are solid at room temperature.
   2. Alkanes are nonpolar and, hence, insoluble in water but they dissolve other nonpolar organic substances such as fats, oils, and waxes.
   3. Alkanes are less dense than water, so they float on water.
   4. Generally, alkanes are not particularly reactive; their primary reaction is combustion. Among their most important uses is as fuels. When they burn, they release a great deal of heat.
   5. In the lungs, alkanes can lead to chemical pneumonia.
   6. Heavy alkanes (long C chains) act as emollients (skin softeners).

F. Cyclic Hydrocarbons: Rings and Things.
   1. Compounds with closed rings of carbon atoms are called cyclic and are often represented with geometric shapes: Three-carbon ring: cyclopropane: triangle; four-carbon ring: cyclobutane: square; five-carbon ring: cyclopentane: pentagon; six-carbon: cyclohexane: hexagon.
   2. These compounds are named in a similar fashion to the aliphatic hydrocarbons except that each name begins with the prefix "cyclo."

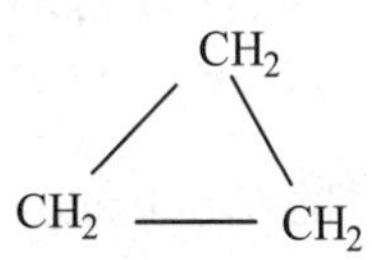

Cyclopropane

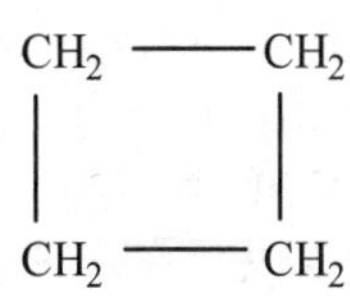

Cyclobutane

3. We can represent these structures with skeletal structures showing the shape of the molecule without drawing in the atoms (see Figure 9.5). Note: In skeletal structures, each line represents a shared electron pair, and there is a C atom at the ends of each line that is bonded to enough H atoms to complete its octet. Any atoms other than C and H must be indicated, and any multiple bonds must also be shown.

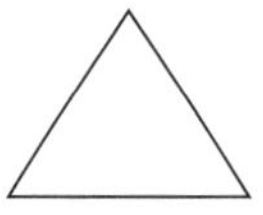

Cyclopropane

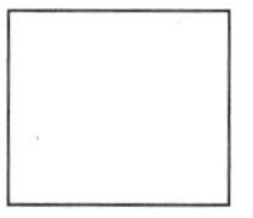

Cyclobutane

G. Unsaturated Hydrocarbons: Alkenes and Alkynes.
   1. An alkene is a hydrocarbon that contains one or more carbon–carbon double bonds (C=C).
   2. An alkyne is a hydrocarbon that contains one or more carbon–carbon triple bonds (C≡C).
   3. Collectively, alkenes and alkynes are called unsaturated hydrocarbons because they can have more hydrogen atoms added to them.

H. Properties of Alkenes and Alkynes.
   1. Unsaturated hydrocarbons burn in the presence of oxygen.
   2. The carbon–carbon double bond in these compounds is reactive; many different small molecules can add to the compound at this site, including, among others, hydrogen (H_2), chlorine (Cl_2), bromine (Br_2), and water (H_2O). These reactions are called addition reactions.

I. Because rotation is restricted at the site of a C=C bond, alkenes can form geometric isomers.
   1. When a C=C occurs other than between the first and second C atoms in a chain, two distinct compounds can form.

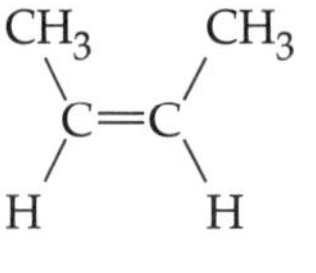

cis-2-butene

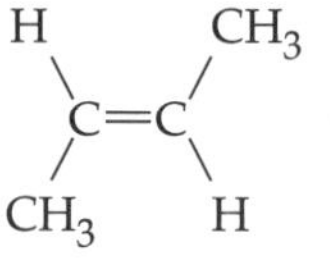

trans-2-butene

   2. In the *cis* isomer, the chain enters and exits the C=C site from the same side of the molecule; in the *trans* isomer, the chain enters and exits the C=C from opposite sides.
   3. For geometric isomers to form, two requirements must be met: (1) Rotation must be restricted in the molecule, and (2) there must be two nonidentical groups on each of the doubly bonded C atoms.

Answers to Self-Assessment Questions

1. d Organic chemistry is the study of the compounds of carbon.
2. d Hydrocarbons have the simplest composition because they are formed from only C and H atoms. As you will see, in addition to C and H atoms, alcohols and carbohydrates contain O atoms, and amino acids contain N, O, and sometimes S atoms in addition to C and H atoms.
3. b Each successive member of the alkane series differs by CH_2.
4. c Isomers have different structures but the same chemical formula.
5. b Boiling points of alkanes increase as the molecular weights increase. Among other things, this is because the extent to which dispersion forces occur between molecules increases as their molecular masses increase.
6. c The prefix "cyclo" indicates a cyclic compound, "hex" indicates six carbons, and the ending "ane" indicates that the compound is a hydrocarbon.
7. c Alkynes are identified as compounds containing one or more C≡C bond.

8. a Because the carbon atoms in alkanes are already bonded to four different atoms, it is not possible to insert any more atoms into those structures. The carbon atoms involved in the C=C bond of alkenes are bonded to three different atoms, so it is possible to break one of the electron pairs in the C=C bond and attach each of the carbons to one more atom.
9. b. Geometric isomers occur when rotation is restricted and there are two nonidentical groups attached to the C atoms in the restricted area.

9.2 Aromatic Hydrocarbons: Benzene and Its Relatives

Learning Objectives: • Define aromatic compound and recognize the structural feature such compounds share. • Name simple aromatic hydrocarbons.

A. Benzene was first isolated in 1825. It has the molecular formula C_6H_6.
B. In 1865, Kekulé proposed a cyclic structure with alternating single and double carbon–carbon bonds for benzene. However, it has been determined that all of the bonds in benzene are identical and that there are no isolated double bonds.
   1. Benzene doesn't react by addition reactions the way unsaturated hydrocarbons do, but rather by substitution reactions, like saturated hydrocarbons.
C. Benzene and compounds containing the benzene-type ring structure are called aromatic compounds. Many have pleasant aromas; others stink.
   1. Today aromatic means a compound with rings of electrons and properties similar to benzene. It no longer refers to odor.
   2. Monosubstituted benzenes are compounds where one of the H atoms is replaced with another atom or group called a substituent.

Answers to Self-Assessment Questions

1. a The formulas for alkanes follow the pattern C_nH_{2n+2}. Alkenes and alkynes have fewer H atoms because of the C=C or C≡C bonds; thus, at the very most, the formula for an unsaturated compound follows the pattern C_nH_{2n}. Answer a is the pattern for an unsaturated compound (an alkane).
2. c Chlorobenzene is a ring structure (benzene in which one of the H atoms is replaced by a Cl atom). Cyclohexene is a six-membered ring with just one C=C. The other two choices are not ring structures.
3. b Aromatic hydrocarbons have benzene-like structures.
4. d Naphthalene is an aromatic compound with alternating single and double bonds in two fused rings.
5. d The reactivity of the benzene ring reflects the spreading of six electrons over all six carbon atoms, so it does not have isolated double bonds.

9.3 Halogenated Hydrocarbons: Many Uses, Some Hazards

Learning Objective: • Name a halogenated hydrocarbon given its formula, and write the formula for such a compound given its name.

A. Methane derivatives.
   1. Substitution of a chlorine atom for one of the H atoms in an alkane occurs in the presence of ultraviolet light.
   2. Methyl chloride (CH_3Cl): used in making silicones.
   3. Methylene chloride (CH_2Cl_2): solvent, paint remover.
   4. Chloroform ($CHCl_3$): early anesthetic; industrial solvent.
   5. Carbon tetrachloride (CCl_4): dry-cleaning solvent.

B. Properties of chlorinated hydrocarbons.
   1. Only slightly polar, thus insoluble in water.
   2. These compounds dissolve fats, greases, and oils, thereby making them useful as solvents.
   3. Some chlorinated compounds (DDT, PCBs, etc.) are stored in fatty animal tissue.

C. Chlorofluorocarbons and fluorocarbons.
   1. CFCs: Carbon compounds containing fluorine, as well as chlorine.
   2. Typical compounds are gases or low-boiling liquids at room temperature.
   3. They are insoluble in water and inert toward most chemical substances.
      a. Used as propellants (for aerosol spray cans).
      b. Their inertness leads to stability in nature, meaning they break down slowly. Diffusion into the stratosphere may damage the ozone layer.
   4. Perfluorinated compounds have been used as blood extenders (they dissolve large amounts of oxygen) and in Teflon.
   5. The prefix "per" means all the H atoms have been replaced.

Answers to Self-Assessment Questions

1. a $CHCl_3$ is named trichloromethane and commonly known as chloroform.
2. a CF_2Cl_2 is a chlorofluorocarbon because it contains both Cl and F atoms.
3. c 1-chloropropane, $CH_3CH_2CH_2Cl$, is an isomer of $CH_3CHClCH_3$, which is 2-chloropropane.

9.4 The Functional Group

Learning Objectives: • Classify an organic compound according to its functional group(s), and explain why the concept of a functional group is useful in the study of organic chemistry. • Recognize and write the formulas of simple alkyl groups.

A. Organic compounds can be divided into groups or families based on groups of atoms in their structures that account for their chemical and physical properties.
   1. A functional group: an atom or group of atoms that gives a family of organic compounds its characteristic chemical and physical properties. See Table 9.4, which summarizes the functional groups' names, structure, and general formulas. Note: It is very useful for you to become familiar with these functional groups so that when you encounter the structure of a molecule you can identify its functional group(s) and anticipate the compound's chemical and physical properties

2. In many simple molecules, functional groups are attached to hydrocarbon stems called alkyl groups.
   a. Because the chemical properties of a substance are usually determined by reactions of the functional group, it often doesn't matter what alkyl groups are attached. The letter "R" represents alkyl groups in general structures.

B. Alkyl groups can be derived from alkanes by removing an H atom from the alkane. They are named by replacing "-*ane*" with "-*yl*."

methyl $CH_3–$ propyl $CH_3–CH_2–CH_2–$
ethyl $CH_3–CH_2–$ isopropyl $CH_3–CH–CH_3$ (with a bond | from the central CH)

Answers to Self-Assessment Questions

1. c A functional group is a specific arrangement of atoms that gives characteristic properties to an organic molecule.
2. a $CH_3CH_2CH_2$— is a three-carbon alkyl group.
3. a There are two carbon atoms in an ethyl group.
4. a The formula $CH_3CH_2CH_2CH_2$— represents a butyl group.

9.5 Alcohols, Phenols, and Ethers

Learning Objectives: • Recognize the general structure and properties for an alcohol, a phenol, and an ether. • Name simple alcohols, phenols, and ethers.

A. An alcohol has a hydroxyl group (OH) substituted for an H of the corresponding alkane.
   1. The names for alcohols are based on those of alkanes, with the ending changed from *–e* to *–ol*. A number designates the carbon to which the —OH group is attached: $CH_3CH_2CH_2CH_2OH$ is 1-butanol and $CH_3CH_2CHOHCH_3$ is 2-butanol.
   2. Many alcohols are also known by common names.

B. Methyl alcohol (methanol) is the first member of the family.
   1. Methanol is called wood alcohol and once was made by the destructive distillation of wood.
   2. Methanol, a valuable industrial solvent, is now made from carbon monoxide and hydrogen.

$$CO(g) + 2\,H_2(g) \rightarrow CH_3OH(l)$$

C. Ethyl Alcohol (Ethanol).
   1. Ethyl alcohol (ethanol) is also called grain alcohol.
   2. Most ethyl alcohol is made by the fermentation of grain.
   3. Denatured alcohol, ethanol to which toxic substances have been added, is not fit to drink.

D. Toxicity of Alcohols.
   1. Methyl alcohol can cause blindness or death.
   2. Ethyl alcohol is less toxic than methyl alcohol.
   3. "Proof" of an alcoholic beverage is twice the percent of alcohol in it, by volume.

E. Multifunctional Alcohols.
   1. Some alcohols have more than one hydroxyl (OH) group.
      a. Ethylene glycol (an ingredient in some antifreeze).

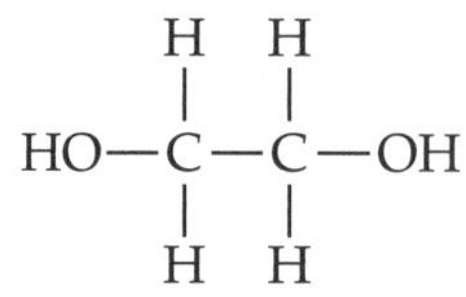

Ingestion leads to kidney damage and possibly death.

b. Glycerol is used in lotions and for making nitroglycerin.

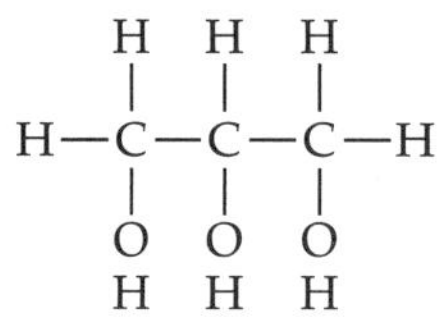

Glycerol

F. Phenols.
 1. Phenols have an —OH group attached directly to a benzene ring.
 2. Phenols are widely used as antiseptics.

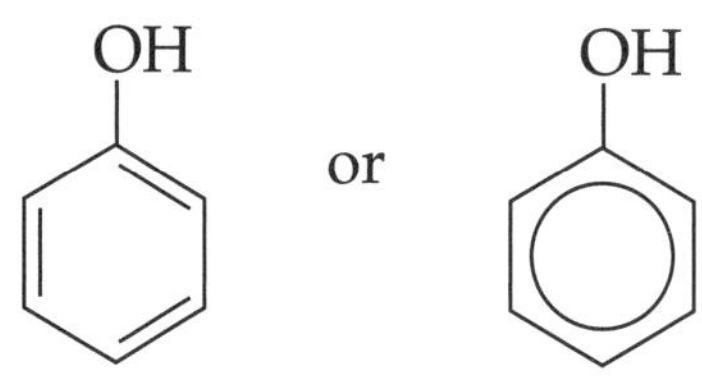

Phenol, also known as carbolic acid

G. Ethers
 1. Ethers have two hydrocarbon groups attached to the same oxygen atom (R—O—R′).
 2. Diethyl ether, once used as an anesthetic, is an important solvent. It is highly flammable.
 3. Methyl *tert*-butyl ether is sometimes used as an octane booster in gasoline.

Answers to Self-Assessment Questions

1. a Only the first formula contains an OH group attached to a carbon atom.
2. b There are three —OH groups in a glycerol molecule.
3. d The —OH groups in ethylene glycol are on separate carbon atoms.
4. a Alcohols are compounds with an —OH group attached to an aliphatic (nonaromatic) carbon atom.
5. d Compounds with the smallest structures and least amount of intermolecular interaction have the lowest boiling point. All of the choices are alcohols, so they can experience hydrogen bonding. The number of carbon atoms in propanol (3) is smaller than the number of C atoms in any of the other choices, so the dispersion forces felt between propanol molecules would be the least consequential and its boiling point would be lower.
6. b Ethers have two hydrocarbon groups attached to the same oxygen atom.
7. d The two aliphatic groups attached to the oxygen atom are both ethyl groups.
8. d One of the aliphatic groups attached to the oxygen has four carbons (butyl), and the other has two carbons (ethyl).
9. d Isomers have the same chemical formulas but different structures. Both $CH_3CHOHCH_2CH_3$ and $CH_3CH_2CH_2CH_2OH$ are forms of butanol. The first is 2-butanol, while the compound in answer d is 1-butanol.

9.6 Aldehydes and Ketones

Learning Objective: • Name simple aldehydes and ketones, and list their important properties.

A. Both aldehydes and ketones have a carbonyl functional group.

B. Aldehydes

1. Aldehydes have at least one H atom attached to the carbonyl carbon atom.

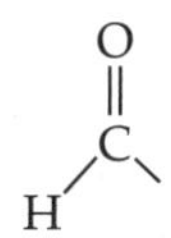

2. Many aldehydes have common names. The systematic name of an aldehyde is based on the name of the alkane with the same number of carbon atoms, to which the ending *–al* is added. For example, CH_3CHO is ethanal.
3. The simplest aldehyde is formaldehyde, made by the oxidation of methanol. It is used as a preservative and to make plastics.

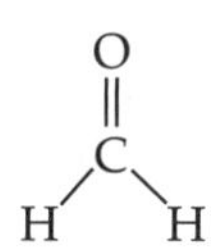

4. The next higher aldehyde is acetaldehyde, made by the oxidation of ethanol.

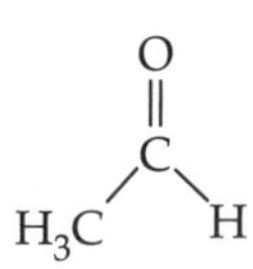

5. Benzaldehyde has an aldehyde group attached to a benzene ring. It is called oil of almond and is used in flavors and perfumes.

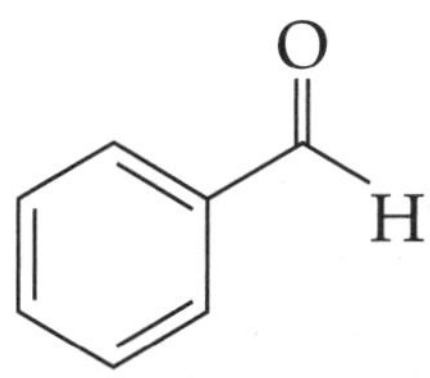

C. Ketones have their carbonyl carbon atom joined to two other carbon atoms. Acetone, a common solvent, is an example.

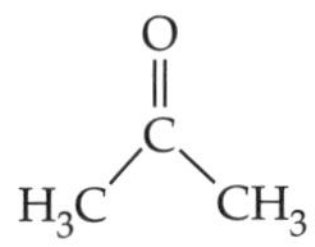

1. The systematic names for ketones are based on those of alkanes, with the ending changed from *-e* to *-one*. When necessary, a number is used to indicate the location of the carbonyl group. For example, $CH_3COCH_2CH_2CH_3$ is 2-pentanone.

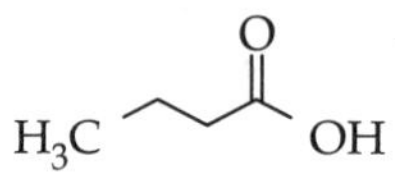

Answers to Self-Assessment Questions

1. a The aldehyde functional group has an H atom attached to a carbonyl group.
2. c The ketone functional group has a carbonyl group attached to two carbon atoms.
3. c R—CO—R′ is a ketone, with two alkyl groups attached to the carbonyl carbon atom.
4. b The carbonyl carbon atom is doubly bonded to one oxygen atom and singly bonded to the other. 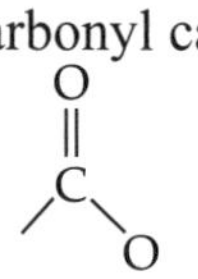
5. a $CH_3COCH_2CH_3$ is a four-carbon ketone, so it is butanone ("but" indicates 4 carbons, "one" indicates the ketone functional group).
6. a Formaldehyde has one carbon and is produced from CH_3OH, which has one carbon.
7. d Acetone has two carbon atoms on each side of the central oxygen atom and is produced from the oxidation of $CH_3CHOHCH_3$.

9.7 Carboxylic Acids and Esters

Learning Objectives: • Name simple carboxylic acids and esters, and list their important properties.

A. Organic or carboxylic acids have the carboxyl group as their functional group.
   1. Carboxyl group.

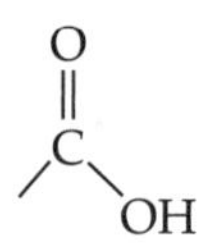 or –COOH

   2. Carboxylic acids are weak acids.

B. Common carboxylic acids include

1.	HCOOH	formic acid	in ant bites, bee stings
2.	CH_3COOH	acetic acid	in vinegar
3.	CH_3CH_2COOH	propionic acid	salts used as preservatives
4.	$CH_3CH_2CH_2COOH$	butyric acid	rancid butter (stench)
5.	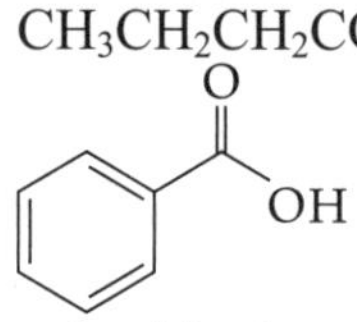	benzoic acid	salts used as preservatives

B. Carboxyl acid salts are used as food preservatives (calcium propionate, sodium benzoate).

C. Esters: The Sweet Smell of RCOOR'
   1. Esters are derived from the reaction of a carboxylic acid with an alcohol or phenol.
   2. General formula.

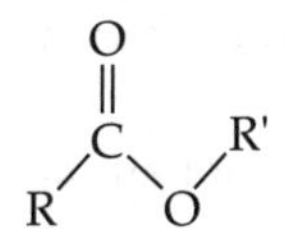

   3. Although derived from carboxylic acids with unpleasant odors, many esters have pleasant odors.
   4. They are used in fragrances and flavors.

D. Green Chemistry—The Art of Organic Synthesis: Green Chemists Find a Better Way.
   1. Most drugs are synthesized by combining smaller, simpler molecules, often using petroleum-based solvents which are flammable, toxic, and/or present disposal problems.
   2. New methods are being developed that require no solvent at all or that can be carried out in renewable solvent.
   3. The water insolubility of many organics is being overcome by "on water" reactions.
   4. Pfizer was awarded a Green Chemistry Challenge Award in 2002 when it was able to replace the four solvents originally used in the manufacture of Zoloft with ethanol, with a 10-fold reduction in the total volume of solvent required.
   5. Dow Chemical revised its synthesis of epichlorohydrin to better fit with green chemistry principles.
   6. As a means of reducing the energy demands of some synthetic reactions, microwave radiation is replacing standard heating processes.

E. Salicylates: Pain Relievers Based on Salicylic Acid.
   1. Salicylic acid, isolated from willow bark in 1838, was used as a fever reducer but was sour and irritating when taken orally.
   2. After trying several structural modifications that were not satisfactory, salicylic acid was reacted with acetic acid, which produced acetylsalicylic acid, the compound we now know as aspirin.

Answers to Self-Assessment Questions

1. c The COOH in $CH_3CH_2CH_2COOH$ is the carboxylic acid group.
2. d $CH_3CH_2CH_2COOH$ is a carboxylic acid with four carbon atoms.
3. d The reaction products of the oxidation of secondary alcohols are ketones.
4. a Ethanol with two carbon atoms would be oxidized to ethanoic (acetic) acid or CH_3COOH.
5. c RCOOR′ is the general formula for an ester.
6. c $CH_3CH_2CH_2CH_2COOCH_3$ is an ester that was formed from the reaction of a five-carbon carboxylic acid (pentanoic acid) with a one-carbon alcohol (methanol). The compound name is methyl pentanoate.
7. d $CH_3CH_2COOCH_3$ is an ester formed from the reaction of a three-carbon carboxylic acid (propanoic acid) with a one-carbon alcohol (methyl alcohol), so its name is methyl propanoate.

9.8 Nitrogen-Containing Compounds: Amines and Amides

Learning Objectives: • Name and write formulas of simple amines and amides. • Recognize the structure and characteristics of a heterocyclic compound.

A. Amines are basic compounds derived from ammonia (NH_3) by replacing one or more of the hydrogen atoms with alkyl or aromatic groups.
   1. RNH_2, R_2NH, and R_3N are amines.
   2. Amines are denominated by naming the alkyl groups first and adding the ending "-amine."
      CH_3NH_2 — Methylamine; $CH_3CH_2NH_2$ — Ethylamine; CH_3NHCH_3 — Dimethylamine
   3. The simplest aromatic amine has the special name aniline.

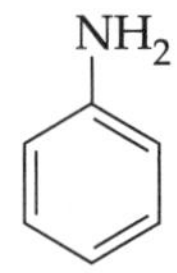

   4. The —NH_2 group is called an amino group. An example of the use of this term is with amino acids, molecules that contain both a carboxylic acid and an amino group and are the structural units for proteins.

B. Amides have a nitrogen atom attached to a carbonyl carbon atom.

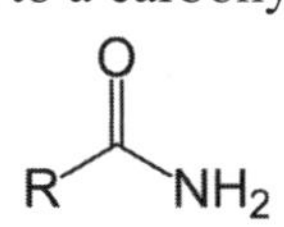

   1. Urea (H_2NCONH_2), the compound that helped change the understanding of organic chemistry, is an amide.
   2. Nylon, silk, and wool are amides.
   3. Names for simple amides are derived from the corresponding carboxylic acids or based on those of alkanes with the ending changed from *–e* to *–amide*.

C. Heterocyclic Compounds: Alkaloids and Others.
   1. Heterocyclic compounds are those having one or more atoms other than C in a ring structure.
   2. These compounds usually contain N, O, or S in the ring.
   3. Pyrimidine and purine, found in DNA, are examples.

D. Alkaloids are amines, often heterocyclic (with N), that occur naturally in plants.
   1. Morphine, nicotine, caffeine, and cocaine are examples.

Answers to Self-Assessment Questions

1. c In primary amines there are two H atoms bonded to the N atom, which is the case for isopropyl amine.
2. b $CH_3CH_2NHCH_3$ has an ethyl group and a methyl group attached to the N atom. Notice that the two alkyl groups are named in alphabetical order.
3. a CH_3CONH_2 is acetamide (a common name) or ethanamide.
4. b Amides have a carbonyl group attached to a nitrogen atom.
5. c Heterocyclic compounds contain at least one atom in the ring that is not a carbon atom.
6. a Alkaloids are cyclic organic compounds found in plants that have at least one nitrogen atom in the ring.

Green Chemistry: The Art of Organic Synthesis: Green Chemists Find a Better Way

Learning Objective: • Identify greener solvents that can replace those from nonrenewable fossil fuels, methods for incorporating renewable resources into organic synthesis, and efficient energy sources for enhancing chemical reactions.

A. Many medicines are synthesized through chemical reactions rather than extracted from natural products.
   1. Most of these synthesis reactions involve solutions with solvents that will dissolve the starting materials, many of which are nonpolar.
   2. Typically used nonpolar solvents are toxic and flammable, so green chemists are designing synthesis reactions using water or more environmentally friendly organic solvents, or using no solvent at all.
   3. Green chemical processes are also being facilitated through the use of especially designed microwave ovens to speed reactions and improve their yields.

LEARNING OBJECTIVES

You should be able to . . .

1. Define *hydrocarbon* and recognize the structural features of alkanes, alkenes, and alkynes. (9.1)
2. Identify hydrocarbon molecules as alkanes, alkenes, or alkynes, and name them. (9.1)
3. Define *aromatic compound*, and recognize the structural feature such compounds share. (9.2)
4. Name simple aromatic hydrocarbons. (9.2)
5. Name a halogenated hydrocarbon given its formula, and write the formula for such a compound given its name. (9.3)
6. Classify an organic compound according to its functional group(s), and explain why the concept of a functional group is useful in the study of organic chemistry. (9.4)
7. Recognize and write the formulas of simple alkyl groups. (9.4)
8. Recognize the general structure and properties for an alcohol, a phenol, and an ether. (9.5)
9. Name simple alcohols, phenols, and ethers. (9.5)
10. Name simple aldehydes and ketones, and list their important properties. (9.6)
11. Name simple carboxylic acids and esters, and list their important properties. (9.7)
12. Name and write the formulas of simple amines and amides. (9.8)
13. Recognize a structure and characteristics of a heterocyclic compound. (9.8)

14. Identify greener solvents that can replace those from nonrenewable fossil fuels, methods for incorporating renewable resources into organic synthesis, and efficient energy sources for enhancing chemical reactions.

DISCUSSION

Understanding the chemistry of carbon and the hydrocarbons is essential for comprehending our energy problems and associated pollution problems. Perhaps equally important, the chemistry of the hydrocarbons is the basis for most of the organic and biological chemistry that we will encounter in subsequent chapters. Organic chemistry is very extensive because carbon bonds to itself so well and in so many combinations. Focus on the characteristics of the classes (alkanes, alkenes, etc.) of organic compounds to organize your study.

Once again, you may find flash cards helpful in learning the names of the hydrocarbons. Just write the name on one side of a note card and the structure on the other. Look at the name and see if you can write the structure (or vice versa). Then flip the card to see if you got it right.

Adapting the principles of green chemistry to a wide variety of organic reactions has greatly improved the speed, cost, and safety of these processes. In addition, the amount of hazardous waste produced in these reactions has been greatly reduced.

ANSWERS TO ODD-NUMBERED REVIEW QUESTIONS AND SOLUTIONS FOR ODD-NUMBERED END-OF-CHAPTER PROBLEMS

Review Questions

1. Three characteristics of the carbon atom that make possible millions of organic compounds are:
    a. Carbon atoms can bond strongly to each other;
    b. Carbon atoms can bond strongly to other atoms;
    c. Carbon atoms can form chains, rings, and other kinds of structures.

3. Isomers have the same general formula but different structures. If you can draw two or more structures of compounds with the same general formula, they are isomers.

5. The circle represents the six delocalized electrons from the three double bonds and indicates that the ring is aromatic.

7. Methane through butane (C_1 through C_4) are gases.
Pentane through hexadecane (C_5 through C_{16}) are liquids.
Heptadecane (C_{17}) and larger hydrocarbons are solids.

9. a. Grain alcohol is ethyl alcohol or ethanol.
    b. Rubbing alcohol is isopropyl alcohol or 2-propanol.
    c. Wood alcohol is methyl alcohol or methanol.

11. Diethyl ether was used as an anesthetic but is now used primarily as solvent.

13. a. ROH is the general formula for alcohols.
    b. RCOR′ is the general formula for ketones.
    c. RCOOR′ is the general formula for esters.
    d. ROR′ is the general formula for ethers.
    e. RCOOH is the general formula for carboxylic acids.
    f. RCHO is the general formula for aldehydes.

Problems

15. Compounds a and d are organic because they contain carbon. Compounds b and d do not contain carbon and, therefore, are not considered to be organic.

17. a. Hexane has six carbon atoms.
    b. Nonane has nine carbon atoms.
    c. Cyclopentane has five carbon atoms.
    d. 2-pentene has five carbon atoms.

19. a. The general formula for alkanes is C_nH_{2n+2}. C_8H_{18} is the molecular formula of an alkane with eight carbon atoms.
    b. The molecular formula of an alkane with 14 carbon atoms is $C_{14}H_{30}$.

21. a. $CH_3CH_2CH_2CH_3$ is butane.
    b. $CH_2{=}CHCH{=}CH_2$ is 1,3-butadiene or butadiene.
    c. C_2H_2 is ethyne and is commonly known as acetylene.

23. a. The molecular formula for pentane is C_5H_{12}, and the condensed formula is $CH_3CH_2CH_2CH_2CH_3$.
    b. The molecular formula for decane is $C_{10}H_{22}$, and the condensed formula is $CH_3CH_2CH_2CH_2CH_2CH_2CH_2CH_2CH_2CH_3$.

25. a. CH_3— is a methyl group.
    b. CH_3CH_2CH— is a *sec*-butyl group.

           |
           CH_3

27. a. CH_3CH_2OH is ethanol.
    b. $CH_3CH_2CH_2OH$ is 1-propanol.
    c. The structure of methanol is CH_3OH.
    d. The structure of 2-heptanol is $CH_3CH_2CH_2CH_2CH_2CH(OH)CH_3$.

29. The structure for phenol is

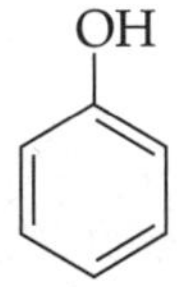

or

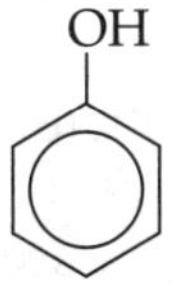

a. The structure of 2-methyl phenol is:

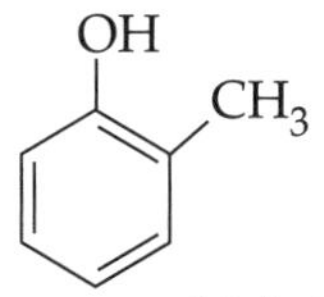

b. The structure of 4-iodophenol is:

31. a. The structure of dipropyl ether is $CH_3CH_2CH_2OCH_2CH_2CH_3$.
    b. The structure of butyl ethyl ether is $CH_3CH_2CH_2CH_2OCH_2CH_3$.

33. a. The structure of butanone is $CH_3CH_2COCH_3$.
    b. The structure of formaldehyde is HCHO.
    c. The structure is butanal.
    d. The structure is butyl ethyl ketone or 3-heptanone.

35. a. CH_3CH_2COOH is propanoic acid.
    b. $CH_3CH_2CH_2CH_2COOH$ is pentanoic acid.
    c. The structure of formic acid is HCOOH.
    d. The structure of benzoic acid is C_6H_6.

37. a. The structural formula for methyl acetate is CH_3COOCH_3.

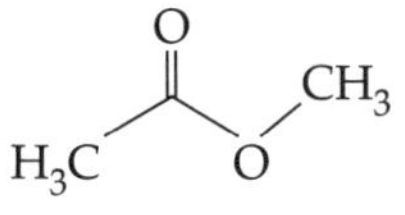

    b. The structural formula for ethyl butyrate is $CH_3CH_2CH_2COOCH_3$.

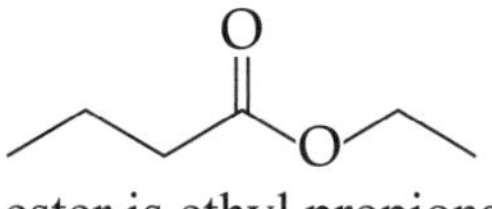

    c. This ester is ethyl propionate.

39. a. The structural formula for triethylamine is $(CH_3CH_2)_3N$.
    b. The formula for ethylmethylamine is $CH_3CH_2NHCH_3$.
    c. $CH_3CH_2CH_2NH_2$ is propylamine.
    d. $CH_3CH_2NHCH_3$ is ethylmethylamine.

41. a. These are the same compound, ethane, drawn in different directions.
    b. These are the same compound, propane, drawn in different ways.
    c. These are different isomers. The first is 2-methylpentane, and the second is 3-methylpentane.

43. a. $CH_3CH_2CH_3$ and $CH_3CH_2CH_2CH_3$ are homologs as they differ by only one CH_2 unit.
b. These compounds are none of these because one is a cyclic compound and the other is not.

45. a. This is unsaturated and an alkene.
b. This is unsaturated and an cyclic alkene.

47. Malic acid has one alcohol group (–OH) and two carboxylic acid groups (–COOH).

49. a. This is a heterocyclic compound and an amine.
b. This is a cycloalkane substituted with an amino group (not heterocyclic).

Additional Problems

51. a. The molecular formula of gingerol is $C_{17}H_{26}O_4$.
b. The molecular formula of shogaol is $C_{17}H_{24}O_3$.
c. The structure of gingerol includes a phenol (an aromatic alcohol), an ether, a ketone, and an aliphatic alcohol.
d. The structure of shogaol includes a phenol (an aromatic alcohol), an ether, a ketone, and an alkene (C=C).
e. In order to create shogaol from gingerol, the aliphatic alcohol group and one H on an adjacent C atom are removed. This is a dehydration reaction because the elements of water are removed.

53. a. $CH_3CH_2C{\equiv}CCH_3 + 2\ H_2 \rightarrow CH_3CH_2CH_2CH_2CH_3$
b. $CH_3CH{=}C(CH_3)CH_2CH_2CH_2CH_3 + H_2 \rightarrow CH_3CH_2CH(CH_3)CH_2CH_2CH_2CH_3$

55. a. $CH_3CH_2CH_2CHO$, butanal, is produced by the oxidation of $CH_3CH_2CH_2CH_2OH$, 1-butanol.
b. $CH_3CH_2COCH_2CH_3$, 3-pentanone, is produced by the oxidation of 3 pentanol, $CH_3CH_2CH(OH)CH_2CH_3$.
c. $CH_3COCH(CH_3)_2$, 3-methyl-2-pentanone, is produced by the oxidation of 3-methyl-2-pentanol, $CH_3CHOHCH(CH_3)_2$.
d. C_6H_5CHO, benzaldehyde, is produced by the oxidation of benzyl alcohol, $C_6H_5CH_2OH$.

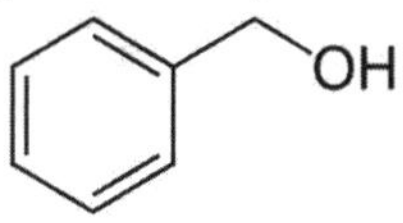

benzyl alcohol

57. These compounds are all isomers—they all have the same chemical formula ($C_4H_{10}O$), but the arrangement of the atoms in each of the molecules is different. Two of the compounds are alcohols and two are ethers, and neither of the alcohols is the same, nor are the ethers the same.

59. The balanced reaction is $CH_2{=}CH_2 + H_2O \rightarrow CH_3CH_2OH$ (it's sometimes easier to change from structures to formulas to make it easier to balance equations: $C_2H_4 + H_2O \rightarrow C_2H_6O$).
The molar masses of ethylene ($CH_2{=}CH_2$) and ethanol (CH_3CH_2OH) are 28.0 g/mol and 46.0 g/mol, respectively.

$$\text{mass of } C_2H_6O \text{ formed} = (445\ \text{g}\ C_2H_4)(1\ \text{mol}\ C_2H_4/28.0\ \text{g}\ C_2H_4)(1\ \text{mol}\ C_2H_6O/1\ \text{mol}\ C_2H_4)(46.0\ \text{g}\ C_2H_6O/\text{mol}\ C_2H_6O) = 731\ \text{g}\ C_2H_6O$$

61. a. The oxygen–carbon bonds in dimethyl ether (CH_3OCH_3) are polar covalent bonds, allowing for relatively weak dipole-dipole intermolecular forces between the molecules.
b. Both the C-O and O-H bonds in ethanol, CH_3CH_2OH, are polar, and the O-H bond allows for hydrogen-bonding interactions. Hydrogen bonds are much stronger than dipole-dipole forces, which is the reason ethanol is a liquid at room temperature while dimethyl ether is a gas.

63. a. The structure of hydroquinone:

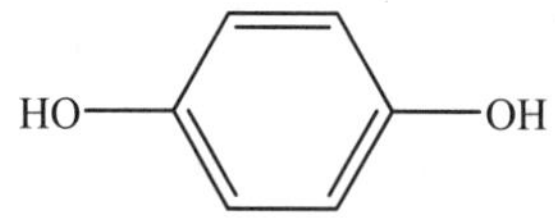

b. The structure of para-benzoquinone

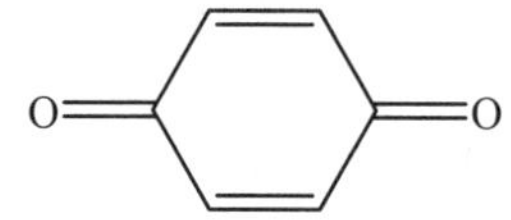

65. Density = mass/volume. The ideal gas law, PV = nRT, can be rewritten as PV = [(mass)RT]/(molar mass). Rearranging the ideal gas law and solving for density, d = (P)(molar mass)/(RT). The molar masses of the first four gases in Table 9.3 are: methane, 16.0 g/mol; ethane, 30.0 g/mol; propane, 44.0 g/mol; and butane, 58.0 g/mol. Substituting the molar mass of each gas into the equation, we can calculate the density of each gas.

a. Methane: d = [(1 atm)(16.0 g/mol)]/[(0.0821 L atm/mol K)(293.1 K)] = 0.655 g/L
Converting this density, in g/mL d = (0.655 g/L)(1 L/1000mL) = 6.55×10^{-4} g/mL

b. Ethane: d = [(1 atm)(30.0 g/mol)]/[(0.0821 L atm/mol K)(293.1 K)] = 1.25 g/L
Converting this density, in g/mL d = (1.25 g/L)(1 L/1000 mL) = 1.25×10^{-3} g/mL

c. Propane: d = [(1 atm)(44.0 g/mol)]/[(0.0821 L atm/mol K)(293.1 K)] = 1.83 g/L
Converting this density, in g/mL d = (1.83 g/L)(1 L/1000 mL) = 1.83×10^{-3} g/mL

d. Butane: d = [(1 atm)(58.0 g/mol)]/[(0.0821 L atm/mol K)(293.1 K)] = 2.41 g/L
Converting this density, in g/mL d = (2.41 g/L)(1 L/1000 mL) = 2.41×10^{-3} g/mL

Expressing the density of gases in units of g/L is much more convenient than expressing those densities in units of g/mL.

67. The answer is b. Because many organic molecules are nonpolar, they have limited solubility in polar solvents. Therefore, water is not always a good solvent for reactions involving those compounds.

69. First, the solvents originally used in the synthesis of sertraline, toluene, tetrahydrofuran, dichloromethane, and hexane are compounds derived from petroleum, a nonrenewable resource. In addition, these solvents are difficult to dispose of safely. The synthesis has been made much greener by substituting ethanol, a solvent that can be prepared from plant materials, for these four petroleum-based solvents. Second, the volume of solvent used in the synthesis has been reduced from 60,000 gallons to 6000 gallons per ton of sertraline prepared, which greatly reduces the disposal problems as well as the cost of the synthesis.

71. The advantage of using microwaves to heat reactions is that less energy is used in the process than is used when other forms of heat are employed. Less energy is used because the molecules absorb the heat directly with less loss to the surroundings, so reactions proceed more rapidly and are completed much more quickly than is the case when more traditional heating methods are used.

CHAPTER

10

Polymers

Giants Among Molecules

CHAPTER SUMMARY

10.1 Polymerization: Making Big Ones Out of Little Ones

Learning Objectives: • Define *polymer* and *monomer*. • List several natural polymers, including a chemically modified one.

A. Polymer comes from the Greek term for "many parts."
   1. Polymers are macromolecules.
   2. The building blocks of polymers are small molecules called monomers ("one part").
B. Polymerization is the process by which monomers are converted to polymers.
   1. The physical properties of the monomers are very different from those of the polymer formed from the monomer. For example, ethylene, a monomer, is a gas while polyethylene, a polymer of ethylene, is the material from which plastic bags and other familiar objects are made.
C. Natural polymers such as proteins have served humanity for as long as we have existed. Other polymers, such as cotton and silk, have been familiar for centuries.
   1. Starch is a polymer made up of glucose units.
   2. Cotton and wool are also polymers of glucose.
   3. Proteins are polymers of amino acids.
   4. Nucleic acids are polymers in DNA.
D. Celluloid represents the first attempt to improve on natural polymers by chemical modification.
   1. Celluloid, also called cellulose nitrate, is derived from cellulose, a polymer of glucose, treated with nitric acid.
   2. Celluloid was used as a substitute for ivory in billiard balls and as a means of permanently stiffening shirt collars.
   3. Because it is flammable, celluloid was removed from the market when safer substitutes became available.
   4. The first truly synthetic polymers were phenol-formaldehyde resins first made in 1909. Bakelite is an example.

Answers to Self-Assessment Questions

1. d   Cellulose is a polymer composed of a long chain of glucose molecules.

2. b Glucose, $C_6H_{12}O_6$, is a monomer that forms polymers such as starch and cellulose.
3. b Celluloid is a modification of the naturally occurring polymer cellulose that has been treated with nitric acid.
4. a Bakelite, formed from phenol and formaldehyde, is not a naturally occurring polymer.

10.2 Polyethylene: From the Battle of Britain to Bread Bags

Learning Objectives: • Classify a particular change within a redox reaction as either oxidation or reduction. • Identify an oxidation–reduction reaction.

A. Polyethylene is made from ethylene ($CH_2{=}CH_2$), a hydrocarbon derived from petroleum.

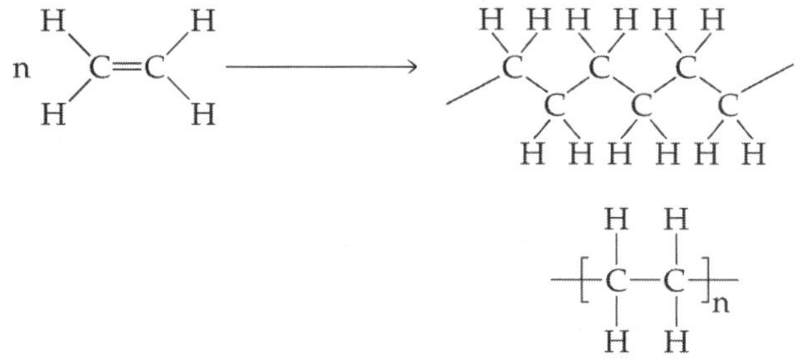

B. There are three major types of polyethylene plastics.
   1. High-density polyethylene (HDPE) is made from closely packed linear polymer strands that assume a fairly well-ordered, crystalline structure. HDPEs are relatively dense and tend to be rigid, have great tensile strength, and are not deformed when placed in boiling water.
      a. Used in threaded bottle caps, toys, bottles, and milk jugs.
   2. Low-density polyethylene (LDPE) is made from branched chains that cannot pack as tightly, thus producing a more flexible, noncrystalline material that is less dense than HDPE and that is seriously deformed in boiling water.
      a. Used in plastic bags, plastic film, squeeze bottles, and electric wire insulation.
   3. Linear low-density polyethylene (LLPE) is a copolymer (formed of two different monomers) of ethylene and a higher branched-chain alkene such as 4-methyl-1-pentene.
      a. Used to make plastic films for use as landfill liners, trash cans, tubing, and automotive parts.

C. The Many Forms of Carbon.
   1. Carbon forms a variety of kinds of structures.
      a. Diamonds are pure crystalline carbon in which each carbon atom is bonded to four other carbon atoms.
      b. Graphite (used in pencil lead, for example) is formed from carbon where each carbon atom is bonded to three other carbon atoms.
         i. Graphite is formed from many stacked layers of graphene (below).
      c. Graphene is a one-atom-thick planar sheet of carbon atoms lifted from graphite and is useful in electronics, sensing devices, and touch screens.
      d. Geim and Novoselov were awarded the 2010 Nobel Prize in Physics for their work on graphene.
      e. Buckminster fullerene, also called buckyballs, is a spherical collection of hexagons and pentagons, like the pattern on a soccer ball, with the formula C_{60}.
      a. The structure resembles a geodesic dome.

C. Thermoplastic and Thermosetting Polymers.
   1. Polyethylene is a thermoplastic material; it can be softened by heat and then re-formed.

2. Some plastics are thermosetting plastics, or plastics that harden permanently when formed, and cannot be softened by heat and remolded.
   a. Permanent hardness of thermosetting plastics is due to cross-linking (side-to-side connections of the polymer chains).
   b. When heated, thermosetting polymers discolor and decompose.

Answers to Self-Assessment Questions

1. b The building blocks that join to form a polymer (many parts) are called monomers (single parts).
2. d Because there are more branches in LDPE chains than in HDPE chains, the LDPE chains cannot pack as tightly together, resulting in materials with lower densities and lower melting points than HDPEs.
3. c Plastic grocery bags are made from polyethylene. The other choices, pre-1900 electrical wire insulation, foamed coffee cups, and plastic plumbing pipes are all made of polymers but just not polyethylene.
4. d Copolymers are made from the polymerization of two different monomers.
5. d Because of the extensive cross-linking between chains in thermosetting polymers, they harden permanently when formed and cannot be melted and reshaped.
6. a Buckyballs are structures consisting of sixty carbon atoms arranged in a roughly spherical pattern that resembles the geodesic dome structures first designed by architect R. Buckminster Fuller. Therefore, the molecule is called buckministerfullerene. There are several big molecules of this sort, all of which are referred to as 'buckyballs'.

10.3 Addition Polymerization: One + One + One + ... Gives One!

Learning Objective: • Identify the monomer(s) on an addition polymer, and write the structural formula for a polymer from its monomer structure(s).

A. There are two general types of polymerization reactions.
   1. Addition polymerization, also called *chain-reaction polymerization*, uses an addition reaction in such a way that the polymeric product contains all of the atoms of the starting monomers.
   2. Condensation polymerization results in a product where some part of each monomer molecule is not included in the polymer chain.

B. Polypropylene.
   1. Formed from the addition polymerization of propylene, C_3H_6.

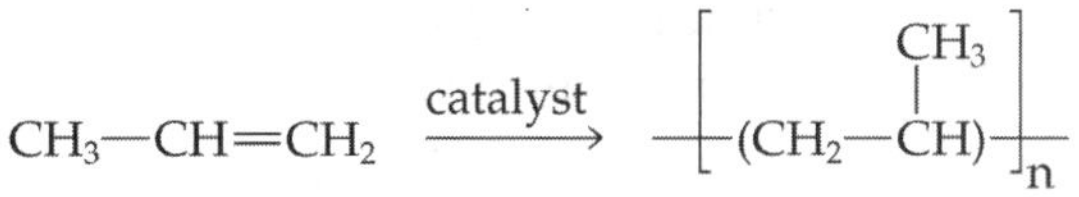

   2. The chain of carbon atoms is called the *backbone*, while groups attached to the backbone (CH_3- in the case of polypropylene) are called *pendant groups*.
   3. Polypropylene is tough and resists moisture, oils, and solvents.
      a. It is molded into hard-shell luggage, battery cases, and some appliance parts.
      b. Polypropylene is also used to make packaging material, fibers for textiles such as upholstery fabrics and carpets, and ropes that float.
      c. Because the melting point of polypropylene is 121 °C, it is used to make objects that can be sterilized with steam.

C. Polystyrene.

1. Formed from the addition polymerization of styrene, C_8H_8, ethylene with a benzene pendant group,

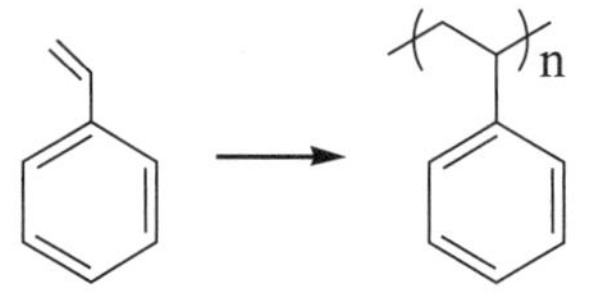

2. Used to make transparent disposable drinking cups and, when a gas is blown into polystyrene liquid, it foams and hardens into Styrofoam® used for ice chests and disposable coffee cups.
3. Polystyrene can be formed into shapes as packing material for shipping instruments and appliances, and is also widely used as insulation.

D. Vinyl Polymers.

1. Formed from the addition polymerization of vinyl chloride, C_2H_3Cl, an ethylene molecule in which one of the H atoms is replaced by a chlorine atom, resulting in polyvinyl chloride (PVC).

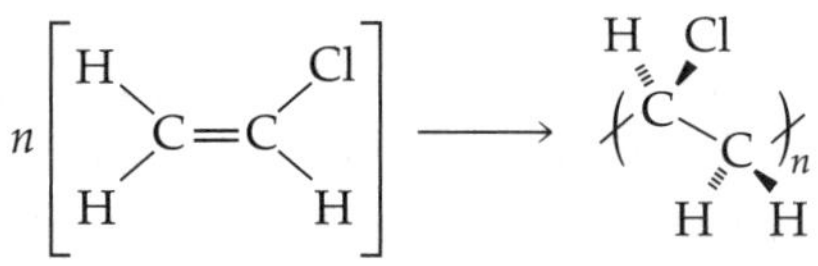

2. Clear, transparent polymer is used in plastic wrap and clear plastic bottles. Adding color and other ingredients produces artificial leather.
3. Most floor tile and shower curtains are made from vinyl plastics, which are also widely used to simulate wood in home siding panels and window frames.
4. Vinyl chloride, the monomer from which vinyl plastics are made, is a carcinogen.

E. PTFE: The Nonstick Coating.

1. Made from the polymerization of tetrafluoroethylene, $CF_2{=}CF_2$.

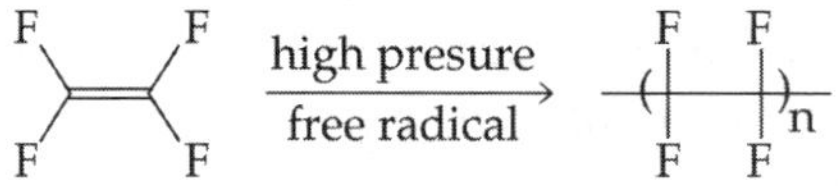

2. Called Teflon, PTFE is tough, unreactive, nonflammable, strong, and heat- and chemical-resistant.
3. Used to make electric insulation, bearings, and gaskets, as well as to coat surfaces of cookware to eliminate sticking of food.

F. Processing Polymers.

1. Many polymers are called plastics.
2. Plastic materials:
   a. Can be made to flow under heat and pressure.
   b. Are often made from granular polymeric material by *compression molding* or *transfer molding*.
   c. In compression molding, heat and pressure are applied directly to the granular material in the mold cavity.
   d. In transfer molding, the polymer is softened by heating before being poured into the molds to harden.
   e. Other methods of molding molten polymers are *injection molding* (plastic is melted in a heating chamber and then forced by pressure into a cold mold) and *extrusion molding* (the melted polymer is extruded through a die in continuous form to be cut into lengths or coiled).

G. Conducting Polymers: Polyacetylene.

1. Acetylene molecules have triple bonds and can undergo addition reactions.

$$H-C\equiv C-H$$

a. The result is a polymer with a backbone, with alternating carbon–carbon double and carbon–carbon single bonds.

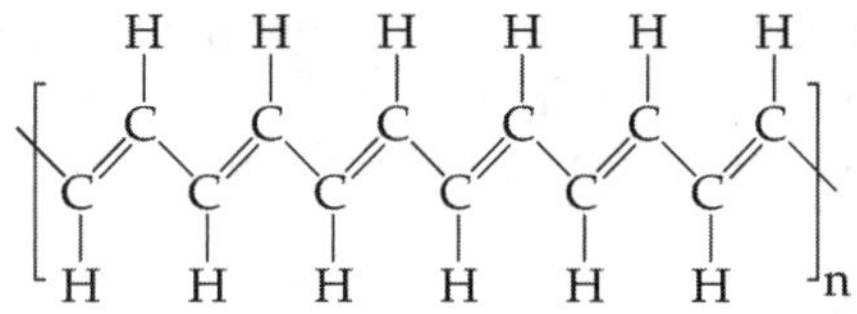

b. The alternating single and double bonds in this polymer, called a conjugated system, enable the polymer to conduct electricity.

Answers to Self-Assessment Questions

1. d Addition polymers form when an unpaired electron pairs up with one of the electrons in the C=C double bond of a monomer.
2. c Polystyrene has the CH_2CH– group as a backbone with C_6H_6 as pendants.
3. d Polyvinyl chloride, the polymer of vinyl chloride, is frequently used to make water pipes, bottles, and raincoats.

10.4 Rubber and Other Elastomers

Learning Objectives: • Define cross-linking and explain how it changes the properties of a polymer.

A. The plastics industry grew out of the need for a substitute for natural rubber during World War II.
B. Natural rubber is a polymer of isoprene.

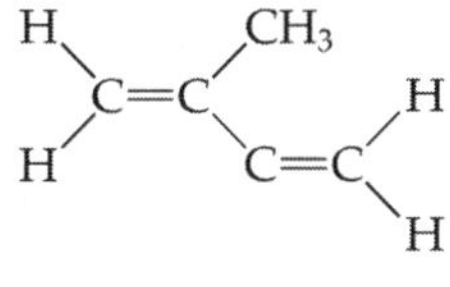

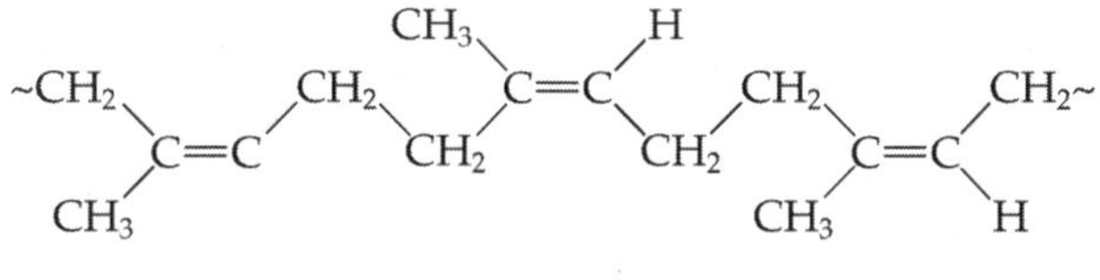

C. Vulcanization is a process in which natural rubber is heated with sulfur, resulting in cross-linking of the hydrocarbon chains with sulfur atoms.
   1. Vulcanized rubber is harder than natural rubber.
   2. Elastomers are stretchable materials that return to their original structure.
D. Synthetic Rubber.
   1. Some synthetic elastomers are polybutadiene, polychloroprene (Neoprene), and styrene-butadiene rubber (SBR).
      a. Neoprene is more resistant to oil and gasoline than other elastomers are, so it is used to make gasoline hoses.

2. Styrene-butadiene rubber (SBR) is a copolymer of styrene (~25%) and butadiene (~75%).
   a. SBR is more resistant to oxidation and abrasion than natural rubber, but its mechanical properties are less satisfactory.
   b. SBR can be cross-linked by vulcanization.
   c. SBR is used mainly for making tires.

E. Polymers are added to paint to help harden it into a continuous surface coating.
   1. This polymer is called a binder or resin.
   2. Paints made with elastomers are resistant to cracking.

Answers to Self-Assessment Questions

1. c Natural rubber is a polymer of isoprene.
2. d Isoprene has the structure $CH_2=C(CH_3)—CH=CH_2$.
3. a A cross-linking agent forms covalent bonds between polymer chains, which makes them stronger or harder.
4. d Charles Goodyear discovered that when natural rubber was heated with sulfur, the natural rubber became cross-linked.
5. a Styrene–butadiene rubber is a copolymer, a polymer formed from two different monomer units, styrene and butadiene.
6. a Paint binders are polymers.

10.5 Condensation Polymers

Learning Objectives: • Differentiate between addition and condensation polymerization. • Write the structures of the monomers that form polyesters and polyamides.

A. In condensation polymerization, also called step-reaction polymerization, a small molecule, often water (sometimes methanol, ammonia, or HCl), is formed as a by-product of the reaction.

B. Nylon and Other Polyamides.
   1. Nylon-6 is formed by splitting out water to join 6-aminohexanoic acid molecules into long polyamide chains.
   2. An amide bond is formed between the carbonyl of one monomer and the amine group of another.
   3. Water molecules are the by-products.
   4. Nylon is a polyamide—amide linkages hold the molecule together.
   5. Most nylon is made into fibers, some silk-like, some wool-like, and much of it used in carpeting.

C. Polyethylene Terephthalate and Other Polyesters.
   1. A polyester is a condensation polymer made from molecules with alcohol and carboxylic acid functional groups.
   2. The most common polyester is made from ethylene glycol and terephthalic acid and is called polyethylene terephthalate (PET).

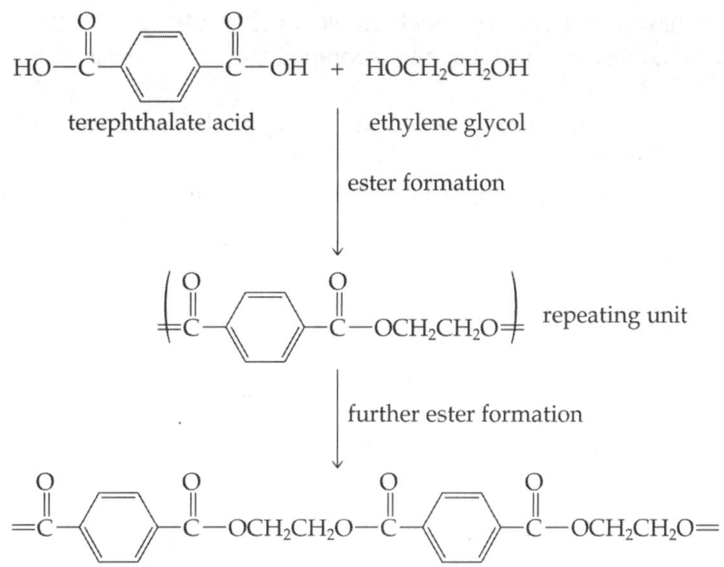

polyethylene terephthalate (polyester)

a. PET can be molded into bottles, made into film used to laminate documents, and to make packaging tape.

3. Polyester finishes are used on musical instruments and on the interiors of vehicles and boats. Polyester fibers are strong, quick-drying, and mildew-resistant, so they are used in home furnishings (carpets, curtains, sheets and pillow cases, and upholstery). They do not absorb water, so they are used for insulation in boots and sleeping bags. One type of polyester is the Mylar® used in balloons.

D. Phenol-Formaldehyde and Related Resins.
   1. Bakelite, a condensation polymer of phenol and formaldehyde, is a stable, thermosetting, condensation polymer that won't soften on heating.
   2. It is a three-dimensional polymer composed of phenol units, each with three formaldehyde cross-links.
   3. Thermosetting resins harden permanently; they cannot be softened and remolded.
   4. Bakelite is used to bind wood chips together in particle board.
   5. Melamine-formaldehyde resins are used in plastic (Melmac) dinnerware and in laminate countertops.

E. Other Condensation Polymers.
   1. Polycarbonates—"tough as glass" polymers.
      a. Used in bullet-proof windows, protective helmets, safety glasses, clear plastic water bottles, baby bottles, and dental crowns.
      b. One polycarbonate is made from bisphenol-A (BPA) and phosgene.
   2. Polyurethane
      a. Used in foam rubber in cushions, mattresses, and padded furniture, as well as for skate wheels, running shoes, and protective gear for sports activities.
   3. Epoxies
      a. Used in surface paints and coatings and on the inside of metal cans to protect from rusting.

F. Composite Materials.
   1. High-strength fibers (glass, graphite, or ceramics) are held together with a polymeric matrix (usually a thermosetting condensation polymer).

2. Composites have the strength of steel but weigh only a fraction of the weight.
   a. Powerful adhesives form when two components are mixed and polymer chains become cross-linked.
   b. Widely used in boat hulls, molded chairs, automobile panels, and sports gear such as tennis rackets.

G. Silicones—Polymers of Alternating Silicon and Oxygen Atoms.

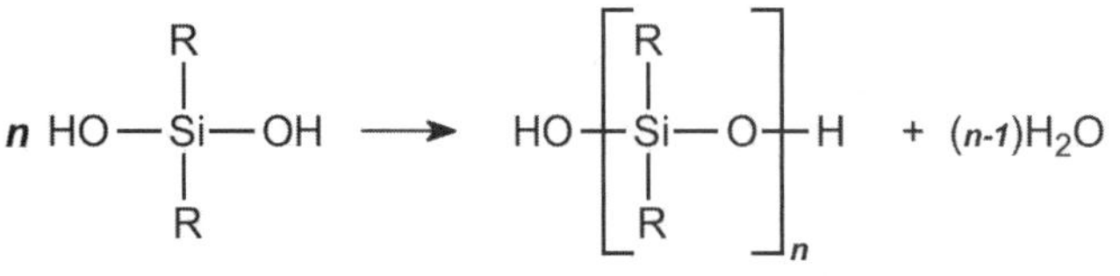

1. Can be linear, cyclic, or cross-linked networks.
2. Are very heat-stable and resistant to most chemicals.
3. Used as waterproofing materials, "Silly Putty," and synthetic body parts.

Answers to Self-Assessment Questions

1. c Water is usually the co-product with a polymer in condensation polymerization reactions between molecules with two functional groups each. Under these circumstances, the individual monomers can attach to the chain and still have an unreacted functional group to which another monomer can attach..
2. d Polyesters are the concentration products of carboxylic acids and alcohols. $HOOCCH_2CH_2CH_2CH_2COOH$ has carboxylic acid groups on both ends and $HOCH_2CH_2CH_2CH_2OH$ has hydroxyl groups on both ends, so these molecules can form condensation products proceeding from both ends of the molecules.
3. c $HOOCCH_2CH_2CH_2CH_2CH_2NH_2$ has both the needed amine and carboxylic acid groups (right and left ends of molecule, respectively) to make a polyamide.
4. c Thermosetting polymers harden permanently when they are first formed, so they cannot be melted and remolded.
5. c The backbone of a silicone polymer chain is repeating units of –SiO–.

10.6 Properties of Polymers

Learning Objectives: • Understand the concept of the glass transition temperature. • Explain how crystallinity affects the physical properties of polymers.

A. Polymers differ from substances consisting of small molecules in three main ways.
   1. The long chains can tangle with one another, lending strength to many plastics and elastomers.
   2. The magnitude of intermolecular forces is greater in polymers than in small molecules because of the extent to which the forces exist, giving strength to the polymeric materials.
   3. Large chains have less motion than small molecules, so their solutions are usually very viscous.

B. Crystalline and Amorphous Polymers.
   1. Crystalline polymers: Molecules line up to form long fibers of high tensile strength.
   2. Amorphous polymers: Randomly orientated molecules tangle with one another providing good elasticity.
   3. Two molecular structures can be grafted onto one polymer chain resulting in both sets of properties—flexibility and rigidity.

C. Glass Transition Temperature (T_g).
  1. Polymers are rubbery above their glass transition temperature (T_g) and are hard and brittle, like glass, below their T_g.

D. Fiber Formation.
  1. Three-fourths of fibers and fabrics used in the United States are synthetic.
     a. Synthetic fibers can mimic silk and wool, or enhance the properties of cotton by making it wrinkle-resistant.
     b. Microfibers can be made from almost any fiber-forming polymer.

Answers to Self-Assessment Questions

1. d Many fibers are composed of crystalline polymers.
2. c The glass transition temperature (T_g) is the temperature below which a polymer is hard and brittle and above which the same polymer is soft and stretchy.

10.7 Plastics and the Environment

Learning Objectives: • Describe the environmental problems associated with plastics and plasticizers. • Name two types of sustainable, nonpetroleum-derived polymers and give their sources.

A. Most plastics break down slowly in the environment, leading to litter and solid-waste problems.
  1. When disposed of in oceans, plastics pose a danger to fish by clogging their digestive tracts.
  2. The volume of plastics being disposed of puts stresses on the available space in landfills.
  3. While incinerating (burning) plastics is a source of energy, the process can cause environmental problems, such as the production of toxic gases like HCl when PVC is incinerated.

B. Degradable Plastics.
  1. One approach to solving the problem is to use biodegradable and/or photodegradable polymers in the manufacture of disposable items.

C. Recycling.
  1. Only two kinds of plastics are recycled on a large scale: PET (28% recycled) and HDPE (29% recycled).
  2. Many plastics can be recycled, but they must be separated according to type.
  3. Code numbers specifying the type of plastic and stamped on the bottom of plastic containers help in the separation process.
     a. Only about 7% of waste plastic was recycled in 2007.

D. Plastics and Fire Hazards.
  1. Because polymers are flammable and are used in so many fabrics, fire retardants, usually containing bromine and/or chlorine atoms, are incorporated into the polymeric fibers.
  2. Burning plastics often produce toxic gases, so taking safety precautions is required when anyone works with materials that could combust.
  3. Green chemists are working to make new kinds of polymers that don't burn or that don't generate toxic chemicals when they burn.

E. Plasticizers and Pollution.
   1. Plastics are made more flexible and less brittle by adding plasticizers to lower their glass transition temperatures, T_gs.
      a. Plasticizers are liquids of low volatility that are generally lost by diffusion and evaporation as a plastic article ages, at which point the plastic becomes brittle and cracks and breaks.
      b. Polychlorinated biphenyls (PCBs), which resemble DDT in structure, were once used as plasticizers but are now banned because they bio-concentrate in the food chain and have the same general physiological effect as DDT.
      c. The most widely used plasticizers for vinyl plastics are phthalate esters which have low acute toxicity and are considered to be generally safe.

F. Plastics and the Future.
   1. In medicine, body replacement parts made from polymers have become common.
   2. PVC water pipes, siding and window frames, plastic foam insulation, and polymeric surface coatings are now used in home construction, as are lumber and wall panels fabricated from artificial wood made from recycled plastics.
   3. Synthetic polymers are used in airplane interiors. The bodies and wings of some planes and some automobile parts are made of composites.
   4. Electrically conducting polymers are used in lightweight batteries for electric automobiles.
   5. Several new types of plastics are made from nonpetroleum-based monomers such as polylactic acid and polyhydroxybutyrates.

G. Green Chemistry: Life-Cycle Impact Assessment of New Products.
   1. Life-cycle impact assessment (LCIA) is a way of "grading" a new compound or product in terms of its environmental footprint.
      a. The first step is to outline the product's full life cycle, including manufacturing of monomer and product all the way to the recycling or disposal of the item.
   2. Once the life cycle is outlined, the inventory phase of the LCIA totals all of the inputs (mass, water, energy) and outputs for each step, along with any waste that is emitted to air, land, and water.
   3. Next, inventory is converted in one of several ways into impact.
      a. In midpoint analysis (widely used in North America), the potential for a given chemical to cause harm is calculated for each impact category by comparing the chemical with a compound whose behavior is well studied.
   4. The final "report card" shows whether or not the process creates any unfortunate trade-offs (for example, a drop in fossil fuel use and an increase in algae bloom potential in estuaries).

Answers to Self-Assessment Questions

1. b Generally, incineration is the worst way to dispose of plastics because of the hazards associated with the gaseous combustion products of many polymers.
2. d Many plastics produce toxic gases (for example, HCl from burning polyvinyl chloride) when burned.
3. c PVCs have structures very similar to DDT, which is known to have negative environmental effects.
4. a Many flame-retardant fabrics incorporate Br and Cl atoms.
5. c Most synthetic polymers are made from petroleum and natural gas.
6. c Polylactic acid is made from corn, which is a renewable resource.

LEARNING OBJECTIVES

You should be able to . . .

1. Define *polymer* and *monomer*. (10.1)
2. List several natural polymers, including a chemically modified one. (10.1)
3. Describe the structure and properties of the two main types of polyethylene. (10.2)
4. Use the terms *thermoplastic* and *thermosetting* to explain how polymer structure determines properties. (10.2)
5. Identify the monomer(s) of an addition polymer, and write the structural formula for a polymer from its monomer structure(s). (10.3)
6. Define *cross-linking*, and explain how it changes the properties of a polymer. (10.4)
7. Differentiate between addition and condensation polymerization. (10.5)
8. Write the structures of the monomers that form polyesters and polyamides. (10.5)
9. Understand the concept of the glass transition temperature. (10.6)
10. Explain how crystallinity affects the physical properties of polymers. (10.6)
11. Describe the environmental problems associated with plastics and plasticizers. (10.7)
12. Name two types of sustainable, nonpetroleum-derived polymers and their sources. (10.7)
13. Explain how life-cycle impact analysis can be used to identify the environmental benefits and effects of newly designed products.

DISCUSSION

The age we're living in can be called the Plastic Age (as contrasted to the Stone Age or Iron Age). Humanity has used natural polymers (rubber, wood, cotton) for millennia. Since the early 1900s, chemists have been able to produce polymers and tailor their properties by modifying their structure. Since polymers ("many units") are made out of monomers ("one unit"), the structure of the polymer is determined by the structure of the monomer.

We classify polymers by several methods or types:

1. Thermoplastic—can be re-formed by softening with heat.
   Thermosetting—once made, these cannot be re-formed.
2. Addition polymers—monomer needs a double (or triple) bond.
   Condensation polymers—monomer needs two functional groups, and a small molecule is "condensed" out.
3. Low-density polymers—loosely packed polymer strands; soft and flexible.
   High-density polymers—tightly packed polymer strands; harder and rigid.

You should concentrate on relating the monomer structure to the polymer it makes and the properties of the polymer. The properties of the polymers are determined by their monomers but can be modified by changing their physical properties by introducing additives such as plasticizers, flame retardants, and fibers (to make composites). While polymers have many advantages, fire hazards and disposal issues are problems. The research in polymers is very active, with new developments occurring frequently.

ANSWERS TO ODD-NUMBERED REVIEW QUESTIONS AND SOLUTIONS FOR ODD-NUMBERED END-OF-CHAPTER PROBLEMS

Review Questions

1. The monomer in polyethylene is ethylene, $H_2C{=}CH_2$, while that in polyvinyl chloride is vinyl chloride, $H_2C{=}CHCl$. Structurally, vinyl chloride is an ethylene molecule with one of the hydrogen atoms replaced with a chlorine atom. Polyethylene is used for, among other things, toys, plastic bags, and trash cans.

3. Addition polymerization is the polymerization in which all the atoms in the monomer molecules are included in the polymer. Most addition polymers are made from alkenes which contain carbon–carbon double bonds. The monomers vary from one another because of the pendant groups they contain.

5. Polystyrene is used to make disposable foamed drinking cups. Polystyrene is made from styrene.

7. Celluloid is a semisynthetic material made from natural cellulose (perhaps from cotton and wood) that has been treated with nitric acid.

9. Synthetic fibers are used more than natural fibers because they are cheaper and have a greater range of properties.

11. In order to be recycled, plastics must be collected, sorted, chopped, melted, and then remolded. The sorting process is simplified by the code numbers stamped on plastic containers.

Problems

13. The polymer chains in LDPE have many branches, so the chains cannot pack closely together; thus, LDPE has a more amorphous structure than HDPE has. As a result, LDPE is less dense and more flexible than HDPE.

15. a. The structure of polyethylene monomer is $H_2C{=}CH_2$
    b. The structure of the monomer used to make polystyrene is $H_2C{=}CHC_6H_5$

17. a. Polyacrylonitrile chloride has the structure

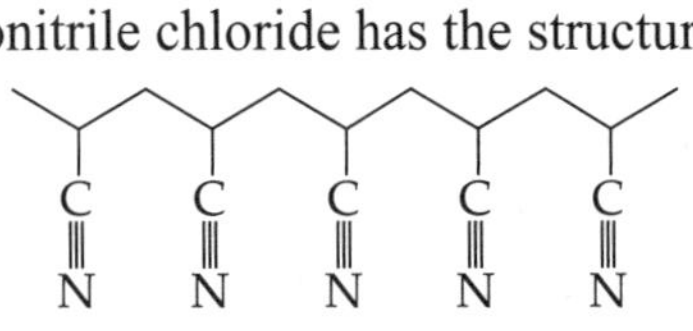

    b. Vinylidene has the structure -{CH_2CF_2- CH_2CF_2- CH_2CF_2- CH_2CF_2}-

19. a. 1-pentene gives a polymer with the structure
-$CH_2CH(A)CH_2CH(A)CH_2CH(A)CH_2CH(A)$-] where A = ($CH_3CH_2CH_2$)-

b. Methyl cyanoacrylate gives a polymer with the structure

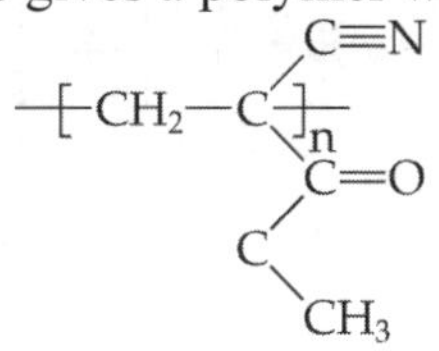

21 The structure of isoprene is:

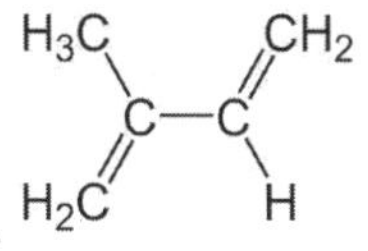

23. The long chains of polymerized isoprene are coiled and twisted and intertwined with one another, making the material elastic. When stretched, rubber's coiled molecules are straightened. When released, the molecules coil again. Many other polymers are rigid because their chains are cross-linked or are strongly attracted to each other through intermolecular attractions.

25. SBR, styrene-butadiene rubber, is a copolymer prepared from ~25% styrene and ~75% butadiene.

27. $+CO(CH_2)_6\,CONH(CH_2)_8NHCO(CH_2)_6\,CONH(CH_2)_8NH+$

29. $+OCH_2COOCH_2COOCH_2COOCH_2CO+$

31. Atoms are the smallest units of matter. Elements are either individual atoms (as in the case of the noble gases) or two or more of the same kind of atoms bonded together (as in the case of O_2, N_2, and Br_2, for example). Compounds can be formed from as few as two atoms (for example, CO) but often contain three or more atoms connected with ionic or covalent bonds (for example, $CaCO_3$, CO_2, $C_6H_{12}O_6$). Polymers are long chains of monomers joined together with covalent bonds. In terms of relative sizes then, atoms < elements < compounds < polymers.

33. The glass transition temperature, T_g, is the temperature at which the properties of the polymer change from its being hard, stiff, and brittle to rubbery and tough. Rubbery materials such as automotive tires should have a low T_g, while glass substitutes should have a high T_g.

35. The monomer unit (a) is CH_2=CHF and is the monomer from which this addition polymer is formed. The four-unit polymer segment (labeled c) is -$CH_2CHF\ CH_2CHF\ CH_2CHF\ CH_2CHF$ with a repeating unit (b) of CH_2CHF. This is an addition polymer because all parts of the monomer are incorporated into the polymer.

37. The monomer structures are $CH_2=CCl_2$ and $CH_2=CHCl$.

39.

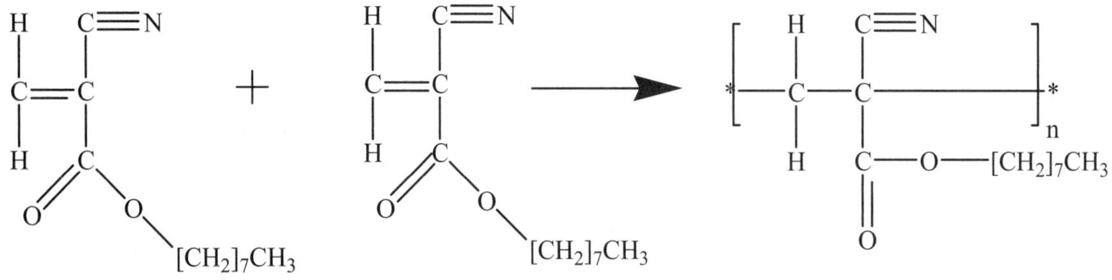

41. {-CH_2-$C(CH_3)_2$-CH_2-$C(CH_3)_2$--CH_2-$C(CH_3)_2$-CH_2-$C(CH_3)_2$-}

43. The monomer 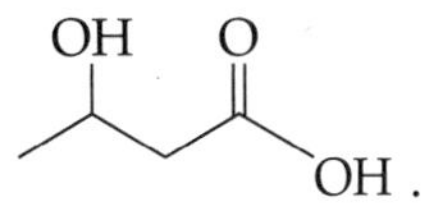

45. The student's structure appears to have three carbon atoms in the backbone of the monomer unit. The correct structure of propylene has two carbon atoms in the backbone and a methyl pendant group. The structure of the polymer should look like this:

$$CH_3—CH=CH_2 \xrightarrow{\text{catalyst}} \left[(CH_2—\overset{\overset{CH_3}{|}}{C}H) \right]_n$$

47. All three balls are made of elastomers or elastic polymers. Of the three balls, the golf ball exhibits the property to the greatest degree, and the SuperBall is a close second.

49. Because 1-butanol has an OH group on one end of the molecule only, the reaction cannot continue after the first ester forms on both ends of the terephathalic acid. The final product will be limited to a trimer of butanol-terephthalic acid—butanol or $CH_3(CH_2)_3OC(O)C_6H_5C(O)O(CH_2)_3CH_3$. A polymer cannot form.

51. a. The structure of the polymer is as follows.

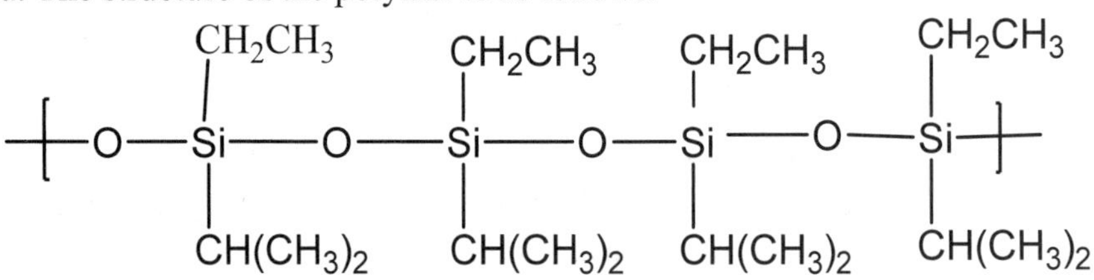

53. a. $+CH_2CH_2OCH_2CH_2OCH_2CH_2OCH_2CH_2O+$

b. $+CH_2CH_2SCH_2CH_2SCH_2CH_2SCH_2CH_2S+$

55. The answer is b. Polylactic acid (PLA) is a polymer made from corn. The other three examples are made from monomers distilled from petroleum.

57. The answer is c. Only two kinds of plastics are recycled on a large scale: PET (28% recycled) and HDPE (29% recycled).

CHAPTER

11

Nuclear Chemistry

The Heart of Matter

CHAPTER SUMMARY

11.1 Natural Radioactivity

Learning Objectives: • Identify the sources of the natural radiation to which we are exposed. • List the sources and dangers of ionizing radiation.

A. Many nuclei are unstable and undergo radioactive decay. These nuclei are called radioisotopes.
B. Background radiation: radiation from cosmic rays, and natural radioactive isotopes found in air, water, soil, and rocks.
   1. Harmful effects arise from the interaction of radiation with living tissue.
   2. Radiation that has sufficient energy to knock electrons from atoms and molecules, converting them into ions, is called ionizing radiation.
C. Radiation causes damage to cells.
   1. Radiation can break up molecules and convert them into free radicals (species with unpaired electrons).
      a. White blood cells, responsible for fighting infection, are especially vulnerable.
      b. Leukemia can be induced by exposure to radiation.
   2. Radiation causes changes in DNA that produce mutations in offspring.

Answers to Self-Assessment Questions

1. d According to Figure 11.1, 82% of background radiation comes from natural sources.
2. c According to Figure 11.1, the largest artificial source of background radiation is from radon.
3. b Of the choices, medical X-rays are most likely to cause cancer.

11.2 Nuclear Equations

Learning Objectives: • Balance nuclear equations. • Identify the types of products formed by various nuclear decay processes.

A. Nuclear equations differ in two ways from chemical equations.

1. While chemical equations must have the same elements on both sides of the arrow, nuclear equations rarely do.
2. While we balance atoms in ordinary chemical equations, we balance the nucleons (protons and neutrons) in nuclear equations.

B. There are three common types of nuclear reactions.

1. Alpha Decay
   a. An alpha particle, $^{4}_{2}He$, is a helium nucleus.
   b. The product when a Ra-226 nucleus releases an alpha particle is Rn-222. We know the equation is balanced when the sum(s) of the mass numbers on both sides of the equation are equal (226 on the left, 222 + 4 = 226 on the right) and the sum(s) of the atomic numbers on both sides of the equation are equal (88 on the left, 86 + 2 = 88 on the right).

$$^{226}_{88}Ra \longrightarrow ^{222}_{86}Rn + ^{4}_{2}He$$

2. Beta Decay
   a. A beta particle, $^{0}_{-1}e$, is identical to an electron.
   b. The product when C-14 releases a beta particle is N-14. In beta decay, a neutron in the nucleus is converted into a proton (which remains in the nucleus) and an electron (which is ejected). Notice the sums of the atomic numbers (7 + (-1) = 6).

$$^{14}_{6}C \longrightarrow ^{14}_{7}N + ^{0}_{-1}e$$

3. Gamma Decay
   a. Gamma decay (v) is different from alpha and beta decay because gamma radiation has no mass or charge. Neither the nucleon number nor the atomic number of the emitting atom is changed: the nucleus simply becomes less energetic.
   b. The penetrating power of gamma rays is extremely high.

C. Positron emission and electron capture are two more types of nuclear reactions.

1. A positron, $^{0}_{+1}e$, is a particle equal in mass but opposite in charge to the electron. An example of positron emission:

$$^{11}_{6}C \longrightarrow ^{0}_{+1}e + ^{11}_{5}B$$

2. Electron capture is a process in which a nucleus absorbs an electron from an inner electron shell, usually the first or second. When an electron from a higher shell drops to the level vacated by the captured electron, an X-ray is released. Once inside the nucleus, the captured electron combines with a proton to form a neutron. An example of electron capture:

$$^{195}_{79}Au + ^{0}_{-1}e \longrightarrow ^{195}_{78}Pt$$

D. Some Differences between Chemical and Nuclear Reactions.

1. Atoms in chemical reactions retain their identity but may change in nuclear reactions.
2. Chemical reactions involve electrons (usually only valence electrons), whereas nuclear reactions involve mainly protons and neutrons.
3. Chemical reaction rates can be changed by varying the temperature, while nuclear reaction rates are unaffected by temperature changes.
4. The energy absorbed or released by chemical reactions is rather small when compared with the energy associated with nuclear reactions.
5. Mass is conserved in chemical reactions, while mass and energy are conserved in nuclear reactions.

Answers to Self-Assessment Questions

1. c A positron has no mass and a charge of +1.
2. a An alpha particle, ${}^{4}_{2}\text{He}$, is needed to balance the atomic numbers (lower left) and nucleon numbers (upper left).
3. b Beta particle emission occurs when a neutron in the nucleus is converted into a proton plus an electron. The beta particle is ejected, but the proton remains in the nucleus, increasing the atomic number by one unit.
4. c Gamma radiation has no charge and no mass; it is pure energy. The number of protons and neutrons in an atom is the nucleon number.
5. c Electron capture involves the absorption of an electron, usually from the first or second shell, into the nucleus.
6. a An alpha particle, ${}^{4}_{2}\text{He}$, is the largest of the particles emitted by a radioactive nucleus.

11.3 Half-Life and Radioisotopic Dating

Learning Objectives: • Solve simple half-life problems. • Use the concept of half-life to solve simple radioisotopic dating problems.

A. Radioactivity is dependent on the isotope involved but is generally independent of any outside influence such as temperature or pressure.
B. Radioactivity is a random process. Large numbers of atoms have a predictable half-life characteristic of each radioisotope.
   1. The half-life is a property of a radioactive isotope and represents the period of time required for one-half of the radioactive atoms to undergo decay.
      a. Half-lives can vary enormously in length.
      b. The fraction of the original radioactive sample remaining after n half-lives is given by $1/2^n$, or 1/2, 1/4, 1/8, 1/16, and so forth.
C. Radioisotopic Dating.
   1. The half-lives of radioisotopes can be used to estimate the age of rocks and artifacts, such as the Shroud of Turin, dated to between AD 1260 and 1340, and the Dead Sea Scrolls, dated 2000 years ago.
   2. Many artifacts are dated by their carbon-14 content.
   3. Tritium dating is used to date items up to 100 years old.
   4. The variance in mass among isotopes can affect their relative rates of reaction. This difference can lead to a measurable characteristic called the isotopic signature in a material.

Answers to Self-Assessment Questions

1. a 36 h is 6 half-lives. This gives the fraction $1/2^6 = 1/64$. (48 mg)(1/64) = 0.75 mg.
2. b 6.25 g is one-eighth of 50 g or $1/2^3$. This decay occurred over three half-lives, so one half-life is 55.7 d/3 = 17.9 d.
3. d 285 years is 10 half-lives. This gives the fraction $1/2^{10}$ or 1/1024 = 0.000977. (64.0 g)(0.000977) = 0.0625 g.
4. c The decline from 16 Bq to 2 Bq represents 3 half-lives (2 Bq is one-eighth of 16 Bq or $1/2^3$). (3)(22.26 y) = 67 y.
5. b 3.9 counts per minute is one-fourth of 15.6 counts per minute or $1/2^2$, so two half-lives have elapsed since the live wood was cut and ceased to metabolize. (2)(5730 y) = 11,460 y.

11.4 Artificial Transmutation

Learning Objective: • Write a nuclear equation for a transmutation, and identify the product element formed.

A. Transmutation involves converting one element into another, which requires changing the number of protons in the nucleus. This cannot be accomplished chemically.
B. Nuclear changes can also be brought about by the bombardment of stable nuclei with subatomic particles such as alpha particles, beta particles, or neutrons.

$$^{9}_{4}Be + ^{4}_{2}He \longrightarrow ^{12}_{6}C + ^{1}_{0}n$$

   1. A nuclear change whereby one element is changed into another is called transmutation.
C. Transmutations carried out by Rutherford with alpha particles led to the discovery of fundamental particles such as protons.

Answers to Self-Assessment Questions

1. b The nucleon number must be 14 and the nuclear charge must be 6 to balance the mass and charge. The new isotope is carbon-14.
2. c The nucleon number must be 6 and the nuclear charge must be 3 to balance the mass and charge. The new isotope is lithium-6.
3. a The nucleon number must be 30 and the nuclear charge must be 15 to balance the mass and charge. The new isotope is phosphorus-30.

11.5 Uses of Radioisotopes

Learning Objective: • List some applications of radioisotopes.

A. Radioisotopes in industry and agriculture.
   1. Radioactive isotopes can substitute chemically for their nonradioactive counterparts but have the advantage of being easily detected; they act as tracers.
      a. Radioactive I-131, with a short half-life (8.04 days), is used to detect leaks in underground pipes.
      b. Beta-emitters are used to measure the thickness of sheet metal during production.
      c. Carbon-14 is used to determine frictional wear in piston rings.
      d. Phosphorus-32, a beta-emitter with a half-life of 14.3 d, is used to study the uptake of phosphorus and its distribution in plants.
   2. Radioisotopes are used to study the effectiveness of weed killers, compare the nutritional value of feeds, determine optimal methods for insect control, and monitor the fate and persistence of pesticides in soil and groundwater.
   3. Radioisotopes can be used to induce heritable genetic alterations known as mutations into seeds.
   4. Gamma-ray emitting radioisotopes are used in the irradiation of food to kill microorganisms.
B. Radioisotopes in medicine.
   1. Nuclear medicine involves two distinct uses of radioisotopes: therapeutic and diagnostic.
   2. An example of a therapeutic use is radiation therapy: Radiation from radioisotopes is used to treat cancer.
   3. Radioisotopes are used for diagnostic purposes to provide information about the functioning of some part of the body or about the type or extent of an illness.

Isotope	Half-life	Use
Carbon-11	20.39 min	Brain scans
Chromium-51	27.8 d	Blood volume determination
Cobalt-57	270 d	Measuring vitamin B_{12} uptake
Cobalt-60	5.271 y	Radiation cancer therapy
Gadolinium-153	242 d	Determining bone density
Gallium-67	78.1 h	Scan for lung tumors
Iodine-131	8.040 d	Thyroid diagnoses and therapy
Iridium-192	74 d	Breast cancer therapy
Iron-59	44.496 d	Detection of anemia
Phosphorus-32	14.3 d	Detection of skin cancer or eye tumors
Plutonium-238	86 y	Provision of power in pacemakers
Radium-226	1600 y	Radiation therapy for cancer
Selenium-75	120 d	Pancreas scans
Sodium-24	14.659 h	Locating obstructions in blood flow
Technetium-99m (m means metastable)	6.0 h	Imaging of brain, liver, bone marrow, kidney, lung, or heart
Thallium-201	73 h	Detecting heart problems during treadmill stress test
Tritium (H-3)	12.26 y	Determining total body water
Xenon-133	5.27 d	Lung imaging

4. Computer-aided medical imaging methods such as positron emission tomography (PET) scans use of positron-emitters to generate gamma rays inside the body and measure dynamic processes such as blood flow or the rate at which oxygen or glucose is being metabolized.

Answers to Self-Assessment Questions

1. a I-131's short half-life makes it ideal for detecting leaks in underground pipes.
2. c Gamma radiation is used to irradiate foodstuffs to extend shelf life.
3. a Cobalt-60 is used in radiation therapy.
4. b Gadolinium-153 is used in determining bone density.
5. d Thallium-201 is used to detect heart problems during treadmill stress tests.
6. c Iodine-131 is used to treat thyroid disorders.
7. c Positrons are used in PET scans.

11.6 Penetrating Power of Radiation

Learning Objective: • Describe the nature of materials needed to block alpha, beta, and gamma radiation.

A. The penetrating power of different types of radiation varies widely with gamma > beta > alpha owing in part to their masses as well as to their charges.
   1. Alpha particles outside the body do little damage because they can't penetrate the skin. Inside the body, alpha particles inflict great damage because they are trapped in a small area.
   2. Outside the body, beta particles are somewhat more penetrating than alpha particles; while inside the body they do their damage over a somewhat larger area because they travel further.
   3. Gamma radiation is the most penetrating of all, both inside and outside the body.

B. Protection from radiation.
   1. Move away from the source.
   2. Use shielding between you and the source of the radiation.

Answers to Self-Assessment Questions

1. a The least penetrating type of radiation are alpha particles, and the most penetrating type of radiation is gamma radiation.
2. b Beta particles cannot penetrate a thin sheet of aluminum but, unlike alpha particles which are stopped by fabric, skin, and the like, beta particles are much smaller and require more of a barrier than alpha particles require.
3. a The reduction in Geiger counter activity indicates that most of the radiation was blocked by the piece of paper. Alpha particles are easily blocked.

11.7 Energy from the Nucleus

Learning Objectives: • Explain where nuclear energy comes from. • Describe the difference between fission and fusion.

A. Albert Einstein (1905) derived a relationship between matter and energy, $E = mc^2$, where E is energy, m is mass, and c is the speed of light, as a part of his theory of relativity.
B. Binding energy, the energy that holds the nucleons together in the nucleus, is equivalent to the difference in mass between the individual protons and neutrons and the mass of the nucleus they form. Binding energy can be calculated using the equation $E = mc^2$ where m is the difference in the calculated and actual nuclear masses, called the mass defect.
   1. Elements with the highest binding energy per nucleon are the most stable.
   2. In nuclear fission reactions, one nucleus splits into two or more smaller nuclei with higher binding energies and, therefore, with greater stability than the original nucleus that split.
   3. In nuclear fusion reactions, a great deal of energy is released when small atoms with lower binding energies combine to form a larger atom with more binding energy. This is what happens when a hydrogen bomb explodes and is also the source of the sun's energy.
C. Nuclear Fission.
   1. In the process of bombarding U-238 nuclei with neutrons, Fermi and others discovered but couldn't explain the presence of smaller nuclei in the reaction mixture.
      a. They concluded that the uranium nucleus split into fragments.
D. Nuclear Chain Reaction.
   1. Szilard discovered that when a uranium nucleus was bombarded with neutrons it split into two smaller fragments and, at the same time, gave off three neutrons, which then could trigger the fission of other uranium atoms in a chain reaction.
E. Thermonuclear Reactions.
   1. The thermonuclear reactions that take place in the sun required enormously high temperatures (millions of degrees) to initiate them.
   2. The intense temperatures and pressures in the sun cause small nuclei to fuse into larger ones.

Answers to Self-Assessment Questions

1. b Helium is more stable than H and Li because of its higher binding energy per nucleon.
2. c Iron nuclei are the most stable nuclei because they have the highest binding energy per nucleon.

3. c Nuclei with nucleon numbers above that of iron are less stable than iron because their binding energies per nucleon are lower.
4. c $E = mc^2$ relates the mass defect in a fission reaction with the energy released by the reaction.
5. c Neutrons carry out the chain reaction.
6. d In order to balance the reaction, three neutrons are needed.
7. c The fusion process involves combining two small nuclei into one bigger nucleus, a process that is accompanied by the release of energy.
8. b. A neutron is needed to balance the nuclear reaction.

11.8 Nuclear Bombs

Learning Objectives: • Describe the goals of the Manhattan Project and how uranium and plutonium bombs were made. • Identify the most hazardous fallout isotopes, and explain why they are particularly dangerous.

A. The Manhattan Project was the research effort launched in 1939 by President Roosevelt to study atomic energy.
   1. The program involved four research teams studying how to:
      a. sustain the nuclear fission reaction;
      b. enrich uranium so that it contained ~90% U-235;
      c. make plutonium-239, another fissionable isotope;
      d. construct a bomb based on nuclear fission.

B. Sustainable Chain Reaction.
   1. Studied by Enrico Fermi's group at the University of Chicago.
   2. One goal was to determine the critical mass; the amount of U-235 needed to sustain the fission reaction.
   3. On December 2, 1942, Fermi's group achieved the first sustained nuclear fission reaction using 16 kg of uranium enriched to about 94% U-235.

C. Isotopic Enrichment.
   1. Natural uranium, U-238, does not undergo fission.
   2. U-235 makes up only 0.72% of natural uranium.
   3. Chemical separation was almost impossible because the two isotopes behave almost identically.
      a. Converted U to gaseous uranium hexafluoride, $^{235}UF_6$ and $^{238}UF_6$.
      b. After traveling long distances, $^{235}UF_6$ ultimately moved farther than $^{238}UF_6$ because it was slightly lighter, allowing separation of the isotopes.

D. Plutonium Synthesis.
   1. Effort led by Glenn Seaborg's group at Oak Ridge (Tennessee).
   2. Found that, although U-238 does not fission, it does decay into Pu-239, which is fissionable.
   3. A group of large reactors were built near Hanford, Washington, to produce Pu.

E. Bomb Construction.
  1. Construction of atomic bombs based on U-235 and Pu-239 was done at Los Alamos, New Mexico, under the direction of J. Robert Oppenheimer.
  2. The synthesis of Pu turned out to be easier than the separation of U isotopes.
  3. The first atomic bomb (a Pu device) was tested in the desert near Alamogordo, New Mexico, on July 16, 1945.
  4. On August 6, 1945, President Truman ordered the dropping of the U bomb "Little Boy" on Hiroshima, causing more than 100,000 casualties. Three days later, another plutonium bomb, "Fat Man," was dropped on Nagasaki. World War II ended with the surrender of Japan on August 14, 1945.

F. Radioactive Fallout.
  1. The primary fission products are radioactive, as are many of the daughter isotopes.
     a. Sr-90 (half-life = 28.5 y) reaches us primarily through dairy products and vegetables and is incorporated into bone, becoming an internal radioactive source.
     b. I-131 (half-life = 8 days) is transferred up the food chain and incorporated in the thyroid gland. For healthy individuals, the presence of I-131 has damaging effects.
     c. Cs-137 (half-life = 30.2 y) is similar to potassium and is taken up as part of body fluids.
  2. Concern over radiation damage from nuclear fallout led to a movement to ban atmospheric testing.
     a. Nobel laureate Linus Pauling was an instrumental spokesperson for banning atmospheric tests.
  3. In 1963, a nuclear test ban treaty was signed by the major nations, with the exception of the People's Republic of China.
     a. Pauling was awarded the Nobel Peace Prize in 1962.

Answers to Self-Assessment Questions

1. c Of the uranium isotopes, only U-235 undergoes fission at a level that can be used in a bomb.
2. a Less than 1% of uranium ore is U-235.
3. d U-238 bombarded with a neutron produces Np-239, which quickly decays into Pu-239.
4. a Calcium and strontium are in the same group in the periodic table and have similar chemical properties.
5. d Giving those individuals exposed to nuclear fallout large doses of KI was done as a means of "diluting" the iodine-131 to reduce absorption of radioactive iodine by their thyroid glands.
6. c Of the isotopes listed, iron-59 poses the least hazard.
7. b Iodine-131 is concentrated in the thyroid gland.

11.9 Uses and Consequences of Nuclear Energy

Learning Objective: • List some uses and consequences of nuclear energy.

A. Nuclear energy accounts for one-fifth of the energy produced in the United States.
  1. 80% of the energy in France comes from nuclear power plants.
  2. Belgium, Spain, Switzerland, and Sweden generate about one-third of their power from nuclear reactors.

B. One of the main problems with nuclear power production arises from disposal of nuclear waste and the possibility of transforming this waste into weapons-grade material.
  1. Discarded fuel rods can be used to construct plutonium bombs.
     a. Several countries (for example, North Korea and Iran) have facilities and materials for constructing these bombs.

2. The amount of nuclear waste from reactors and from isotopic enrichment is huge, and ways to deal with the waste and to minimize the amount of material have not been well developed.
3. Recovery of unused uranium and plutonium can result in 25–30% more energy coming from the original fuels.
4. Interest in recovering actinides with plutonium has grown.
   a. This would lower the amount of long-term radioactivity in the waste as well as the risk of having plutonium used for illegal activities..
   b. The U.S. government has prohibited reprocessing by civilian companies.

C. The Nuclear Age.
1. See Figure 11.15 for examples of productive uses of nuclear energy.

D. Green chemistry: Can nuclear power be green?
1. The primary challenges of nuclear power include:
   a. The need for specialized spent fuel handling and waste storage.
   b. Base load mismatch to peak demand.
   c. Safety: Potential for environmental contamination and broad-scale lethal exposure.
   d. Security: Potential use in weapons proliferation.
   e. High capital costs.
2. Fuel Handling and Waste Storage
   a. If all of the nuclear fuel used in the United States to date were fully reprocessed and useful isotopes were recycled, the remaining radioactive waste would fit in a building about one-third the size of a small house.
   b. Also, it is possible to convert that waste into harmless isotopes, and future technology may make that affordable.
3. Base-Load Mismatch
   a. A way to store energy when it's not needed should be devised so that more can be available at times of greater need.
4. Safety
   a. Chemical processes during accidents include the buildup of pressure, formation of hydrogen, combustion of hydrogen with air, and reactions and transport of radioactive fission products and other substances. Use of green adsorbents that will remove these radioactive components from the vented gases is a green chemistry solution.
5. Security
   a. It is difficult to make a nuclear bomb from low-enriched uranium (used in all power reactors) because the process involves separating uranium isotopes. It is easier to separate plutonium from the spent fuel (mainly uranium-238) because plutonium is a different element with different chemical properties, and hence plutonium should be removed from the spent fuel for recycling or safeguarding.
   b. The process for this separation (known as PUREX) is a first-generation commercial process. The objective is to remove the plutonium from spent fuel and to use it in new fuel rods, thereby minimizing the amount on site; this represents a "recycle to extinction process.
6. High Costs
   a. Solutions to higher plant costs are not so much a matter of green chemistry but rather of good practices that improve the economics.

LEARNING OBJECTIVES

You should be able to . . .

1. Identify the sources of the natural radiation to which we are exposed. (11.1)
2. List the sources and dangers of ionizing radiation. (11.1)
3. Balance nuclear equations. (11.2)
4. Identify the products formed by various nuclear decay processes. (11.2)
5. Solve simple half-life problems. (11.3)
6. Use the half-life concept to solve simple radioisotope dating problems. (11.3)
7. Write a nuclear equation for a transmutation, and identify the product element formed. (11.4)
8. List some applications of radioisotopes. (11.5)
9. Describe the nature of materials needed to block alpha, beta, and gamma radiation. (11.6)
10. Explain where nuclear energy comes from. (11.7)
11. Describe the difference between fission and fusion. (11.7)
12. Describe the goals of the Manhattan Project and how uranium and plutonium bombs are made. (11.8)
13. Identify the most hazardous fallout isotopes, and explain why they are particularly dangerous. (11.8)
14. List some uses and consequences of nuclear energy. (11.9)
15. Explain how applications of green chemistry principles can be applied to handling radioactive isotopes in spent fuel rods.

DISCUSSION

The nucleus is 1/100,000 the size of the atom, yet it contains all the mass and all its positive charge. Nuclear symbols summarize the information needed to describe an isotope: the number of protons (atomic number or nuclear charge) and the number of neutrons (mass number). In a nuclear equation, the atomic numbers and mass numbers are conserved; that is, they are equal before and after the reaction. The half-life of each isotope is characteristic of that isotope and a constant that cannot be changed. A half-life is the time it takes for half of the material to decay. Transmutations can be accomplished by bombarding nuclei with subatomic particles. Radioisotopes are used for their penetrating power and the amount of change they can cause in medical, food, and industrial applications. The ages of rocks and archaeological artifacts are determined using half-lives. In a nuclear reaction, mass is converted into energy, producing a million times the energy of an equivalent chemical reaction. This energy is used in fission (splitting nuclei) processes such as the A-bomb and in current nuclear power plants. These processes require sophisticated technology to increase ("enrich") the amount of fissionable material and have hazards associated with radioactivity. The reaction of the sun and H-bomb is thermonuclear fusion—union of nuclei.

EXAMPLE PROBLEMS

1. Thorium-232 ($^{232}_{90}Th$) undergoes alpha decay. What new element is formed?

   Mass and charge are conserved. Alpha particles are helium nuclei and thus carry away 4 mass units and 2 units of charge. The new nuclei must have a mass of 232 - 4 = 228 and a nuclear charge of 90 - 2 = 88. The nuclear charge (atomic number) identifies the new element as radium. The equation for the process is

$$^{232}_{90}Th \longrightarrow ^{4}_{2}He + ^{228}_{88}Ra$$

2. Iodine-131 ($^{131}_{53}I$) undergoes beta decay. What new element is formed?

   The beta particle takes away essentially no mass (the nucleon number remains unchanged) and a charge of -1. Subtracting 1 increases the nuclear charge by one. The new element has a nuclear charge of 54, identifying it as xenon. The equation is

$$^{131}_{53}I \longrightarrow ^{0}_{-1}e + ^{131}_{54}Xe$$

3. Radioactive nitrogen-13 has a half-life of 10 minutes. After an hour, how much of this isotope would remain in a sample that originally contained 96 mg?

   One hour is 60 minutes or 6 half-lives (n = 6) and m_0 = 96 mg.
   Mass remaining, mg $= (1/2^6)(96 \text{ mg}) = (1/64)(96 \text{ mg}) = 1.5 \text{ mg}$

4. Radioactive thalium-154 has a half-life of 5 seconds. After 10 seconds, how many milligrams of this isotope remain in a sample that originally contained 160 mg?

   Ten seconds is 2 half-lives (n = 2), and m_0 = 160 mg.
   Mass remaining $= (1/2^2)(160 \text{ mg}) = (1/4)(160 \text{ mg}) = 40 \text{ mg}$

ADDITIONAL PROBLEMS

1. Complete the following equations by supplying the missing component.

   a. $^{234}_{90}Th \longrightarrow ^{0}_{-1}e + ?$

   b. $^{222}_{86}Rn \longrightarrow ^{4}_{2}He + ?$

   c. $^{56}_{26}Fe + ^{2}_{1}H \longrightarrow ^{54}_{25}Mn + ?$

2. Protactinium-234 has a half-life of 1 minute. How much of a 400-μg sample of protactinium would remain after 1 minute? After 2 minutes? After 4 minutes?

3. The half-life of plutonium-239 is 24,300 years. About 8 kg of this isotope is released in a nuclear explosion. How many years would pass before the amount was reduced to 1 kg?

ANSWERS TO ADDITIONAL PROBLEMS

1. a $^{234}_{90}\text{Th} \rightarrow\ ^{0}_{-1}e +\ ^{234}_{91}\text{Pa}$

   b. $^{222}_{86}\text{Rn} \rightarrow\ ^{4}_{2}He +\ ^{218}_{84}\text{Po}$

   c. $^{56}_{26}\text{Fe} +\ ^{2}_{1}\text{H} \rightarrow\ ^{54}_{25}\text{Mn} +\ ^{4}_{2}\text{He}$

2. At the end of 1 min, half of the original Pa-234 would have decomposed, leaving 200 μg remaining. After 2 min, half of the remaining 200 μg would decompose, leaving 100 μg. after four half-lives only $1/2^n$ or $1/2^4 = 0.0625$ of the sample, or (400 μg)(0.0625) = 25 μg would remain.
3. 1 kg is one-eighth of the original sample or $1/2^3$, so the reduction would occur over three half-lives, which would be equivalent to (3)(24,300 years) = 72,900 years.

ANSWERS TO ODD-NUMBERED REVIEW QUESTIONS AND SOLUTIONS FOR ODD-NUMBERED END-OF-CHAPTER PROBLEMS

1. d. Radon comes from the natural decay of U-238 in the rocks and soil (rocks and soil themselves are not radioactive).

3. (i) The answer is b. New compounds are formed as the result of chemical reactions.
   (ii) The answer is a. New elements are formed as the result of nuclear transformations.
   (iii) The answer is c. The size, shape, appearance, or volume of a substance changes without changing its composition as the result of a physical change.

5. a. Alpha particle emission ($^{4}_{2}\text{He}$) results in a loss of 4 in the nucleon number and a loss of 2 in the atomic number.
   b. Gamma ray emission (ν) does not change either the nucleon number or the atomic number of the product. Gamma rays are bundles of energy.
   c. Proton particle emission ($^{1}_{1}\text{H}$) results in the loss of one in the nucleon number and one in the atomic number.

7. The statement is false. While it is correct that a half-life is the time it takes for 50% of a radioactive sample to decay, in the period of a second half life, half of the remaining radioactive sample decays, leaving, in this case, 25% of the original radioactive sample.

9. After 33.0 s a 5.00 μg F-19 sample with a half-life of 11.0 s has gone through 3 half-lives, the amount of F-19 remaining in the sample is 1/23 = 1/8 = 0.125 of the original amount.
   Amount of F-19 remaining = (5.00 μg)(0.125) = 0.625 μg.

11. Technetium-99m produces gamma rays that can help with diagnosis. What is especially desirable about this isotope is that it does not release alpha, beta, or positron particles that could do internal harm and it has a short half-life (6.0 hr), so it does not linger in long in the body of the person being tested.

13. Two kinds of protection for workers dealing with radioactive materials are making certain the materials are kept and used in shielded containers (or areas), and increasing as much as possible the distance between the worker and the radioactive source(s).

15. The four primary objectives of the Manhattan Project were to (1) develop the conditions for a sustained nuclear fission reaction, (2) develop a process to enrich uranium to produce a sample containing 90% U-235, (3) to find an effect way of producing plutonium-239, which is another fissionable isotope, and (4) to construct a bomb based on a nuclear fission reaction.

17. Uncontrolled nuclear fission reactions can be used to cause destruction as in the case of bombs. Second, the same basic reactions can be controlled and used instead of fossil fuels to generate power for the general population. Third, fission reactions can be controlled and used to power sea-going vessels.

Problems

19. The answer is b. Of the four types of electromagnetic radiation listed, the other three are each energetic enough to ionize atoms and molecules. Microwave radiation is very much less energetic and is considered safe for home use in microwave ovens.

21. a. $^{250}_{98}\text{Cf} \rightarrow {}^{4}_{2}\text{He} + {}^{246}_{96}\text{Cm}$

b. $^{210}_{83}\text{Bi} \rightarrow {}^{210}_{84}\text{Po} + {}^{0}_{-1}\text{e}$

c. $^{117}_{53}\text{I} \rightarrow {}^{0}_{+1}e + {}^{117}_{52}\text{Te}$

23. a. $^{179}_{79}\text{Au} \longrightarrow {}^{175}_{77}\text{Ir} + {}^{4}_{2}\text{He}$

b. $^{12}_{6}\text{C} + {}^{2}_{1}\text{H} \longrightarrow {}^{13}_{6}\text{C} + {}^{1}_{1}\text{H}$

c. $^{154}_{62}\text{Sm} + {}^{1}_{0}\text{n} \longrightarrow 2\,{}^{1}_{0}\text{n} + {}^{153}_{62}\text{Sm}$

25. $^{99m}_{43}\text{Tc} \rightarrow {}^{99}_{43}Tc + \gamma$ The reaction emits gamma radiation.

27. The original nucleus must have had a nucleon number of 233+ 4 or 237 and an atomic number of 91 + 2 or 93. The balanced equation for this reaction is $^{237}_{93}\text{Np} \rightarrow {}^{4}_{2}He + {}^{233}_{91}Pa$

29. Reaction for the first step: $^{107}_{47}\text{Ag} + {}^{1}_{0}\text{n} \rightarrow {}^{108}_{47}Ag$

Reaction for the second step: $^{108}_{47}\text{Ag} \rightarrow {}^{0}_{-1}e + {}^{108}_{48}\text{Cd}$

31. The niobium-94 sample must have gone through 5 half-lives in order to have decayed from 2.00 μg to 0.0625 μg. We can calculate this either by going through the process (2.00 μg → 1.00 μg → 0.500 μg → 0.250 μg → 0.125 μg → 0.0625 μg) or we can solve the equation $(2.00\ \mu g)(1/2^n) = 0.0625\ \mu g$ for n. The length of time required to go through 5 half-lives is (5)(20,000 years) = 100,000 years.

33. The answer is b. A sample with a very short half-life (for example, 122 ms), will go through many half-lives in a very short time, dramatically reducing the amount of radioactivity coming from the sample very quickly.

35. A reasonable estimate of the half-life of sulfur-35 is 90 days. The activity rate dropped from 1000 cps to 525 cps (slightly more than half the original) in 80 days; and to 446 cps (slightly less than half) in 100 days. The half-life must be between 80 and 100 days, so 90 days would be a good estimate. (The actual half-life of S-35 is 87 or 88 days.)

37. It will take 2 half-lives for the count to drop from 20,000 cpm to 5000 cpm. We calculate this as follows: (20,000 cpm)($1/2^n$) = 5000, so $1/2^n = 0.250$ and $2^n = 4.00$; thus, n, or the number of half-lives, is 2. According to Table 11.6, the half-life of Ga-67 is 78.1 hours, so the time required is:
Time = (2)(78.1 hr) = 156 hr.

39. The mass of K-40 in a banana is (600 mg K)(0.0117 mg K-40/100 mg K) = 0.0702 mg K-40. If 1.00 g of K-40 has an activity of 2.63×10^5 Bq, the activity of the K-40 in the banana is:
activity = (0.0702 mg K-40)(1 g/1000 mg)(2.63×10^5 Bq/1.00 g K-40) = 18.5 Bq.

41. It would require 3 half-lives for a radioactive sample to decay from 10.0 to 1.25 mg. If this process required 75 days, then 75 days must represent 3 half-lives. One half-life must be equal to 25 days.

43. ${}^{48}_{20}\text{Ca} + {}^{247}_{97}\text{Bk} \rightarrow {}^{293}_{117}X + 2{}^{1}_{0}\text{n}$; Two neutrons were released.
${}^{48}_{20}\text{Ca} + {}^{247}_{97}\text{Bk} \rightarrow {}^{294}_{117}X + {}^{1}_{0}\text{n}$; One neutron was released.

45. a. ${}^{64}_{28}\text{Ni} + {}^{238}_{92}\text{U} \rightarrow {}^{302}_{120}\text{X}$
b. ${}^{58}_{26}\text{Fe} + {}^{244}_{94}\text{Pu} \rightarrow {}^{302}_{120}\text{X}$
c. ${}^{54}_{24}\text{Cr} + {}^{248}_{96}\text{Cm} \rightarrow {}^{302}_{120}\text{X}$

47. a. Gamma radiation is the most penetrating, followed by beta radiation and then alpha radiation.

49. ${}^{235}_{92}\text{U} + {}^{1}_{0}\text{n} \rightarrow {}^{236}_{92}\text{U}^* \rightarrow {}^{92}_{36}Kr + {}^{141}_{56}\text{Ba} + 3\,{}^{1}_{0}\text{n}$
The chain reaction continues when each of the three neutrons emitted by the fission of U-236 strikes another U-235 nucleus in what is called a chain reaction.

51. a. In order to apply Einstein's mass-energy equation, we need to change the mass of the sample from grams to kilograms as part of the calculation.
: $E = (1\text{ g})(1\text{ kg}/1000\text{ g})(3.00 \times 10^8\text{ m/s})^2 = 9 \times 10^{13}\text{ kg}\cdot\text{m}^2/\text{s}^2 = 9 \times 10^{13}\text{ J}$
equivalent number of calories = (9×10^{13} J)(1 cal/4.184 J) = 2×10^{13} cal
equivalent number of kilocalories = (2×10^{13} cal)(1 kcal/1000 cal) = 2×10^{10} kcal
b. number of bowls of corn flakes = (9×10^{10} kcal)(1 bowl/110 kcal) = 2×10^8 bowls.

53. Synthesizing plutonium-239 from uranium-238 is a three-step process:
${}^{238}_{92}\text{U} + {}^{1}_{0}n \rightarrow {}^{239}_{92}\text{U}$
${}^{239}_{92}\text{U} \rightarrow {}^{239}_{93}\text{Np} + {}^{0}_{-1}\text{e}$
${}^{239}_{93}\text{Np} \rightarrow {}^{239}_{94}\text{Pu} + {}^{0}_{-1}e$

55. The original U-235 that undergoes fission forms two smaller nuclei, both of which are also radioactive. Note that the exact identity of these smaller nuclei can vary (see examples shown in Figure 11.10).

57. Many apply: Power plants control the rate of the fission reaction, so it is slow, continuous, and controlled. The bomb's fission reaction rates are almost instantaneous; all the energy is released at once and not controlled. Power plants use control rods to regulate the amount of neutrons and how much material is above the critical mass. Bombs have no control rods, and it is all brought above the critical mass at once. People can be nearby and live during the operation of a nuclear power plant, but you need to be a long ways away during the operation of a nuclear bomb! Another difference is that

the level of enrichment of U-235 required for use of a uranium sample in a power plant is only 3–5%, while that required for use in a bomb is 90%. Another difference is that the rate of the fission reaction in a nuclear power plant is very carefully controlled by the use of neutron-absorbing control rods, while the fission reaction in a bomb is not controlled at all.

59. The decay of polonium-210 added to Alexander Litvinenko's tea is: $^{210}_{84}\text{Pu} \rightarrow {}^{4}_{2}He + {}^{206}_{82}\text{Pb}$. The reaction products are lead-206 and an alpha particle.

61. $^{223}_{88}\text{Ra} \rightarrow {}^{219}_{86}\text{Rn} + {}^{4}_{2}\text{He}$

$^{223}_{88}\text{Ra} \rightarrow {}^{209}_{82}\text{Pb} + {}^{14}_{6}\text{C}$

63. $^{9}_{4}\text{Be} + {}^{4}_{2}\text{He} \rightarrow {}^{12}_{6}C + {}^{1}_{0}\text{n}$

This was the first experimental detection of the neutron.

65. For a 100 g sample of a radioactive element to decay to 1 g would require slightly more than 66.6 half-lives. (100 g)($1/2^n$) = 1 g so n = 6.65. The time required for the radium to decay is (6.65)(11.4 days) = 76 days. Because this equation provides an estimated result, it is safe to say that the isotope would go through 7 half lives and (7)(11.4 days) = 80 days.

67. An activity of 980 cph is one-fourth of the original activity of 3920 counts/hour, so the sample has gone through two half-lives. The half-life of C-14 is 5730 years. This means that a period of (2)(5730 years) = 11,460 years has elapsed between the time the tree died and the present time.

69. Two half-lives are required for the activity of a sample to drop from 56 ppm/m^3 to 14 ppm/m^3. The amount of time required is (2)(12.7 h) = 25.4 hr, or a little more than 25 hr. Thus, if the sample was taken on Monday, the water should be safe on Tuesday.

71. a. If the half-life of K-40 is 1.2 billion years, then 3.6 billion years represents three half-lives. The amount of sample remaining after 3 half-lives is $1/2_3$ = 0.125 or 12.5%.
b. Uranium-238 is more useful than C-14 for determining the age of rocks because there is very little carbon in the rocks and the much shorter half-life of C-14 means that much less of it would remain to be analyzed after very long time periods.

73. activity = (0.25 mg Am-241)(1 g/1000 mg)(1.26×10^{11} Bq/g Am-241) = 3.2×10^7 Bq

75. d = m/V so V = m/d

Volume occupied by plutonium = (16,300 g)/19.1 g/cm^3)
= 853 cm^3

Volume of coated plutonium = (2500 g) /(19.1 g/cm^3)
= 131 cm^3

Comparing the size of this mass (from above, 131 cm^3) to that of a baseball, a volleyball, and a basketball:
The equation for a sphere is $4/3\pi r^3$, the radius of a baseball is 3.56 cm, that of a volleyball is 10.54 cm, and that of a basketball is 12.17 cm.
volume of baseball = $(4/3)(3.14)(3.56\text{ cm})^3 = 189\text{ cm}^3$
volume of volleyball = $(4/3)(3.14)(10.54\text{ cm})^3 = 4900\text{ cm}^3$
volume of basketball = $(4/3)(3.14)(12.17\text{ cm})^3 = 7550\text{ cm}^3$
The volume of a critical mass of coated plutonium is smaller than a baseball.

77. Using the formula provided and information about the isotopes remaining in the control rod, we calculate the concentration of remaining isotope:
For I-131: concentration = (initial concentration)$(0.5^{3650\text{ d}/8\text{ d}})$
= essentially zero
For Tc-99: concentration = (initial concentration)$(0.5^{10/211,000})$
= 0.9999(initial concentration) so essentially all the initial concentration remains after 10 years
For Pu-239: concentration = (initial concentration)(0.510/24,100)
= 0.9997(initial concentration) so essentially all the initial concentration remains after 10 years.
In terms of disposal, for I-131, allow natural decay at the reactor site, which is green chemistry principle 10 on design for degradation. For Tc-99, separate and set aside for long-term storage (note that although radioactive, it is not very radioactive and thus not highly dangerous), which should use green principles 3 and 12 on the recycle chemistry. For Pu-239, separate and use as fuel in a nuclear reactor, which, if recycling is considered as part of the overall process, is an application

CHAPTER

12

Chemistry of the Earth

Metals and Minerals

CHAPTER SUMMARY

12.1 Spaceship Earth: Structure and Composition

Learning Objectives: • Describe the structure of Earth and the regions of Earth's surface. • List the most abundant elements in Earth's crust.

A. Structurally, the Earth is divided into three main regions: the core, the mantle, and the crust.
   1. The core is thought to be mainly iron and some nickel. Its diameter is ~3500 km. It has two physical states: solid (inner; ~1300 km in diameter) and liquid (outer; ~ 2200 km in thickness).
   2. The mantle is believed to consist mainly of silicates (sulfur, oxygen, and metal compounds). It is ~2900 km thick and consists of a lower (~2000 km thick), transition (~500 km thick), and upper (~360-km-thick) mantle.
   3. The crust is the outer shell of the Earth, which is between 8 km (oceans) and 40 km (continents) thick. The crust is mainly composed of oxygenated Si (~60%) and Al (~15%) compounds: silica, quartz, silicates, silicoaluminates, and others (e.g., metal oxides).
   4. The crust is divided into three parts.
      a. The lithosphere is the solid part and is about 35 km thick under the continents and 10 km thick under the oceans.
      b. The hydrosphere is the oceans, lakes, rivers, and so on.
      c. The atmosphere (air) is the gaseous part.

B. The most abundant element in the Earth's crust by both mass and atom percent is oxygen, followed by silicon and hydrogen.
   1. In the atmosphere, oxygen occurs primarily as molecular oxygen, O_2. In the lithosphere, it occurs primarily in combination with silicon (sand is SiO_2) and some other elements, while in the hydrosphere, oxygen occurs primarily as water, H_2O.

C. The Lithosphere: Composed of Organic and Inorganic Substances.
   1. The lithosphere is composed of three equally important components: inorganic (rocks and minerals), organic (humic materials), and biological (flora and fauna) aspects.
   2. The predominant rocks and minerals in the lithosphere are silicates, carbonates, oxides, and sulfides.

3. A much smaller (in quantity) organic portion includes all living creatures, their waste and decomposition products, and fossilized minerals.
    a. Organic material always contains carbon, nearly always contains hydrogen, and often contains oxygen, nitrogen, and other elements.

D. The Lithosphere's Bounty: From Materials to Energy.
1. Over the centuries, people have learned to convert natural materials into products with superior properties.
2. Fire was one of the earliest agents of chemical change.

Answers to Self-Assessment Questions

1. b The regions of the Earth from the center to outside are the core, the mantle, and the crust.
2. b The Earth's core is mostly iron.
3. c The lithosphere can be divided into inorganic, organic, and biological materials.
4. d The mantle is the middle region of the Earth, not part of the crust.
5. d Oxygen is the most abundant element, followed in decreasing order by silicon and hydrogen.

12.2 Silicates and the Shape of Things

Learning Objectives: • Describe the arrangement of silicate tetrahedral in common silicate minerals. • Describe how glass differs in structure from other silicates.

A. The basic silicate unit is the SiO_4 tetrahedron.
1. Silicates can exist singly or be arranged linearly in fibers, in planar sheets, or in complex three-dimensional arrays.
2. Zircon exists as simple silicate anions (SiO_4^{4-}).

B. Quartz is pure silicon dioxide, SiO_2; however, each silicon atom in the crystal is surrounded by four oxygen atoms.
1. Amethyst, citrine, rose, and smoky quartz contain impurities, which give them their characteristic color.

C. Micas are composed of sheets of SiO_4 tetrahedra, giving a two-dimensional structure.

D. Asbestos is a generic term for fibrous silicates.
1. Chrysotile is a form of asbestos composed of a double chain of SiO_4 tetrahedra bonded to magnesium ions.
2. Inhalation of asbestos fibers over 10 to 20 years causes asbestosis, a severe respiratory disease. After 30 to 45 years, mesothelioma or cancer may develop.
3. Cigarette smoking and inhalation of asbestos fibers act synergistically to greatly increase the risk of lung cancer.

E. Aluminum Silicates: From Clays to Ceramics.
1. Clay is formed when two-dimensional sheets of connected SiO_4 tetrahedra are stacked to produce a three-dimensional mineral phase
2. There are two major classes of clays, based on how the tetrahedral silicate sheets are assembled with the octahedral, mainly aluminum hydroxide, sheets.
    a. 1:1 clays: In these clays, a tetrahedral silicate sheet and an octahedral aluminum hydroxide sheet are connected by oxygen atoms bound to both the Si atoms of the tetrahedral sheet and the Al atoms of the octahedral sheet, to form a layer. These 1:1 layers are then connected by hydrogen bonding to each other, alternating the silicate and aluminum hydroxide sheets as they pile one on top of the other. It also means that these clays do not expand or swell. A well-known example of this type of clay is a kaolinite.

b. 2:1 clays: These clays consist of layers composed of one octahedral aluminum hydroxide sheet sandwiched between two tetrahedral silicate sheets, bridged by oxygen atoms. If the space between the 2:1 layers contains H2O, this type of clay contracts when it is dried and expands when it is wetted. Expanding clays have found applications in drilling muds, in clumping cat litter, and in pollutant cleanup, to mention just a few.

3. Ceramics (modified sand, clays, and limestone) have been developed with specialized properties such as high heat resistance, magnetic properties, and computer memory capabilities.
   a. Bricks and pottery are examples of ceramics.

F. Glass.
1. Glass is a noncrystalline solid resulting from heating a mixture of sand, "soda" (sodium carbonate, Na_2CO_3), and limestone (calcium carbonate, $CaCO_3$).
   a. When heated, glass softens and can be blown, rolled, pressed, or molded.
   b. Properties result from an irregular three-dimensional arrangement of SiO_4 tetrahedra connected by chemical bonds of varying strength.
2. Basic ingredients in glass can be mixed in different proportions.
3. Glass manufacture uses no vital raw materials, but the furnaces require a great deal of energy to reach high temperatures.
4. Glass is easily recycled.

G. Ceramics
1. Ceramics are inorganic nonmetallic solids that are usually crystalline or partly crystalline. The heating techniques used in the manufacturing of ceramics along with the flexibility of the materials used to make them means that ceramics can be tailored to specific uses. Many ceramics have been made from kaolinite clays and aluminum oxide

H. Cement and Concrete.
1. Cement is a complex mixture of calcium and aluminum silicates.
   a. Raw materials are finely ground limestone and clay, heated to 1500 °C.
2. The finished product is mixed with sand, gravel, and water to form concrete.
   a. Concrete is widely used in construction because it is inexpensive, strong, chemically inert, durable, and tolerant of a wide range of temperatures.
   b. Its production is expensive, in terms of both environmental damage and cost.
   c. Concrete can be recycled as rock fill.

Answers to Self-Assessment Questions

1. d All silicate minerals contain oxygen.
2. c Quartz is pure SiO_2.
3. c In a silicon tetrahedron, Si is at the center of the tetrahedron, with the four oxygen atoms at the corners.
4. a Asbestos has double chains of silicon and oxygen atoms.
5. b Mica has sheets of silicon tetrahedra.
6. d Quartz has a three-dimensional array of silicon and oxygen atoms.
7. c Zircon has a structure of SiO_4^{4-} anions (see Table 12.3).
8. f Clays are formed from stacked sheets of SiO_4 tetrahedra.
9. e Glass is an amorphous substance created when crystalline materials are broken down by heating.
10. a 2:1 clays consist of layers composed of one octahedral aluminum hydroxide sheet sandwiched between two tetrahedral silicate sheets, bridged by oxygen atoms.
11. d The gold color in glass can be produced by nanoparticles of gold.
12. a Ceramics are hard and durable but break easily (brittle).

13. d Ordinary glass is made of mostly sand, sodium carbonate, and limestone.
14. a The main ingredients of ordinary cement are clay and limestone.

12.3 Metals and Ores

Learning Objectives: • List the most important metals with their principal ores, and explain how they are extracted. • Describe some of the environmental costs associated with metal production.

A. Copper and Bronze.
   1. Copper usually occurs with sulfur from which it can be separated by heating.
   2. Bronze, a copper-tin alloy (~ 90% Cu, ~ 10% Sn) is harder than copper.
      a. An alloy is a mixture of two or more elements, at least one of which is a metal.

B. Iron and Steel.
   1. Carbon reduces iron oxides to iron metal. (The actual reducing agent is carbon monoxide.)
   2. Iron is converted to steel (an alloy of carbon with other metals) by reacting oxygen with the impurities in iron and adjusting the carbon content.
      a. High-carbon steel is hard and strong.
      b. Low-carbon steel is ductile and malleable.
   3. Iron reacts with atmospheric oxygen, forming a porous, flaky coating that flakes off, allowing further oxidation.

C. Aluminum: Abundant and Light.
   1. Unlike iron, aluminum forms an oxide film on its surface that protects the metal from further corrosion.

D. The Environmental Costs of Iron and Aluminum.
   1. Aluminum is the most abundant metal in the Earth's crust, occurring as aluminum oxide (bauxite).
   2. Making an aluminum can from bauxite requires 6.3 times as much energy as making a steel can.
   3. When aluminum is recycled, only a fraction of the energy needed to make a steel can is required.

E. Other Important Metals.
   1. Table 12.6 in your text lists other technologically important metals such as indium, lithium, and palladium.
   2. Although atoms are conserved globally, use of metals scatters these atoms throughout the environment.

Answers to Self-Assessment Questions

1. b Copper is the element most likely to be found as a free element.
2. d SO_2 is produced when Cu is produced from CuS.
3. c Bronze is an alloy of Cu and Sn (tin).
4. a Carbon serves both to reduce the iron in the ore and to harden the resultant steel.
5. Electrolysis (c) is used to reduce aluminum when it is isolated from its ore. Heat (b) is used to reduce copper from its ore. Carbon (a) is used to reduce iron from its ore.
6. a The ore composed of primarily Al_2O_3 is called bauxite.
7. d The oxidation number in all of the compounds of Al in all the reactions shown is +3. The other common oxidation number of Al is zero when it is in its elemental form, Al.
8. c Recycling an aluminum can requires only 5% of the energy that would be required to make one from raw materials.

9. b Classic bronze contains 90% Cu and 10% Sn. Mild bronze, which is softer than classic bronze, contains 94% Cu and 6% Sn.

12.4 Earth's Dwindling Resources

Learning Objectives: • List the main components of solid waste. • Name and describe the three Rs of garbage.

A. Most of the high-grade ores in the United States are gone.
   1. It takes more energy to obtain metals from low-grade ores than from high-grade ores.
B. Land Pollution: Solid Wastes.
   1. Once discarded, most solid goods are sent to municipal landfills. This material is called municipal solid waste (MSW).
   2. MSWs are handled in three principal ways.
      a. In 2010, 33.8% of MSW was recovered and recycled or composted.
      b. 11.9% was burned at combustion facilities (incinerators).
      c. 54.3% was disposed of in sanitary landfills.
   3. In sanitary landfills, trash is compacted and covered over.
      a. Some landfills leak and contribute to groundwater contamination.
      b. Materials in landfills decompose slowly.
      c. Land for landfills is increasingly scarce.
   4. Incineration can reduce the volume of trash, but it can lead to air pollution.
C. The Three Rs of Garbage: Reduce, Reuse, Recycle.
   1. Reduce amount of throwaway material produced.
   2. Reuse materials: Durable objects can be used repeatedly and require less energy than recycling. Plastic bags are not as good as paper bags.
   3. Recycle materials: This requires less energy than making objects from ores.
D. How crowded is our spaceship?
   1. Due to medical advances, our death rate has been lowered while our birth rate has stayed the same.
   2. As a result, our birth and death rates are out of balance and our population is increasing at a very fast rate.
      a. This creates a strain on our resources.
   3. We're going to need a bigger Earth.
      a. We need to consider the atom economy of the processes used to produce goods—for example, replace steel with aluminum.
      b. We should consider the environmental friendliness of the materials we use—for example, aluminum is greener than steel. Although it takes more energy to produce, recycling aluminum saves 95% of the energy used to produce the metal.
      c. We should use nondepleting raw materials.

Answers to Self-Assessment Questions

1. c Paper is the largest component of the garbage put into landfills.
2. c Recycling aluminum requires only 5% of the energy to produce aluminum from bauxite, so, from an energy perspective, we could make 20 cans by recycling for every can we made starting with the ore.
3. d The three Rs of garbage are reduce, reuse, and recycle.

Green Chemistry: Critical Supply of Key Elements

A. We are rapidly depleting the reserves of many of the elements we require for our modern day life.
   1. Although we recycle many things, some critical elements are difficult to recycle.
B. The solution to the problem will be finding green chemistry solutions for achieving the same technological goals using more readily available and easily recyclable materials.

LEARNING OBJECTIVES

You should be able to . . .

1. Describe the structure of Earth and the regions of Earth's surface (12.1)
2. List the most abundant elements in Earth's crust. (12.1)
3. Describe the arrangement of silicate tetrahedra in common silicate minerals. (12.2)
4. Describe how glass differs in structure from other silicates. (12.2)
5. List the most important metals with their principal ores, and explain how they are extracted. (12.3)
6. Describe some of the environmental costs associated with metal production. (12.3)
7. List the main components of solid waste. (12.4)
8. Name and describe the three Rs of garbage. (12.4)
9. Identify the green chemistry principles that can guide the determination of which metal, Fe or Al, is greener.
10. Describe how recycling supports sustainable use of natural resources.

DISCUSSION

This chapter describes the resources of planet Earth that are easily accessible to us. By far the most abundant atom is oxygen, followed by silicon and hydrogen. Combinations of silicon and oxygen (the formulas for sand and silicates are SiO_2 and SiO_4, respectively) form an important part of the materials used in our everyday life. Quartz, mica, and asbestos are different forms of pure silicates; ceramics (made by heating clays) are complex silicates; glass is sand mixed with various inorganic salts; cements and concretes are aluminum silicates mixed with limestone. Most metals are found combined with oxygen and have to be reduced to produce the pure metal. Copper was the first metal to be widely used, followed by iron and steel. Although the most abundant metal, aluminum, is difficult to separate from oxygen, it is still cheaper to recycle aluminum than it is to process its ore. Since the Earth neither gains nor loses appreciable mass, and since mass is conserved, the effectiveness with which we manage our resources determines whether or not we have readily available resources and an unpolluted environment. Another factor to consider in the availability of resources and quality of the environment is the world's population.

ANSWERS TO ODD-NUMBERED REVIEW QUESTIONS AND SOLUTIONS FOR ODD-NUMBERED END-OF-CHAPTER PROBLEMS

Review Questions

1. A synergistic effect is one that is greater than the sum of expected effects. The especially adverse health effects of coupling smoking and asbestos exposure is an example of a synergistic effect.

3. Montmorillonite is added to cat litter so that it will clump in the presence of moisture, making it easier to separate the litter that has absorbed moisture from that which has not. The use of this kind of litter accomplishes two purposes: it makes cleaning the boxes easier, and it reduces the amount of "good" litter that would otherwise be discarded.

5. Cement production requires extensive mining; the rotary kiln used in its production consumes fossil fuels; and particular matter from the crushing operations, along with smoke and sulfur dioxide from the burning of fossil fuels, make air pollution by cement plants especially serious. The technology for manufacture of goods from copper and bronze (Cu + Sn) preceded that for iron and steel for several reasons. Copper and tin are easier than iron to extract from ores (copper is often found in its native state while iron reacts so readily with oxygen and sulfur that it is not found uncombined in nature), coupled with the fact that iron is much higher melting than copper or bronze.

7. The technology for manufacture of goods from copper and bronze (Cu + Sn) preceded that for iron and steel for several reasons. Copper and tin are easier to extract from ores than is iron (copper is often found in its native state, while iron reacts so readily with oxygen and sulfur that it is not found uncombined in nature), coupled with the fact that iron is much higher melting than copper or bronze. Aluminum ores are thinly distributed in clays rather than localized.

9. Aluminum ores are thinly distributed in clays rather than localized.

11. The three methods of solid waste disposal are storing them in sanitary landfills, incinerating them, or recycling them. Some sanitary landfills have issues with leakage which can result in groundwater contamination, and because the materials are covered, many decompose very slowly due to the lack of oxygen. Incinerators can be poorly designed and generate smoke and odors. Recycling requires energy and some of the material is lost.

13. Metals are fairly easily recycled because most of the energy required to produce them initially is not required in the recycling process. We can run out of metal if goods produced from metal are discarded into landfills rather than being recycled. The problem is keeping the metal in a usable form and not dispersed through the environment.

Problems

15. The answer is d. The Earth's surface consists of the solid lithosphere, the hydrosphere (water), and the atmosphere (air).

17. The lithosphere is composed of an inorganic component (rocks and minerals), an organic component (humic materials), and a biological component (flora and fauna).

19. In terms of atom percent, the abundances of the listed elements are Si (15.9%) > H (15.1%) > Na (1.8%) > Fe (1.5%) = Ca (1.5%) > Mg (1.4%) > K (1.0%) (see Table 12.1).

21. The rankings in questions 19 and 20 are different because of the way in which the percentages were calculated. Atom percent reflects the relative numbers of individual kinds of atoms, while mass percentages reflect the relative masses contributed by each element. Because some atoms are much heavier than others, the two kinds of percentages do not agree.

23. The chemical formulas of hematite and magnetite are Fe_2O_3 and Fe_3O_4 respectively. The oxidation number of Fe in Fe_2O_3 is +3. The oxidation number of Fe in Fe_3O_4 is +8/3 or +2 2/3.

25. When people learned to control fire, they learned that cooking improved the flavor and digestibility of meat and grains; that heating adobe bricks made stronger ceramic bricks; and that firing clay at high temperatures produced ceramic pots, making cooking and storage of food much easier. As they became able to produce higher temperatures, people learned to make glass and to extract metals from ores. They made tools from bronze and, later, from iron.

27. The ratio of silicon to oxygen atoms in silica is 1:2. However, because each silicon atom is surrounded by *four* oxygen atoms, the basic unit of quartz is the SiO_4 tetrahedron.

29. The Lewis structure of the SiO_4^{4-} ion is:

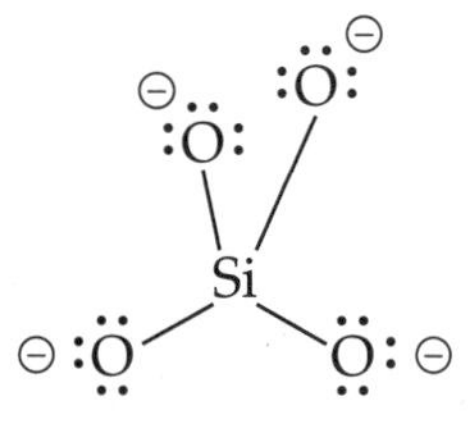

31. Micas are composed of SiO_4 tetrahedra arranged in two-dimensional sheet-like arrays. Micas are easily cleaved into thin, transparent sheets. The sheets are linked by bonds between O atoms and cations, mainly Al^{3+}. The bonding within the sheets is stronger than the attractions from sheet to sheet.

33. a. There are two major classes of clays, based on how the tetrahedral silicate sheets are assembled with the octahedral, mainly aluminum hydroxide, sheets.

35. b. The composition of ordinary glass windows is 75% SiO_2, 15% Na_2O, and 10% CaO. In terms of atom percent, oxygen is the most prevalent. Oxygen is likely the most abundant in terms of mass percent as well. The two elements that would compete for top ranking are Si and O, which appear in a 1:2 ratio in SiO_2 and oxygen appears in the other two major components as well. If the only component that contained Si and O was SiO_2 then oxygen would have a higher mass percentage because the masses of O and Si in SiO_2 are (2)(16.0) = 32 and 28.1, respectively.

37. In the production of glass, crystalline materials are heated to a temperature at which they start to lose their repeating arrangement of atomic units (SiO_4) because the bridging interactions within the crystalline matrix break leading to an irregular three-dimensional arrangement of the SiO_4 tetrahedra,

39. Oxides of various metals can be substituted for some or all of the sand, soda, or limestone. Adding extra elements to glass changes its color.

41. Ceramics are inorganic nonmetallic solids that are usually crystalline or partly crystalline. Heating in the manufacturing of ceramics along with the flexibility of the materials used to make them means that ceramics can be tailored to specific uses. Many ceramics have been made from kaolinite clays and aluminum oxide. The more advanced ceramics originate from silicon carbide as well as tungsten carbide. Ceramic materials are particularly useful when high temperatures are present

43. Cement is a complex mixture of calcium and aluminum silicates. The raw materials for the production of cement are limestone and clay which are finely ground, mixed, and roasted at about 1500 °C. The finished product is mixed with sand, gravel, and water to form concrete.

45. a. $2\ Cu_2O \rightarrow 4\ Cu + O_2$: The oxidation number of Cu in Cu_2O is + 1. In Cu the ON of Cu in its elemental form is zero.
    b. $Fe_2O_3 + 3\ CO \rightarrow 2\ Fe + 3\ CO_2$: The ON of Fe ion Fe_2O_3 is +3. In Fe the ON of Fe in its elemental form is zero.
    c. $2\ Al_2O_3 \rightarrow 4\ Al + 3\ O_2$: 'The ON of Al in Al_2O_3 is +3. In Al the ON of Al in its elemental form is zero.

47. Metals usually occur in their ores combined with oxygen (example: Fe_2O_3) or sulfur (example: CuS) in the form of ionic compounds. In order to isolate the metals in their elemental forms, they must be reduced (gain electrons) from the cationic form to the elemental form.
$2\ Cu_2O \rightarrow 4\ Cu + O_2$; $Fe_2O_3 + 3\ CO \rightarrow 2\ Fe + 3\ CO_2$:; and $2\ Al_2O_3 \rightarrow 4\ Al + 3\ O_2$: are examples of reduction reactions in which metals are produced in their elemental forms.

49. a. Adding tin to copper increases the hardness of the resultant alloy so that it can be used in more applications.
    b. Bronze is an alloy containing ~90% Cu and between 6% and 10% Sn.
    c. Bronze containing 10% Sn is harder than bronze containing 6% tin.

51. Copper is used extensively in electrical wires because of its excellent electrical conductivity. It is also used in roofing because of its thermal conductivity and appearance. In addition, it is used in plumbing pipes. Steel is a versatile material whose properties such as hardness, toughness, or corrosion resistance can be varied by adding other metals to the alloy (see Table 12.5). Steel is used in an almost unimaginably wide range of applications including construction, engines, and prosthetic devices.

53. Of Al, Cu, and Fe, Al requires the most energy to extract and purify because Al is more reactive than either of the other two metals.

55. A katana (a samurai sword) is strong with a sharp edge. The temperature of the smelting process is carefully monitored to control the carbon content of the steel between 0.5 and 1.5%. The higher the carbon content is, the harder but less flexible the steel will be. The process results in several layered steels with different carbon concentrations so that the cutting edge is very sharp and hard, while the center is tough enough to adsorb impact without breaking the blade.

57. Balanced equation: 2 ErF_3 + 3 Mg $\rightarrow$ 2 Er + 3 MgF_2
   a. Er in is reduced from the 3+ ion (ON = +3) in ErF_3 to the elemental form (ON = zero).
   b. Mg provides the electrons to Er^{3+} for the reduction so Mg is the reducing agent.
   c. Mg is oxidized, going from ON = zero to ON = +2.
   d. Er^{3+} in ErF_3 is the oxidizing agent because it removes the electrons from Mg.

59. Balanced equation: 2 $EuCl_3$ $\rightarrow$ 2 Eu + 3 Cl_2
   a. Eu in is reduced from the 3+ ion (ON = +3) in $EuCl_3$ to the elemental form (ON = zero).
   b. Cl^- in $EuCl_3$ provides the electrons to Eu^{3+} for the reduction, so Cl^- is the reducing agent.
   c. Cl^- is oxidized, going from ON = – 1 to ON = zero.
   d. Eu^{3+} in $EuCl_3$ is the oxidizing agent because it removes the electrons from Cl^-.

61. Chemical reaction: 2 $InCl_3$ $\rightarrow$ 2 In + 3 Cl_2
   Molar mass of In = 114.8 g/mol; molar mass of $InCl_3$ = 114.9 + (3)(35.5) = 221.3 g/mol
   Mass of $InCl_3$ = (33.2 g In)(1 mol In/114.8 g In)(2 mol $InCl_3$/2 mol In)(221.3 g $InCl_3$/mol $InCl_3$)
   = 70.0 g $InCl_3$

63. (72,000,000 kg Al)(17 kWh/kg Al) = 1,200,000,000 kWh

65. 4 Au + 8 NaCN + 2 H_2O + O_2 $\rightarrow$ 4 $NaAu(CN)_2$ + 4 NaOH
   Molar mass Au = 197.0 g/mol; molar mass NaCN = 49.0
   Mass of NaCN = (10 g Au)(1 mol Au/197.0 g Au)(8 mol NaCN/4 mol Au)(49.0 g NaCN)
   = 5.0 g NaCN

67. Assuming it isn't recycled, according to Figure 12.9, approximately 60–70% of municipal waste could be burned as fuel.

69. The three Rs stand for reduce, reuse, and recycle.

71. The answer is e. All of the items listed should be considered when deciding whether aluminum metal is greener than steel.

CHAPTER

13

Air

The Breath of Life

CHAPTER SUMMARY

13.1 Earth's Atmosphere: Divisions and Composition

Learning Objectives: • List and describe the layers of the atmosphere. • Give the approximate proportion of N_2, O_2, Ar, and CO_2 in Earth's atmosphere.

A. The atmosphere is divided into layers.
   1. The troposphere is the layer next to the Earth's surface where nearly all life exists.
   2. The next layer is the stratosphere, where the Earth's protective ozone layer is located.
   3. Above the stratosphere is the mesosphere, and above that is the thermosphere.

B. Air is a mixture of gases. Dry air, by volume, is:
   1. 78.08% nitrogen (N_2).
   2. 20.94% oxygen (O_2).
   3. 0.93% argon (Ar).
   4. 0.04% carbon dioxide (CO_2), up from 0.028% two centuries ago.
   5. < 0.01% trace gases.

Answers to Self-Assessment Questions

1. d The part of the atmosphere enveloping virtually all human activity is called the troposphere.
2. b The ozone layer is found in the stratosphere.
3. a The atmosphere blends into outer space.
4. d The approximate percentage of nitrogen in the atmosphere is 78%.
5. a After nitrogen and oxygen, the most abundant gas in dry air is argon.
6. b The CO_2 concentration in the troposphere has risen to 400 ppm:
   (0.04 parts/100 parts air)(1,000,000 parts air) = 400 ppm.
7. c The atmosphere from 11 km to 50 km above Earth's surface is called the stratosphere (see Figure 13.1).

13.2 Chemistry of the Atmosphere

Learning Objectives: • Describe the nitrogen and oxygen cycles • Describe the origin and effects of temperature inversions.

A. The Nitrogen Cycle.
   1. Plants need nitrogen as a nutrient, but cannot use nitrogen in the form of N_2 molecules.
   2. Fixed nitrogen is atmospheric nitrogen combined with other elements.
   3. Lightning fixes nitrogen by causing it to combine with oxygen to make nitrogen oxides.
      a. The nitrogen oxides then react with water to form nitric acid.
   4. Nitrogen is also fixed industrially to make nitrogen fertilizers.
   5. Some bacteria fix nitrogen; others convert it back to N_2. This establishes a nitrogen cycle.

B. The Oxygen Cycle.
   1. In the troposphere our supply of oxygen is constantly replenished by green plants and consumed by animals and plants in the metabolism of foods.
   2. In the stratosphere, oxygen is formed by the action of ultraviolet rays on water molecules.
      a. Some oxygen is converted to ozone.
      b. Ozone shields us from harmful ultraviolet radiation.

C. Temperature Inversions.
   1. Lower stagnant cold air is trapped by warmer air above it.
      a. Pollutants in the cold air are trapped near the ground, sometimes for several days.

Answers to Self-Assessment Questions

1. c Lightning can convert N_2 into nitrogen oxides (NO_x).
2. b Nitrogen fixation is the conversion of nitrogen (N_2) into biologically useful forms.
3. a Hydrogen (H_2) and nitrogen (N_2) are used in the industrial process (the Haber–Bosch process) to make ammonia.
4. a Carbon dioxide is removed from the atmosphere by photosynthesis.
5. b The corrosion of metals consumes oxygen (O_2).
6. c Thermal inversions concentrate air pollutants.

13.3 Pollution Through the Ages

Learning Objectives: • List some natural sources of air pollution. • List the main pollutants formed by burning coal, and describe some technologies used to clean up these pollutants.

A. Volcanoes spew ash and sulfur dioxide into the atmosphere.

B. Dust storms add enormous amounts of particulate matter to the air.

C. The Air Our Ancestors Breathed.
   1. Early people made fires, which added smoke to the atmosphere, and cleared land, which made dust storms worse.
   2. Rome was afflicted with stinking air and soot in C.E. 61.
   3. Heavy smoke promoted regulation of emissions from burning coal in England in the thirteenth century.
   4. The Industrial Revolution caused terrible pollution from burning coal in factory towns.
   5. The level of air pollution today is much more complex than at other times in history.

D. Pollution goes global.
   1. Huge urban areas today are afflicted with air pollution that drifts from one area to another.

2. Pollution from Midwestern power plants leads to acid rain in the Northeast.
3. Norway is afflicted with pollution from Germany and England.
4. Most major cities around the world have suffered serious episodes of air pollution.
5. Cities in China, Iran, Mexico, and Indonesia have experienced serious episodes of air pollution.
6. A pollutant is too much of any chemical in the wrong place or at the wrong time.

E. Coal + Fire → Industrial (Sulfurous) Smog.
1. The term *smog* is a contraction of the words *smoke* and *fog*.
2. There are two basic types of smog: industrial smog and photochemical smog.
    a. Polluted air associated with industrial activities is called industrial smog.
    b. Industrial smog consists of smoke, fog, sulfur oxides, sulfuric acid, ash, and soot and derives primarily from burning coal.
    c. When burned, the carbon in coal winds up as carbon dioxide, carbon monoxide, and soot (unburned carbon).
    d. The sulfur in coal is oxidized to sulfur dioxide (an acrid, choking gas) and then to sulfur trioxide, which reacts with water to form sulfuric acid.
3. Particulate matter (solid and liquid particles of greater than molecular size) consists of minerals (fly ash) and soot.
    a. Visible particulate matter consists of dust and smoke.
    b. Invisible particulates are called aerosols (a dispersion of liquid particles in air).
    c. Small particulates, $< 10\ \mu m$ in diameter and called PM10, are especially harmful.
    d. The EPA monitors even smaller PM2.5 particulates which contribute to tens of thousands of premature deaths each year.
    e. Unburned minerals in coal do not burn and become bottom ash.
        i. When bottom ash is carried aloft in smokestacks, it is called fly ash.

F. Health and Environmental Effects of Industrial Smog.
1. Sulfur dioxide and particulates such as ammonium sulfate act synergistically to cause far greater harm than either would alone.
2. Air pollution contributes to the development of respiratory diseases such as emphysema. (Cigarette smoking is a far larger contributor.)
3. Sulfur oxides and sulfuric acid also damage plants, causing crop losses.

G. What to Do about Industrial Smog.
1. There are several ways to remove particulate matter from smokestack gases.
    a. Electrostatic precipitators induce electric charges on the particles, which then are attracted to oppositely charged plates.
    b. Bag filtration works much like a vacuum cleaner to clean the smokestack gases.
    c. Cyclone separators cause the gas to spiral upward; particles hit the walls and settle out.
    d. Wet scrubbers pass the stack gases through water, from which the particulates are removed.
2. Ash removed from stack gases can be used
    a. To replace clay in making cement.
    b. The rest is stored in ponds and landfills.
3. It is difficult to remove sulfur oxides.
    a. Sulfur can be removed from coal before burning
        i. By flotation.
        ii. By gasification or liquefaction.
    b. Sulfur can be removed after burning by scrubbing the stack gases with a suspension of limestone or dolomite.
        i. The calcium sulfite formed can be converted to the more useful calcium sulfate.

Answers to Self-Assessment Questions

1. c The air in thirteenth-century England was smoky from coal burning.
2. a Oxygen is not a product of the combustion of fossil fuels. Rather, it is an essential reactant.
3. c Particulates are described by their sizes, for example, PM10 and PM2.5.
4. d SO_2 and particulates combine synergistically to cause lung damage.
5. c An electrostatic precipitator cleans smoke from flue gases by producing static charges.
6. d Wet scrubbers remove SO_2 from flue gases, forming $CaSO_3$ from the SO_2.
7. b $CaSO_4$ is a common by-product of sulfur removal from flue gases.

13.4 Automobile Emissions

Learning Objectives: • List the main gases in automobile emissions, and describe how catalytic converters reduce these gaseous pollutants. • Explain how carbon monoxide acts as a poison.

A. The main components of automotive exhaust are water vapor, carbon dioxide, and unreacted nitrogen gas (from the atmosphere) that result from the complete combustion of hydrocarbons.

   1. If combustion is not complete, side reactions occur that produce small quantities of harmful products.
      a. Carbon monoxide (CO), a colorless, odorless, poisonous gas formed by incomplete combustion of fuels.
      b. Nitrogen oxides (NO_x), formed when sunlight breaks down NO_2 to form NO and O atoms, contribute to smog and acid rain and cause lung irritation.
      c. Volatile organic compounds (VOCs) come mainly from unburned fuel or evaporating fuel and react with oxygen atoms to form ground-level ozone (O_3), aldehydes, and peroxyacetyl nitrate (PAN).

B. Carbon Monoxide: The Quiet Killer.
   1. Carbon monoxide forms when a hydrocarbon burns in an insufficient amount of oxygen.
      a. Overall, about 30% of the carbon monoxide that we dump into the atmosphere comes from transportation sources (transportation sources in urban areas can be as high as 95%).
   2. Carbon monoxide is odorless, tasteless, and invisible and can only be detected using CO detectors or test reagents.
   3. Carbon monoxide reacts with hemoglobin in the blood, hindering the transport of oxygen.
      a. Chronic exposure to low levels of carbon monoxide adds stress to the cardiovascular system and may increase the chance of a heart attack.
      b. Exposure to higher levels of carbon monoxide can cause drowsiness and death.
      c. The best antidote is the administration of pure oxygen.

C. Nitrogen Oxides: Some Chemistry of Amber Air.
   1. Any time combustion occurs in air, some of the nitrogen combines with oxygen to form nitrogen oxides. The higher the temperature, the more nitrogen oxides are formed.
   2. Nitric oxide is slowly oxidized in air to amber-colored nitrogen dioxide.

$$2\ NO + O_2 \rightarrow 2\ NO_2$$

   3. Nitrogen oxides produce smog and form nitric acid, which contributes to acid rain.

D. Ozone as an Air Pollutant.
   1. Ozone, O_3, is an allotrope (different form of the same element) of oxygen, O_2.
   2. In the troposphere O_3 is a pollutant, while the presence of O_3 in the stratosphere is essential to the existence of life on Earth.

3. Ozone is a powerful oxidant and extremely reactive, making it a severe respiratory irritant, especially for young children.
4. Ozone causes extensive damage to crops and materials such as rubber.

E. Volatile Organic Compounds (VOCs).
   1. VOCs are organic substances that vaporize significantly at ordinary temperatures and pressures and are major contributors to smog formation.
   2. Natural sources such as swamps release hydrocarbons; only 15% of those in the atmosphere are there as the result of human activity.
   3. In urban areas, the processing and use of gasoline contribute substantially to atmospheric hydrocarbons.
   4. Hydrocarbons react with
      a. Atomic oxygen or ozone to form aldehydes.
      b. Oxygen and nitrogen dioxide to form peroxyacetyl nitrate (PAN).

Answers to Self-Assessment Questions

1. a Transportation vehicles are the largest source of CO, carbon monoxide due to incomplete combustion.
2. a Addiction is not a problem caused by NO_x emissions.
3. a VOCs are almost always composed of molecules containing carbon and hydrogen.
4. d VOCs come mainly from natural sources.
5. b Alkenes react with oxygen atoms to form aldehydes.
6. d Carbon monoxide is produced when fossil fuels are burned in an O_2-deficient atmosphere—that is, in an atmosphere containing less than the required amount of O_2 to produce CO_2.
7. d The NO_x and VOCs associated with automobile exhaust react with sunlight to form O_3 and other irritants.
8. a Carbon monoxide replaces oxygen in hemoglobin, severely limiting breathing efficiency.
9. d. Carbon monoxide concentrations have decreased 25%, partly with the help of more efficient catalytic converters on automobiles and trucks.

13.5 Photochemical Smog: Making Haze While the Sun Shines

Learning Objectives: • Distinguish the origin of photochemical smog from the origin of sulfurous smog. • Describe the technologies used to alleviate photochemical smog.

A. Photochemical Smog.
   1. Photochemical smog is produced by a complex series of reactions that starts with nitrogen oxides and unburned hydrocarbons from automobiles.

B. Solutions to Photochemical Smog.
   1. Modified gas tanks and crankcase ventilation systems have reduced evaporative emissions from automobiles.
   2. Catalytic converters, which reduce hydrocarbon and carbon monoxide emissions in automotive exhausts, are the principal approach to reduce photochemical smog.
   3. Lowering the operating temperature of an engine helps reduce NO_x emissions but lowers engine efficiency.
   4. Driving hybrid vehicles saves gasoline and helps the environment.

Answers to Self-Assessment Questions

1. d Nitrogen oxides and hydrocarbons combine in the presence of sunlight to produce photochemical smog.
2. c An amber haze is an indication of NO_2.
3. b Photochemical smog occurs mainly in dry, sunny weather.
4. a The main source of photochemical smog is automobiles.
5. c A reduction catalyst is required to remove NO_x.
6. a According to Figure 13.8, it takes 2.5 hours for the O_3 level to drop to one-half of its maximum value.

13.6 Acid Rain: Air Pollution → Water Pollution

Learning Objectives: • Name the air pollutants that contribute to acid rain. • List the major industrial and consumer sources of acid-rain-producing pollutants.

A. Sulfur oxides become sulfuric acid, and nitrogen oxides become nitric acid. These acids lower the pH of rainwater.
   a. Rain with a pH below 5.6 is called acid rain.
   b. Normal rainwater is slightly acidic due to dissolved CO_2.
B. Good evidence indicates that acid rain comes from sulfur oxides and nitrogen oxides emitted from power plants, smelters, and automobiles.
C. Acid rain corrodes metals and destroys marble and limestone buildings and statuary.

Answers to Self-Assessment Questions

1. c Acid rain has a pH below 5.6.
2. d The oxidation of S ($S + O_2 \rightarrow SO_2$) leads to acid rain.

13.7 The Inside Story: Indoor Air Pollution

Learning Objectives: • List the main indoor air pollutants and their sources. • Explain where radon comes from and why it is hazardous.

A. Home: No Haven from Air Pollution.
   1. Indoor air is often as bad as or worse than that outside.
      a. Gas ranges and kerosene heaters produce nitrogen oxides.
      b. Formaldehyde is slowly released from building materials and new furniture.
B. Cigarette Smoke.
   1. Cigarette smoke is the most prevalent indoor air pollutant.
   2. More than 40 carcinogens have been found in cigarette smoke.
   3. Carbon monoxide levels often exceed standards for ambient air.
   4. Nonsmokers are also exposed to tars, nicotine, and allergy-triggering substances in cigarette smoke.
C. Radon and Its Dirty Daughters.
   1. Radon is a colorless, odorless, tasteless, chemically unreactive, radioactive gas found in rocks (granite and shale) and minerals.
      a. Radon decays to the daughter isotopes polonium-218, lead-214, and bismuth-214, which are trapped in the lungs and damage tissue.
      b. Trapped inside well-insulated houses, radon levels build up and exceed EPA limits.

D. Other Indoor Pollutants.
   1. Unvented natural gas heaters and kerosene heaters can produce carbon monoxide.
   2. Mold can be produced by excessive moisture.
   3. Some electronic air cleaners produce ozone.

E. It's not easy being green.
   1. Green chemistry solutions have to be carefully studied for their environmental effects.
   2. Liquid crystal displays (LCD) replaced cathode ray tube (CRT) displays to remove the need for lead shielding and to reduce the energy required for operation.
      a. Producing LCDs leads to unwanted materials that can harm the environment. For example, ~16% of the nitrogen trifluoride (NF_3) used for LCDs and solar panel manufacture escapes into the environment where it is 17,000 times more efficient at trapping atmospheric heat than CO_2.
      b. Replacing NF_3 with F_2 is problematic because F_2 is highly toxic.

Answers to Self-Assessment Questions

1. b Gas ranges in kitchens are a major source of NO_x.
2. b Ozone can be produced by some electronic air cleaners.
3. d Radon in homes comes from the uranium present in soil and rocks.
4. c Polonium-218 is a radioactive daughter of radon.

13.8 Stratospheric Ozone: Earth's Vital Shield

Learning Objectives: • Explain the link between CFCs and depletion of the ozone layer. • Describe the consequences of stratospheric ozone depletion.

A. The Stratospheric Ozone Shield.
   1. Oxygen in the mesosphere absorbs short-wavelength ultraviolet radiation, breaking the bond between the oxygen atoms.
   2. Oxygen atoms migrate into the stratosphere where they form O_3.
   3. Ozone molecules absorb longer wavelength ultraviolet radiation, which separates an oxygen atom from O_3, producing O_2 and O.
   4. Undisturbed, ozone is formed and destroyed in a cyclic process in the stratosphere. Levels fluctuate, but human activities may contribute to the destructive part of the cycle.

B. Chlorofluorocarbons and the Ozone Hole.
   1. Chlorofluorocarbons are insoluble in water and inert toward most substances. They persist in the environment for a long time.
   2. Chlorofluorocarbons are broken down by ultraviolet light in the stratosphere to fragments including chlorine atoms.
   3. These chlorine atoms catalyze the destruction of ozone.
   4. The U.S. National Research Council predicts a 2–5% increase in skin cancer for each 1% depletion of the ozone layer.
   5. In 1974, Mario Molina and F. Sherwood Rowland proposed a mechanism for the enhanced ozone depletion by CFCs in the stratosphere for which they received the 1995 Nobel Prize.
      a. They found that one CFC molecule can lead to the destruction of thousands of O_3 molecules.
   6. CFCs were used as the dispersing gases in aerosol cans, as foaming agents for plastics, and as refrigerants.

C. International Cooperation.
   1. The United Nations addressed the problem of ozone depletion through the 1987 Montreal Protocol, an international agreement enforcing the reduction and eventual elimination of the production and use of ozone-depleting substances.
   2. Hydrofluorocarbons (HFCs) have been suggested as substitutes.

Answers to Self-Assessment Questions

1. c Ozone is produced from oxygen in the stratosphere.
2. d Ozone is an allotrope of oxygen.
3. d Ozone in the troposphere is harmful, while the O_3 in the stratosphere protects life by helping absorb harmful ultraviolet radiation before it reaches the troposphere and living things on the land.
4. d Oxygen atoms are produced in the mesosphere by the splitting of oxygen molecules by short-wavelength ultraviolet radiation.
5. d Energy with wavelengths in the ultraviolet region of the spectrum are absorbed by ozone.
6. d Ultraviolet radiation reacts with chlorofluorocarbons to produce chlorine atoms that destroy ozone.

13.9 Carbon Dioxide and Climate Change

Learning Objectives: • List the important greenhouse gases, and describe the mechanism and significance of the greenhouse effect. • Describe some strategies for reducing the amount of CO_2 released into the atmosphere.

A. Nearly all combustible processes yield carbon dioxide as one of the reaction products.
   1. Carbon dioxide levels increased nearly 54% between 1990 and 2011; they are increasing at a rate of 2 ppm per year.
   2. Methane also contributes to the greenhouse effect with concentrations rising from 0.7 ppm in 1750 to 1.8 ppm presently.
B. The sun radiates different types of radiation, of which visible, infrared, and ultraviolet are most prominent.
   1. About half of this energy is either reflected or absorbed by the atmosphere.
   2. The light that gets through (primarily visible radiation) acts to heat the surface of Earth.
C. Molecules That Absorb Infrared Energy.
   1. The three main constituents of the atmosphere—N_2, O_2, and Ar—are small, nonpolar molecules and do not absorb much infrared energy.
   2. Carbon dioxide and some other gases produce a greenhouse effect: they let the sun's visible light pass through the atmosphere to warm the surface, but when the Earth radiates infrared energy back toward space, these greenhouse gases absorb and trap the energy.
   3. Some greenhouse effect is necessary to life—without an atmosphere, all the radiated heat would be lost to outer space and Earth would be much colder than it is.
D. Greenhouse Gases and Global Warming.
   1. An enhanced greenhouse effect is caused by increased concentration of carbon dioxide and other greenhouse gases in the atmosphere.
   2. Enhancement could result in an increase in the Earth's average temperature, an effect called global warming.

3. Methane and other trace gases contribute to the greenhouse effect; methane concentrations are rising.
    a. Methane is 20 to 30 times more efficient at trapping heat than CO_2.
    b. Chlorofluorocarbons are 5000–14,000 times more efficient at trapping heat than carbon dioxide, and HCFCs are up to 11,700 times more efficient at trapping heat than CO_2.

E. Predictions and consequences.
    1. Many scientists believe that global warming is melting the ice caps and causing oceans to rise.

F. Mitigation of global warming.
    1. Technologically advanced countries must make quick and dramatic cuts in emissions.
    2. Technology transfer must occur to developing countries so that those countries can continue to develop their economies without heavy use of coal-fired power plants.
    3. More energy-efficient cars, appliances, and home heating and cooling systems need to be designed and used.
    4. Capture of carbon dioxide from smoke stack emissions by carbon sequestration should be explored.

Answers to Self-Assessment Questions

1. b Infrared radiation is trapped on the Earth's surface by the greenhouse effect.
2. b The global average air temperature increased by 0.875 °C during the twentieth century.
3. a The main cause of global warming is CO_2 from factories, power plants, and automobiles.
4. b CO_2 dissolving in the ocean forms an acid (carbonic acid) that dissolves the shells of sea creatures.
5. d Reducing the use of fossil fuels, and thus CO_2 emissions, is a way of combating global warming.
6. a If there were no greenhouse gases, the mean temperature on Earth would be –18 °C, which is below the freezing point of water.
7. c N_2O, a member of the oxides of nitrogen (NO_x) group, is a greenhouse gas.
8. c. See data on Figure 13.13. The data show an increase slightly above 30%.

13.10 Who Pollutes? Who Pays?

Learning Objective: • List the EPA's criteria pollutants and the major air pollutants that come mainly from automobiles and industry.

A. The EPA lists six criteria pollutants, so called because scientific criteria are employed to determine their health effects.
    1. Carbon monoxide.
    2. Nitrogen dioxide.
    3. Ozone.
    4. Particulate matter.
    5. Sulfur dioxide.
    6. Lead.

B. Where does the pollution come from?
    1. Our transportation system accounts for ~85% of urban carbon monoxide emission, 40% of hydrocarbon emissions, and 40% of nitrogen oxide emissions.
    2. ~40% of PM comes from power plants and ~45% from industrial processes.
    3. More than 80% of sulfur oxide emissions come from power plants, with 15% more coming from other industries.

4. Power plants alone contribute ~55% of nitrogen oxide emissions.
5. Carbon monoxide is toxic but is deadly only in concentrations approaching 4000 ppm, so the World Health Organization rates sulfur oxides as the worst pollutants.

C. Paying the Price.
1. Air pollution costs us tens of billions of dollars each year due to health problems, crop destruction, livestock health issues, and its effects on machines and buildings.
2. Costs increase rapidly as we try to remove larger percentages of pollutants.

Answers to Self-Assessment Questions

1. a The EPA is responsible for gathering and analyzing air pollution data.
2. c Increased pollution controls are the reason air pollutants have decreased significantly since 1970.
3. c Motor vehicles are the largest source of carbon monoxide.
4. c Electric power plants are the largest source of sulfur dioxide (SO_2) pollution.
5. d The World Health Organization cites sulfur dioxide as the worst pollutant in terms of health effects.
6. a The largest sources of NO_x pollution are electric power plants and motor vehicles.
7. b The largest sources of particulate pollution are electric power plants and industry.

Green Chemistry: Making CO_2 Work for You

1. To minimize CO_2 emissions, power plants can separate the CO_2 from the waste gas before it is released into the atmosphere.
   a. This is known as scrubbing and involves cooling the gas and treating it with a liquid solvent that binds to CO_2 (such as aqueous solutions of ammonia or ethanolamine).
2. We can use waste CO_2 to make a number of processes more environmentally friendly.
   a. For example, a very new kind of solvent, called switchable-hydrophilicity solvents (SHS), can exist in either of two forms that can be interconverted by the addition of CO_2.
   b. SHS are amines.
   c. In the absence of CO_2, they are hydrophobic solvents, meaning that they won't mix with water.
   d. However, in the presence of CO_2, SHS become hydrophilic, meaning that they do mix with water.
   e. SHS can be used to extract oily materials from insoluble solids such as removing oil from oil sands (a naturally occurring mixture of sand, clay, and heavy oil), extracting vegetable oil from soybeans, and separating and recycling the oil and plastic in used bottles of motor oil.
   f. Switchable-hydrophilicity solvents, because they can be removed from product and recycled without distillation, do not need to be volatile; they therefore offer a potentially safer alternative.

LEARNING OBJECTIVES

You should be able to . . .

1. List and describe the layers of the atmosphere. (13.1)

2. Give the proportions of N_2, O_2, Ar, and CO_2 in Earth's atmosphere. (13.1)

3. Describe the nitrogen and oxygen cycles. (13.2)

4. Describe the origin and effects of temperature inversions. (13.2)

5. List some natural sources of air pollution. (13.3)

6. List the main pollutants formed by burning coal, and describe some technologies used to clean up these pollutants. (13.3)

7. List the main gases in automobile emissions, and describe how catalytic converters reduce these gaseous pollutants. (13.4)

8. Explain how carbon monoxide acts as a poison. (13.4)

9. Distinguish the origin of photochemical smog from the origin of sulfurous smog. (13.5)

10. Describe the technologies used to alleviate photochemical smog. (13.5)

11. Name the air pollutants that contribute to acid rain. (13.6)

12. List the major industrial and consumer sources of acid-rain producing pollutants. (13.6)

13. List the main indoor air pollutants and their sources. (13.7)

14. Explain where radon comes from and why it is hazardous. (13.7)

15. Explain the link between CFCs and depletion of the ozone layer. (13.8)

16. Describe the consequences of stratospheric ozone depletion. (13.8)

17. List the important greenhouse gases, and describe the mechanism and significance of the greenhouse effect. (13.9)

18. Describe some strategies for reducing the amount of CO_2 released into the atmosphere. (13.9)

19. List the EPA's criteria pollutants and the major air pollutants that come mainly from automobiles and mostly from industry. (13.10)

20. Identify that green chemistry assessment includes evaluation of the safety of substances as well as awareness of long-term effects.

21. Explain how the green chemistry principles can be applied in the development of manufacturing processes.

EXAMPLE PROBLEMS

1. There is 0.36 L of carbon dioxide in 1000 L of air. What is the concentration of CO_2 in air in parts per million (by volume)? The term *parts per million* means the number of a particular item (or volume) in a mixture containing one million total items (or volumes).

To find the answer to this problem, multiply both the numerator and denominator of the ratio.
Concentration, ppm = $(0.36\ L\ CO_2/1000\ L\ air)(1000/1000) = 360\ L\ CO_2/1 \times 10^6\ L\ air$
We now have 360 L CO_2 per 1,000,000 L air, or 360 parts CO_2 per million parts air, or simply 360 ppm CO_2.

2. What mass of particulate matter would be inhaled each day by a person breathing 20,000 L of city air containing 230 $\mu g/m^3$ of particulate matter? ($1\ m^3 = 1000\ L$)
Mass of particulate matter = $(2.0 \times 10^4\ L\ air)(230\ \mu g\ particulate\ matter/m^3\ air)(1\ m^3/1000\ L)$
$= 4.6 \times 10^3\ \mu g$

ADDITIONAL PROBLEMS

1. At present, the atmosphere contains about 2.5 quadrillion kg (2.5×10^{15} kg) of carbon dioxide. By burning fossil carbon, we add about 22 trillion kg (22×10^{12} kg) of CO_2 to the atmosphere each year. If half of this CO_2 remains in the atmosphere, how many years will it take, at the present rate, to double the amount of CO_2 in the atmosphere?

2. A supersonic transport (SST) burns 60,000 kg of fuel per hour. What weight of carbon dioxide and of water vapor will be produced per hour? A (representative) equation is

$$C_{15}H_{32} + 46\ O_2 \rightarrow 15\ CO_2 + 16\ H_2O$$

3. In 10,000 L of air there are 3 L of argon. What is the concentration of argon in parts per million (by volume)?

SOLUTIONS TO ADDITIONAL PROBLEMS

1. Doubling the amount of CO_2 in the atmosphere means that the amount will go from the current 2.5×10^{15} kg to $(2)(2.5 \times 10^{15}\ kg) = 5.0 \times 10^{15}$ kg, representing an addition of 2.5×10^{15} kg. If half of the 22×10^{12} kg emitted per year remains, then each year 11×10^{12} kg will be added per year.
Number of years required = $(2.5 \times 10^{15}\ kg)(1\ year/11 \times 10^{12}\ kg) = 230$ years

2. The molar mass of $C_{15}H_{32}$ is $(15)(12.0) + (32)(1.0) = 212.0$ g/mol
Mass of CO_2 produced = $(60{,}000\ kg\ fuel/hour)(1000\ g/kg)(1\ mol\ C_{15}H_{32}/212\ g\ C_{15}H_{32})$
$(15\ mol\ CO_2/1\ mol\ C_{15}H_{32})(44.0\ g\ CO_2/mol\ CO_2)$
$= 1.9 \times 10^8\ g\ CO_2$ (or $1.9 \times 10^5\ kg\ CO_2$)
Mass of H_2O produced = $(60{,}000\ kg\ fuel/hour)(1000\ g/kg)(1\ mol\ C_{15}H_{32}/212\ g\ C_{15}H_{32})$
$(16\ mol\ H_2O/mol\ C_{15}H_{32})(18.0\ g\ H_2O/mol\ H_2O)$
$= 8.2 \times 10^7\ g\ H_2O$ (or $8.2 \times 10^4\ kg\ H_2O$)

3. ppm Ar = $(3\ L\ Ar/10{,}000\ L\ air)(1 \times 10^6\ L\ air) = 300$ ppm Ar

DISCUSSION

While nitrogen makes up almost 80% of the atmosphere, it needs to be combined with other elements (fixed) in order to be useful. The Earth naturally produces a number of pollutants via volcanoes, dust storms, swamps, and marshes. However, since human beings started living in cities, human-made pollution has been a serious problem. Burning coal produces industrial (London) smog, which is a combination of smoke, fog, sulfur oxides, and particulate matter. To reduce industrial smog requires the removal of sulfur from the fuel and smoke that produces particulates. Photochemical (Los Angeles) smog is produced by automobiles and contains carbon monoxide, nitrogen oxides, and unburned hydrocarbons. The best way to reduce photochemical smog is through the use of catalytic converters. Ozone is a pollutant at ground level but is needed as a protective screen against UV rays in the upper atmosphere. CFCs threaten to diminish this protective layer of ozone. Acid rain occurs when sulfur oxides and nitrogen oxides combine with water to lower its pH. The United States successfully removed lead as an air pollutant by banning tetraethyllead as a gasoline additive. Tighter insulation has increased the level of indoor air pollution. Indoor air pollutants include nitrogen oxides from gas ranges, cigarette smoke, and radon. Carbon dioxide produced by burning coal or hydrocarbons may contribute to global warming. "Waste" heat produced as a by-product during the conversion of one type of energy to another may contribute to global warming in the future.

ANSWERS TO ODD-NUMBERED REVIEW QUESTIONS AND SOLUTIONS FOR ODD-NUMBERED END-OF-CHAPTER PROBLEMS

Review Questions

1. CFCs were used as refrigerants and in the molding of plastic foams.

3. Bottom ash is the mineral matter that is left behind when coal is burned in power plants or factories. Fly ash is unburned minerals that are airborne due to the drafts created by the fire and that settle over the surrounding area, covering everything with dust. Fly ash can be used in concrete or as a soil modifier.

5. A greenhouse gas increases the retention of the sun's heat energy. Three examples of greenhouse gases are carbon dioxide (CO_2), methane (CH_4), and water (H_2O). Molecules that are polar or are easily polarized are effective greenhouse gases.

7. a. Ground-level ozone is a powerful oxidizing agent and is highly reactive. It is a severe respiratory irritant and especially harmful to children. Ozone causes rubber to harden and crack, resulting in damage to automobile tires, and it also causes crop damage.
   b. Depletion of stratospheric ozone lessens the ability of the atmosphere to intercept ultraviolet light from the sun. Rays in this part of the electromagnetic spectrum are capable of doing cellular damage and causing cancer.

9. Photochemical smog occurs when unburned hydrocarbons and nitrogen oxides, in the presence of sunlight, undergo a complex series of reactions to produce a brown haze. This type of smog is often associated with warm, sunny climates. Polluted air associated with industrial activities is often called industrial smog. It is characterized by the presence of smoke, fog, sulfur dioxide, and particulate matter such as ash and soot. Most industrial smog occurs as the result of burning coal. The weather conditions associated with industrial smog are cool/cold temperatures, high humidity, and often fog.

Problems

11. Nitrogen fixation is combining nitrogen with another element to make it available to plants. One very useful form of "fixed" nitrogen is ammonia (NH_3). One reason why nitrogen fixation is important is that the process has resulted in a great increase in food production.

13. 1 km = 0.6214 mi, so height reached by aircraft, km = (67 mi)(1 km/0.6214 mi) = 110 km
    According to Figure 13.1, the aircraft reached the thermosphere.

15. a. $4\ Fe + 3\ O_2 \rightarrow 2\ Fe_2O_3$
    b. $4\ Cr + 3\ O_2 \rightarrow 2\ Cr_2O_3$

17. The weather conditions associated with industrial smog are cool/cold temperatures, high humidity, and often fog.

19. $S + O_2 \rightarrow SO_2$

21. $SO_2 + 2\ H_2S \rightarrow 3\ S + 2\ H_2O$

23. The solid lime (CaO) reacts with SO_2 gas, causing it to form a solid product, $CaSO_3$, which is easy to collect. Some scrubbers react the $CaSO_3$ with oxygen to form a more useful product, $CaSO_4$.
    The reactions for this are: $CaCO_3(s) \rightarrow CaO(s) + CO_2(g)$
    $CaO(s) + SO_2(g) \rightarrow CaSO_3(s)$.
    And, in the case of formation of $CaSO_4(s)$: $2\ CaSO_3(s) + O_2(g) \rightarrow 2\ CaSO_4(g)$.

24. The reaction of N_2 and O_2 requires high temperatures. Otherwise, the two main components of the atmosphere would react at ambient temperatures. $N_2 + O_2 \rightarrow 2\ NO$.

27. PAN stands for peroxyacetyl nitrate ($CH_3COOONO_2$). PAN, which is formed from hydrocarbons, oxygen, and nitrogen dioxide, makes breathing difficult and causes eye irritation.

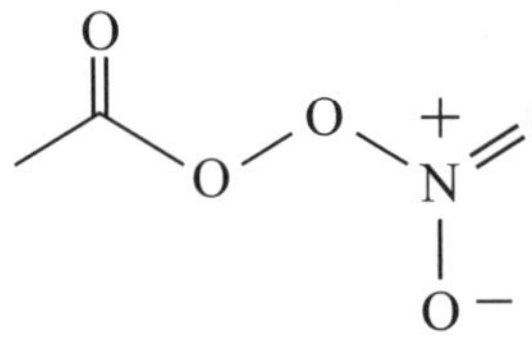

29. The answers are a and c. Consumer products such as paints and aerosol sprays are sources of VOCs, as are petroleum fumes.

31. NO is a free radical. NO has 11 valence electrons, so not all of them can be paired. The other compounds have even numbers of valence electrons, so their valence electrons can occur in pairs.

33. Carbon monoxide binds to hemoglobin, preventing the hemoglobin from carrying O_2.

35. U.S. emissions of CO have decreased from 140 million tons in 1992 to slightly more than 70 million tons in 2013. This decrease is likely due, in part, to improvements in automobile efficiency and to the efficiency of catalytic converters.

37. Allotropes are multiple forms of the same element. In this case the element is oxygen, and the allotropes are O_2 (elemental oxygen) and O_3, ozone.

39. HCFCs are effective greenhouse gases, so leakage of them into the atmosphere contributes to global warming. They are more reactive than CFCs, so they break down while in use and can cause corrosion of the components of compressors and refrigeration coils.

41. In the troposphere, ozone is a pollutant associated with photochemical smog. NO enters the air in automobile exhaust and reacts with the O_2 in the air, forming NO_2. Sunlight splits the NO_2 into NO and O atoms. The O atoms combine with O_2 molecules to form O_3. Chlorofluorocarbons (CFCs) are among the atmospheric components responsible for a decrease in O_3 concentration in the stratosphere. When CFCs interact with UV radiation, the bond between a C atom and a Cl atom breaks, creating a

chlorine radical. This radical enters into the otherwise constant cycle that ensures the constant O_3 concentration, reacting with O_3 to decrease its concentration. One Cl radical can lead to the destruction of thousands of O_3 molecules.

43. Sulfuric acid (H_2SO_4) and nitric acid (HNO_3) are the primary acids responsible for acid rain.

45. $6\ HNO_3(aq) + 2\ Fe(s) \rightarrow 2\ Fe(NO_3)_3(aq) + 3\ H_2(g)$

47. Carbon monoxide is a pollutant both indoors and outdoors. Indoors it results from the use of woodstoves, gas stoves, cigarette smokers, and unvented gas and kerosene space heaters. Outdoors it results primarily from the incomplete combustion of hydrocarbon fuels.

49. In a house built with a crawl space, radon migrating from the nuclear decomposition of rock beneath the house becomes concentrated in the crawl space, which is an uninhabited but usually well-ventilated space in a house. The radon dissipates through vents. In houses built on concrete slabs, the radon migrates through cracks in the slab and into the first floor rooms of the house, posing a greater health risk for the inhabitants of the house.

51. Radon is a radioactive gas that can be breathed into a person's lungs. If the radon nucleus undergoes a nuclear decomposition by alpha emission while it is in a person's lungs, the daughter product, polonium, is a solid that deposits on the lung tissue where it remains while it continues to go through its decomposition sequence.

53. The greenhouse effect is the slow warming of the Earth caused by the trapping of infrared energy.

55. The CO_2 exhaled by humans is part of the natural carbon cycle which begins when photosynthetic organisms store the sun's energy by producing glucose from atmospheric CO_2 and H_2O. Humans fuel their existence by eating food containing the stored energy (and other important elements and compounds). In the broadest of terms, the energy from food is released through essentially the reverse of the photosynthetic reaction, producing the CO_2 and H_2O that we exhale. Thus this cycle is not a big contributor to the atmospheric CO_2 level.

57. According to Figure 13.13, the atmospheric CO_2 concentration in 1880 was approximately 290 ppm while that in 2010 was approximately 400 ppm. This represents a change of 110 ppm or an increase of 40% in 130 years.

59. Zero pollution is not possible for several reasons. There are too many natural pollutants and there are too many people on the planet to eliminate all processes that produce pollutants as a by-product. Also, some waste heat is formed every time energy is transferred from one form into another (the second law of thermodynamics). Finally, the instruments and chemical methods used to quantify pollution are not developed to the point where they can identify extremely minute levels of pollutants.

61. Mass of particulates/day = $(22\ m^3\ air/day)(312\ \mu g\ PM/m^3)(1\ mg/1000\ \mu g) = 6.9\ mg/day$

63. Water vapor will form clouds and may even lead to cooling by reflecting sunlight back to space.

65. According to the data from Problem 64:
Number of mol of molecules in the universe = $(5.2 \times 10^{21}\ g)(1\ mol/29\ g) = 1.8 \times 10^{20}\ mol$
Mass of 1 breath = $(0.50\ L/breath)(1.3\ g/L) = 0.65\ g/breath$

Number of molecules in Buddha's last breath = (0.65 g/breath)(1 mol/29 g)(6.02 x 10^{23} molecules/mol)
= 1.4 x 10^{22} molecules
Number of breaths in atmosphere = (1.8 x 10^{20}mol)(1 breath/0.022 mol) = 8.0 x 10^{21} breaths
There are more molecules from Buddha's last breath than there are breaths in the atmosphere, so, with even distribution, there should be at least one molecule from Buddha's last breath in the breath any one of us just took.

67. $3\ NO_2 + H_2O \rightarrow 2\ HNO_3 + NO$.
$2\ NO + O_2 \rightarrow 2\ NO_2$.

69. Assume that the populations of China and the United States in 2011 were 1.3 billion (1.3 x 10^9) and 311 million (3.1 x 10^8), respectively, when, according to Problem 68, China emitted 2.7 Gt of carbon while 1.6 Gt of carbon was emitted in the United States. From the mass of C emitted, we can determine the equivalent mass of CO_2:
Mass CO_2 emitted in China = (2.7 Gt C)(44.0 Gt CO_2/12.0 Gt C) = 9.9 Gt CO_2
Mass CO_2 emitted in the United States = (1.6 Gt C)(44.0 Gt CO_2/12.0 Gt C) = 5.9 Gt CO_2
On a per capita basis this means that:
Mass of CO_2 emitted/Chinese citizen = (9.9 Gt CO_2/1.3 x 10^9 citizens) = 7.6 x 10^{-9} Gt CO_2/citizen
Mass of CO_2 emitted/U.S. citizen = (5.9 Gt CO_2/3.2 x 10^8 citizens) = 1.9 x 10^{-8} Gt CO_2/citizen
The message to U.S. citizens could be that, individually, they are contributing 40% more to the atmospheric CO_2 concentration than the individuals in China.

71. a. amount of heat absorbed to cause a 1 °C increase = (1.0 kJ/kg °C)(1.0 °C)(5.1 x 10^{18} kg)
= 5.1 x 10^{18} kJ
b. world energy consumption/year, kJ = (5.0 x 10^{20} J)(1 kJ/1000 J)
= 5.0 x 10^{17} kJ
This amount is 10% of the amount of energy needed to cause a 1 ° temperature increase.
c. energy released by hurricane, kJ = (2 x 10^{20} J)(1 kJ/1000 J)
= 2 x 10^{17} kJ
which is about 4% of the amount of energy needed to cause a 1 ° temperature increase.

73. The answer is c.
amount of NF_3 = (20,000 t NF_3 produced)(16 t NF_3 escaped/100 t NF_3 produced)
= 3200 t NF_3

75. The risk phrases for NF_3 indicate the following:
R8: Contact with combustible material may cause fire.
R20: Harmful by inhalation.
R36/38: Irritating to eyes/irritating to skin.
The risk phrases for F_2 indicate the following:
R7: May cause fire.
R8: Contact with combustible material may cause fire.
R26: Very toxic by inhalation.
R35: Causes severe burns.
R41: Risk of serious damage to eyes.
While working with either NF_3 or F_2 requires great care and attention to safety, based on the risk factors, working with F_2 poses the greater hazards. The terms used with NF_3 are "may cause," harmful, and irritating while those associated with F_2 imply a much more immediate risk ("very toxic," "causes," "serious damage").

CHAPTER

14

Water

Rivers of Life, Seas of Sorrows

CHAPTER SUMMARY

14.1 Water: Some Unique Properties

Learning Objectives: • Relate water's unique properties to the polarity and the hydrogen bonding of its molecules. • Explain how water on the surface of Earth acts to moderate daily temperature variations.

A. Water is the only common liquid on the surface of planet Earth.
B. Solid water (ice) is less dense than liquid water.
C. Water has a higher density than most other familiar liquids.
D. Water is polar; it tends to dissolve ionic substances.
E. Water is amphoteric: it can act as an acid or as a base.
F. Water has an unusually high specific heat because of the hydrogen bonding between molecules.
   1. Specific heat is the quantity of heat required to raise the temperature of 1 g of a substance by 1 °C.
   2. The vast amounts of water on the surface of Earth act as a giant heat reservoir to moderate daily temperature variations.
G. Water has an unusually high heat of vaporization (the amount of heat required to evaporate a small amount of water).
H. The properties of water are explained by its structure.
   1. Liquid water is strongly associated through hydrogen bonding, but the molecules are randomly organized.
   2. In ice, molecules have a more ordered arrangement where each water molecule forms hydrogen bonds with four other water molecules, producing a well-defined structure with large hexagonal holes.

Answers to Self-Assessment Questions

1. c The fact that it can exist as a solid, liquid, or gas is not unique to water, as many substances have this property.
2. c Water can form hydrogen bonds with four other water molecules, two through its two H atoms and two more through the two nonbonded electron pairs on the oxygen atom.

3. b The presence of the hydrogen bonds that attract water molecules to each other account for its unusual properties
4. a Hydrogen bonds form between the partially positive H atom in one H_2O molecule and the partially negative O atom in another H_2O molecule.
5. b Frozen water (ice) has an open structure with large hexagonal holes.
6. c amount of heat required, kcal = (1 cal/g °C)(50 g)(50.0 °C – 20.0 °C)
= 1500 cal
7. c amount of heat required, kJ = (0.449 J/g K)(1 kJ/1000 J)((131 g)(368 K – 288 K)
= 4.71 kJ

14.2 Water in Nature

Learning Objectives: • Explain why humans can only make use of less than 1% of all the water on Earth. • Explain the water cycle on Earth.

A. Three-fourths of the surface of Earth is covered with water, but 98% of the water on the surface of the Earth is seawater, which is not suitable for drinking.
B. The Water Cycle and Natural Contaminants.
   1. Although the percentages of water apportioned to the oceans, ice caps, rivers, lakes, and streams remain fairly constant, water is dynamically cycled among these various repositories through the water (hydrological) cycle.
   2. Water evaporates from water and land surfaces, condenses into clouds, and returns to Earth as rain, sleet, and snow.
   3. The precipitation becomes part of the ice caps, runs off in streams and rivers, and fills both lakes and underground pools of water in rocks and sand called aquifers.
C. Gases.
   1. Natural water isn't all H_2O.
      a. Rainwater contains dust, dissolved gases (carbon dioxide, oxygen, nitrogen), and, in thunderstorms, nitric acid.
      b. Carbon dioxide makes natural rainwater slightly acidic because it reacts with water to form carbonic acid (H_2CO_3).
      c. Groundwater also contains the radioactive gas radon.
D. Dissolved Minerals.
   1. Groundwater contains dissolved ions.
      a. The principal positive ions (cations) are sodium, potassium, calcium, magnesium, and sometimes iron.
      b. Calcium, magnesium, and iron ions are responsible for hard water.
   2. The principal negative ions (anions) are sulfate, bicarbonate, and chloride.
E. Organic Matter.
   1. Rainwater dissolves matter from decaying plants and animals.
   2. Organic matter in the form of traces of lubricants, fuels, some fertilizers, and pesticides can contaminate water.
   3. Bacteria, other microorganisms, and animal wastes are potential contaminants of natural waters.

Answers to Self-Assessment Questions

1. c The sun is the source of energy to power the water cycle.
2. d Most of the water on Earth is located in its oceans but, as such, is not useful to humans.
3. a Rainwater becomes acidic when carbon dioxide dissolves in it, forming carbonic acid.

4. d Of the substances listed, radon is not commonly found in rainwater (Rn is found in groundwater, however).
5. c The principal cations in water are Na^+, K^+, Ca^{2+}, and Mg^{2+}.
6. c Potassium ions (K^+) do not contribute to hard water.

14.3 Chemical and Biological Contamination

Learning Objectives: • Describe how human activities affect water quality. • Give an example of a biological water contaminant.

A. Too much organic matter means too little oxygen.
   1. Human and animal wastes provide food for waterborne microorganisms.
      a. If oxygen is required for the degradation, the process is termed anaerobic.
      b. When this process occurs in a body of water, dissolved oxygen is depleted.
      c. Biological oxygen demand (BOD) is a measure of the amount of oxygen required.
      d. If too much oxygen is required, then there isn't enough for the fish and other creatures and they die.
   2. Human and animal waste also add nitrates and phosphates to the water, causing algae to grow and create algae blooms
      a. When the algae die, they contribute to the biomass and BOD in a process called eutrophication.
      b. Eutrophication is enhanced when fertilizers run off of farm fields and enter rivers, contributing to dead zones.
      c. When too much organic material accumulates, the water takes on a bad odor.
B. Chemical Pollution from Waste.
   1. Antibiotics and endocrine disruptors in the water have had negative effects on organisms and contributed to creation of antibiotic-resistant microorganisms.
   2. Overall, our consumer society has polluted natural waters with pesticides, radioisotopes, detergents, drugs, toxic metals, and industrial chemicals.
C. Waterborne diseases (cholera, typhoid fever, dysentery) plagued the entire world until about 100 years ago.
   1. Today in developed nations, unlike in developing nations, chemical treatment has made municipal water supplies generally safe.
   2. The threat of biological contamination has not been totally eliminated from developed nations.
      a. *Cryptosporidium* and *Giardia*, excreted in human and animal feces, resist standard chemical disinfection.
   3. People with waterborne diseases fill half of the world's hospital beds.
   4. The recreational safety of water is also decreased by biological contaminants.
D. Acid Waters.
   1. Sulfur oxides (SO_x) and nitrogen oxides (NO_x) emitted into the air are converted to acids by moisture and come down from the sky as acid rain.
   2. The major sources of sulfur and nitrogen oxides are coal power plants, other industries, and automobile emissions.
   3. Acid rain decreases the viability of life in lakes, dissolves limestone and marble, and ruins the finish of automobiles.
   4. Acids are not a threat to life in lakes and streams where limestone (calcium carbonate) is plentiful because limestone can neutralize excess acid.

Answers to Self-Assessment Questions

1. a Half the world's hospital beds are filled by individuals suffering from diseases caused by waterborne microorganisms.
2. a Acidic lakes are often clear.
3. b The BOD indicates the amount of oxygen required to metabolize the organic material, which, in turn, is an indication of the amount of organic material in the water body.
4. c A high BOD means that the dissolved oxygen concentration will be reduced as the organic material is broken down aerobically.
5. c Anaerobic oxidation (without air) produces nonoxygenated products, CH_4, NH_3, and H_2S as plant materials decay.
6. b Eutrophication in a lake or river is caused by excess nitrates or phosphates.
7. d Antibiotics are released into natural waters from both human and animal waste.
8. a Limestone can neutralize acids that are present in the water.
9. b Municipal water supplies are generally considered to be safe.

14.4 Groundwater Contamination → Tainted Tap Water

Learning Objectives: • List some groundwater contaminants. • Identify the sources of nitrates in groundwater.

A. In the United States surface water supplies about half the drinking water; groundwater supplies the other half.
   1. In rural areas 97% of people drink groundwater.
   2. In addition to withdrawing groundwater faster than it can be replaced, human activity is polluting it.

B. Nitrates.
   1. Nitrates (NO_3^-) in water come from fertilizers used on farms and lawns, decomposition of organic waste during sewage treatment, and runoff from animal feedlots.
   2. Nitrates are highly soluble and difficult to remove from water.
   3. Methemoglobinemia, or blue baby syndrome, is one consequence of consuming NO_3^- containing water.

C. Volatile Organic Chemicals.
   1. VOCs add an undesirable odor to water, and many are suspected to be carcinogens.
   2. Benzene, toluene, chlorinated hydrocarbons (trichloroethylene), carbon tetrachloride, chloroform, and methylene chloride are examples of VOCs.
   3. More recently, VOC-contaminated water is being produced during the fracking process.

D. Leaking Storage Tanks.
   1. Underground storage tanks such as those at gas stations often leak their contents into the groundwater.

Answers to Self-Assessment Questions

1. c Methemoglobinemia, or blue baby syndrome, is caused by elevated nitrate concentrations in drinking water.
2. a Concern is great when more water is being withdrawn from an aquifer than is refilling (recharging) it.
3. b The "V" in VOC stands for volatile.
4. d Toluene is a component of gasoline that can leak into the groundwater from underground storage tanks.
5. b There are VOCs in the liquids used during the fracking process.

14.5 Water: Who Uses It and How Much?

Learning Objective: • List the major uses of water.

A. The average American uses ~400 L of water a day.
   1. It takes several hundred kilograms of steel to produce a typical automobile and 100 metric tons (t) of water to produce 1 t of steel.
   2. It takes 800 L of water to produce 1 kg of vegetables and 13,000 L of water to produce a steak.
   3. We use water for recreation.
   4. For most of these purposes, we need water that is free of bacteria, viruses, and parasitic organisms.

B. Fate of Chemicals in the Water Environment.
   1. To track chemicals scientists and engineers divide the environment into its parts: air, water, and solids. Solids include sediments found at the bottom of rivers, lakes, and oceans, soils, and layers of sand or rock below the Earth's surface.
   2. Chemicals move from one part to another by partitioning.
   3. The amount of a chemical that remains in water or that partitions into the solid depends on the properties of water, the solid, and the chemical.
      a. Chemicals that partition into solids tend to remain in the environment longer than those that partition into water.
   4. Organisms exposed to organic chemicals in the environment may change them or break them down into different substances before they excrete them or they may store the chemicals in their fatty tissue.
      a. Chemical absorption in animals usually leads to the progressive increase in the concentration of some organic chemical in animals that eat other animals, a process called biomagnification.

Answers to Self-Assessment Questions

1. a According to Table 14.4, almost 50% of the water used in the United States is used as a coolant during the production of electricity.
2. c Except for the water that is lost through evaporation, most of the water used in the manufacture of steel is recycled.
3. b About 70% of the freshwater supply is used in agriculture, primarily for irrigation.
4. d The direct average daily water use per person in the United States is 400 L.

14.6 Making Water Fit to Drink

Learning Objective: • Describe how water is purified for drinking and other household uses.

A. Safe Drinking Water Act.
   1. Gives the EPA the power to set, monitor, and enforce national health-based standards for contaminants in municipal water supplies.
   2. Modern analytical techniques allow identification of smaller and smaller concentrations of potentially harmful substances.

3. The number of regulated substances with maximum contaminant levels (MCLs) has increased.

B. Calculations of Parts Per Million and Parts Per Billion.
   1. 1 ppm = 1 g solute/1 x 10^6 g solution or 1 mg solute/ Liter of solution and 1 ppb = 1 g solute/1 x 10^9 g solution.
      Change 4.2 mg Cl^-/L to ppm and ppb.
      Assume density of water = 1.00 g/mL, so that we have 4.2 mg Cl^-/1000 g H_2O.
      (4.2 mg Cl^-/1000 g water)(1 g water/1000 mg water) =
      4.2 mg Cl^-/1,000,000 mg water or 4.2 ppm Cl^-
      1000 ppb = 1ppm so (4.2 ppm)(1000 ppb/1 ppm) = 4200 ppb Cl^-

C. Water Treatment.
   1. Water from reservoirs, rivers, and lakes must be treated to make it safe for drinking.
      a. Water is treated with slaked lime and aluminum sulfate.
         1. The aluminum hydroxide formed carries down dirt and bacteria.
      b. The water is then filtered through sand and gravel.
      c. Sometimes the water is aerated to remove odors and improve the taste.
      d. Sometimes the water is filtered through charcoal to remove colored and odorous compounds.
      e. Finally, the water is chlorinated to kill harmful bacteria.

D. Chemical Disinfection.
   1. In the final step of water treatment, chlorine is added to kill any remaining bacteria, a treatment that can produce some unwanted by-products.
   2. Ozone can also be used to disinfect water through transfer of an oxygen atom to a contaminant making the contaminant less toxic, but it does not provide the residual protection against microorganisms that chlorine provides.

E. UV Irradiation.
   1. UV irradiation can purify water containing *Cryptosporidium* and other microorganisms but leaves no residual protection and does not correct for taste and odor issues.

F. Fluorides.
   1. Fluorides are poisonous in moderate to high concentrations.
   2. In concentrations of 0.7 to 1.0 ppm, fluorides in drinking water lead to a reduction in the incidence of tooth decay (dental caries).
      a. Fluorides strengthen tooth enamel by converting hydroxyapatite to fluorapatite.
      b. Excessive fluorides can cause a mottling of tooth enamel and interfere with calcium metabolism, kidney action, and thyroid function.

Answers to Self-Assessment Questions

1. c About 90 substances are regulated under the U.S. EPA drinking water standards.
2. a A value of 3 ppm of dissolved oxygen means that 3 mg O_2/L H_2O.
3. a One part per 1000 parts represents a higher concentration than either 1 part per million (or billion) parts.
4. a The taste of water is improved by aeration.
5. b Chlorine is added to water to kill bacteria.
6. c Ozonation has the advantage that the oxygenated by-products are generally less toxic than chlorinated by-products.
7. d Ozone disinfects through oxidation.
8. c Fluoridation of drinking water converts hydroxyapatite to fluorapatite.

14.7 Waste Water Treatment

Learning Objective: • Describe primary, secondary, and tertiary treatment of wastewater.

A. Primary sewage treatment plants remove 40–60% of suspended solids as sludge and ~30% of organic matter.
   1. The effluent water has a high biochemical oxygen demand (BOD).
   2. None of the nitrates and phosphates are removed in this process.

B. Secondary sewage treatment plants pass the effluent through sand and gravel filters.
   1. The water is aerated in order to aid the action of aerobic bacteria that metabolize organic contaminants.
   2. BOD is lowered by ~90%, but most nitrates and phosphates remain in the water.

C. The activated sludge method combines primary and secondary treatment.
   1. Sewage is aerated with large blowers, forming large porous floes.
   2. Part of the sludge is recycled.

D. Advanced, or tertiary, treatments are increasingly used but are expensive.
   1. If charcoal filtration is used, charcoal adsorbs certain organic molecules that are difficult to remove otherwise.
   2. If reverse osmosis is used, pressure forces wastewater through a semipermeable membrane, leaving the contaminants behind.
   3. If phytoremediation is used, the effluent is passed into large natural or constructed lagoons for storage, allowing plants such as reeds to remove metals and other contaminants.

E. Sewage effluent is chlorinated to kill pathogenic bacteria.
   1. Excess chlorine provides residual protection.
      a. Chlorine is not effective against some viruses.
      b. Chlorine reacts with organic compounds to form chlorinated hydrocarbons, some of which are carcinogens.

F. Water Pollution and the Future.
   1. Each person in the United States flushes ~35,000 L of drinking-quality water each year.
   2. Nutrients from wastes could be returned to the soil instead of dumped into water.
   3. In many societies, human and animal wastes are returned directly to the soil.
   4. Sludge is used as fertilizer in some parts of the United States.
   5. Toilets are available that use no water; they compost wastes.

G. Some Disadvantages of Using Sludge as Fertilizer.
   1. Pathogenic organisms may survive and spread disease.
   2. The sludge is often contaminated with toxic metal ions.

H. We're the solution to water pollution.
   1. There are between 4000 and 40,000 cases of waterborne illnesses in the United States each year.
   2. Twenty million people have no running water, and more still obtain water from suspect sources.
   3. Ten percent of public water supplies do not meet one or more of the EPA standards.
   4. Thirty million people draw their water from individual wells or springs whose water is often of unknown quality.

Answers to Self-Assessment Questions

1. c Bacteria consume dissolved substances during the secondary treatment of wastewater.
2. b Trihalomethanes are the reaction products when chlorine reacts with dissolved organic compounds.

3. a The charcoal filtration method of advanced treatment can be used to remove trihalomethanes and chloroform from water.
4. c Reverse osmosis uses pressure to force water through a semipermeable membrane.
5. a Some plants can concentrate certain elements and remove them from water in a process called phytoremediation.
6. b weight of sewage = (500,000 gal)(8 lb/gal) = 4,000,000 lb
   16 lb chlorine/4,000,000 lb sewage = 4 ppm Cl_2 in the wastewater
7. c volume of water lost/day, L = (0.5 mL/drop)(1 L/1000 mL)(10 drops/min)
   (60 min/hr)(24 hr/day)
   = 7 L/day

Green Chemistry: Fate of Chemicals in a Water Environment

A. The average American uses ~400 L of water a day.
   1. It takes several hundred kilograms of steel to produce a typical automobile and 100 metric tons (t) of water to produce 1 t of steel.
   2. It takes 800 L of water to produce 1 kg of vegetables and 13,000 L of water to produce a steak.
   3. We use water for recreation.
   4. For most of these purposes, we need water that is free of bacteria, viruses, and parasitic organisms.

B. Fate of Chemicals in the Water Environment.
   1. To track chemicals, scientists and engineers divide the environment into its parts: air, water, and solids. Solids include sediments found at the bottom of rivers, lakes, and oceans, soils, and layers of sand or rock below the Earth's surface.
   2. Chemicals move from one part to another by partitioning.
   3. The amount of a chemical that remains in water or that partitions into the solid depends on the properties of water, the solid, and the chemical.
      a. Chemicals that partition into solids tend to remain in the environment longer than those that partition into water.
   4. Chemical absorption in animals usually leads to the progressive increase in the concentration of some organic chemicals in animals that eat other animals, a process known as biomagnification.
   5. New chemicals that are supposed to benefit humans should be designed to be less persistent in the environment, less likely to absorb to solids, and to biomagnify, and that will either break down to harmless substances when their useful lives are over or can be recycled.

LEARNING OBJECTIVES

You should be able to . . .

1. Relate water's unique properties to the polarity and the hydrogen bonding of its molecules. (14.1)

2. Explain how water on the surface of Earth acts to moderate daily temperature variations. (14.1)

3. Explain why humans can use less than 1% of all the water on Earth. (14.2)

4. Describe how human activities affect water quality. (14.3)

5. Give an example of a biological water contaminant. (14.3)

6. List the major uses of water. (14.4)

7. List some groundwater contaminants. (14.5)

8. Identify the sources of nitrates in groundwater. (14.5)

9. Describe how water is purified for drinking and other household uses. (14.6)

10. Describe primary, secondary, and tertiary treatment of wastewater. (14.7)

11. Connect the properties of chemicals to their fate in the environment.

12. Describe how chemicals migrate between environmental compartments, including how the effects of chemicals can magnify as they travel through food webs.

EXAMPLE PROBLEMS

1. The U.S. Food and Drug Administration (FDA) limit for mercury in food is 0.5 ppm. On a seafood diet, a person might consume 340 g (12 oz) of tuna per day. How much mercury would the person get each day if the tuna contained the maximum of mercury? (One microgram is 1×10^{-6} g.)

   mass of mercury = (340 g tuna)(0.5 g Hg/1×10^6 g tuna)(1 μg/1×10^{-6} g) = 200 μg

   (The lowest level at which toxicity symptoms have been observed is about 300 μg /day.)

2. Every 3 parts of organic matter in water require about 8 parts of oxygen for its degradation. How much organic matter is needed to deplete the 10 ppm of oxygen in (a) 1 L (which weighs 1 kg) of water? (b) A small lake containing 1,000,000,000 kg of water?

   a. mass of organic matter = (1 kg H_2O)(1000 g/kg)(10 g O_2/1×10^6 g H_2O)
   (3 g organic matter/8 g O_2)(1000 μg/g)
   = 4 μg organic matter

   b. mass of organic matter = (1×10^9 kg H_2O)(1000 g/kg)(10 g O_2/1×10^6 g H_2O)
   (3 g organic matter/8 g O_2)
   = 4×10^6 g organic matter

ADDITIONAL PROBLEMS

1. Sewage discharged by each person each day in the United States consumes, on the average, 60 g of oxygen. How many kilograms of water, at 10 ppm O_2, are depleted daily by the raw sewage from a city of 100,000 people?

2. Arizona has estimated recoverable groundwater of 8.7×10^{14} kg. At present, the water is being used at a net rate of 5.0×10^{12} kg/year. At this rate, how long will the water last?

SOLUTIONS TO ADDITIONAL PROBLEMS

1. Mass of H_2O, kg = (100,000 people)(60 g O_2/person)(1 x 10^6 g H_2O/10 g O_2)(1 kg/1000 g)
= 6 x 10^8 kg H_2O

2. Number of years = (8.7 x 10^{14} kg H_2O)(1 year/5.0 x 10^{12} kg)
= 170 years

DISCUSSION

Water has some very unusual properties that make life possible. Although three-fourths of the Earth is covered with water, only 1% of the world's water is fresh water. The water cycle (evaporation and condensation) cleans water. Until 100 years ago, the leading cause of death was from contamination of water by human waste. Even now 80% of the world's sickness is caused by contaminated water. Biochemical oxygen demand (BOD) measures the amount of organic material dissolved. Groundwater can be contaminated by volatile organic chemicals (VOCs) and leakage from underground storage tanks (USTs). Recent abilities to detect lower and lower concentrations have led to overdramatization of pollution problems. Acid rain and runoff from abandoned mines can cause lakes to become acidic. Industry uses large amounts of water in its processes, with multiple chances to pollute. Wastewater can be treated by settling (primary), settling and filtering (secondary), and settling, filtering, and advanced treatment (tertiary) methods. Drinking water is often treated with aeration and chlorination. Fluoridation is used to reduce tooth decay, but the concentrations must be carefully monitored. In rural areas, nitrates from farming operations are a problem.

ANSWERS TO ODD-NUMBERED REVIEW QUESTIONS AND SOLUTIONS FOR ODD-NUMBERED END-OF-CHAPTER PROBLEMS

Review Questions

1. Crude oil will not dissolve in water because gasoline is nonpolar and water is polar. The oil will float because oil is less dense than water.

3. Perhaps the major source of the nutrients that enter the waterways is fertilizer runoff from agricultural fields. Another source is waste runoff from feedlots. Indirectly, the microorganisms that feed off of these nutrients exhaust the oxygen supply in the water and die, becoming additional nutrients themselves.

5. Pathogenic microorganisms are microorganisms that cause disease.

7. Leaking underground storage tanks are a major source of groundwater contamination. Groundwater may also be contaminated with nitrates from farm fertilizer runoff and with volatile organic chemicals, or chemicals buried in dumps that have infiltrated the groundwater.

9. Nutrients and wastes could be returned to the soil rather than allowed to enter the water system; composting toilets could replace traditional toilets; buffer zones around agricultural fields and industrial complexes could be designed to include plants that are capable of phytoremediation of the water.

11. Chlorinated hydrocarbons are unreactive and do not break down easily.

Problems

13. The hydrogen bonding between water molecules in ice create larger holes than exist in the liquid phase, causing water to expand when it freezes. Thus, solid water is less dense than liquid water, so ice floats. Fish and other aquatic species can survive winter in the temperate zones because the ice forms a layer on top of the water rather than sinking to the bottom and allowing the water to freeze solid.

15. Heat required = (750 g H_2O)(45.0 °C – 12.0 °C)(1.00 cal/g °C) = 2.48×10^4 cal

17. The specific heat capacity of a substance is the amount of energy required to raise the temperature of one gram of that substance 1 °C, where temperature is a reflection of the kinetic energy (the energy of motion) of the individual particles. The difference in the specific heat capacities for solid and liquid water is due to the way that the water molecules can convert energy into motion, whether vibrational (in the case of ice) or translational (in the case of liquid water).

19. The high heat of vaporization of water means that large amounts of heat can be dissipated by the evaporation of small amounts of water (perspiration) from the skin.

21. When you step out of a swimming pool into a breeze, water evaporates from your skin. The heat required for the evaporation is removed from your skin, cooling it.

23. Rainwater picks up dust and dissolves atmospheric gases such as carbon dioxide and oxygen as it passes through the atmosphere. During thunderstorms, the water reacts with fixed nitrogen to form nitric acid.

25. Aerobic bacteria require the involvement of O_2 in order to break down nutrients. Anaerobic bacteria can break down nutrients in the absence of O_2.

27. Hardness, mg/L = (3 gr/gal)(64.8 mg/gr)(1 gal/3.79 L) = 50 mg/L

29. Bodies of water gain acidity from acid rain formed when sulfur oxides and nitrogen oxides dissolve in rain, fog, and snow. Acids also flow into streams from abandoned mines.

31. Hard water contains relatively high amounts of Ca^{2+} and Mg^{2+} ions. These ions enter the water when acidic water reacts with limestone (primarily $CaCO_3$ and $MgCO_3$) to produce water-soluble calcium and magnesium salts which contribute to water hardness.

33. The optimal level of fluoride in drinking water is 0.7 to 1.0 ppm (by mass).

35. Yes, ultraviolet radiation can be used to disinfect water because it kills microorganisms.

37. Individuals require 2 L of water per day for drinking.

39. Bottled water is safe and convenient but very expensive relative to tap water. It may not be as safe as municipal water because, although it is considered a food and therefore regulated by the EPA, it is not tested as rigorously as municipal water supplies. Perhaps the biggest drawback of bottled water is the need to manufacture the bottles, a process that requires both petroleum supplies and a great deal of energy. Recycling these bottles would help lessen the disadvantages of this product.

41. Primary sewage treatment allows heavy materials to sink to the bottom of a tank holding the sewage and removing the effluent. This process removes 40%–60% of suspended solids as sludge in settling tanks as well as 30% of organic matter.

43. Nitrates and phosphates are not removed by secondary wastewater treatment.

45. a. Charcoal filtration is part of an advanced (or tertiary) treatment process that removes some organics, including trihalomethanes such as chloroform that are hard to remove by any other method.
 b. Settling ponds are part of the primary sewage treatment where substances more dense than water settle to the bottom of a pond and the effluent is removed for further treatment.
 c. Sand and gravel filters are part of the secondary treatment process, which may also include aeration. Bacteria and microorganisms convert organic material to inorganic material in this process.

47. $2\ HNO_3(aq) + CaCO_3(s) \rightarrow Ca^{2+}(aq) + 2\ NO_3^-(aq) + H_2O(l) + CO_2(g)$

49. $Ca_5(PO_4)_3OH(s) + F^-(aq) \rightarrow Ca_5(PO_4)_3F(s) + OH^-(aq)$

51. a. 11 μg/L = 11 x 10^{-6} g benzene/1 x 10^3 g solution = 11 x 10^{-9} g benzene/g solution.
 ppb benzene = (11 x 10^{-9} g benzene/g solution)(1 x 10^9 g solution) = 11 ppb benzene.
 b. 0.014% $BaCO_3$ = 0.014 g $BaCO_3$/100 g solution = 0.00014 g $BaCO_3$/g solution.
 ppm $BaCO_3$ = (0.00014 g $BaCO_3$/g solution)(1 x 10^6 g solution) = 140 ppm $BaCO_3$.

53. According to Table 14.5, the maximum allowable concentrations for copper and nitrate in water are 1.3 mg/L and 45 mg/L, respectively.
18 μg/2 L = 9 μg copper/L or 9.0 x 10^{-3} mg copper/L
16 mg/2 L = 8 mg nitrate/L
The levels of both substances are lower than the acceptable upper limit.

55. According to Table 14.5, the maximum contaminant levels of each of the four substances in water are: arsenic, 0.010 mg/L; cyanide, 0.02 mg/L; benzene, 0.005 mg/L; and heptachlor, 0.0004 mg/L. Therefore, of the four substances, heptachlor is the most toxic because its contaminant level is the lowest. Benzene would be the next most toxic, followed by cyanide and then arsenic.

57. The answer is d. The hydrogen atoms in water are partially positive and the oxygen atom is partially negative because the electronegativity of oxygen is greater than the electronegativity of hydrogen.

59. The answer is b. According to Figure 14.2, 68.8% of the Earth's fresh water is present in glaciers and ice caps.

61. The answer is d. Gases are more soluble in cold water than in warm water, which is the reason the O_2 supply is reduced when water is warmed through thermal pollution.

63. Chlorine gains electrons (is reduced) in going from Cl_2 to 2 Cl^-; it is the oxidizing agent. Sulfur dioxide is oxidized (gains oxygen) in going from SO_2 to SO_4^{2-}; it is the reducing agent.

65. Radon is a monoatomic gas. Therefore, it is nonpolar; so the strongest interactions a Rn atom can have with water are dispersion forces.

67. a. heat, kcal = (50 gal)(3.8 L/gal)(1000 g/1L)(60 °C – 10 °C)(1.00 cal/g °C)(1 kcal/1000 cal)
$= 9.5 \times 10^3$ kcal

b. mass of CH_4, g = (9.5×10^3 kcal)(4.184 kJ/kcal)(16.0 g CH_4/890 kJ) = 715 g CH_4

c. volume CH_4, ft^3 = (715 g CH_4)(1 mol CH_4/16.0 g CH_4)(22.4 L/mol CH_4)(1 ft^3/28.3 L)
= 35.4 ft^3

d. cost = (10-min)(3.0 gal/min)(9.5×10^3 kcal/50 gal)(715 g CH_4/9.5×10^3 kcal)
(35.4 ft^3/715 g CH_4)(1 therm/100 ft^3)($1.22/therm)

= $0.26

69. Polar substances tend to remain dissolved in polar solvents. Because a larger amount of substance X than of Y remained dissolved in the water, X must be more polar than Y.

71. It would probably be better if the pesticide was polar rather than nonpolar so that it would not tend to dissolve in fatty tissue and biomagnify in the food chain. The best pesticide would break down into nontoxic substances once it enters the environment.

CHAPTER

15

Energy

A Fuel's Paradise

CHAPTER SUMMARY

15.1 The Sun: A Giant Nuclear Power Plant

Learning Objective: • Perform power and energy calculations.

A. Nearly all the energy available to us on Earth comes from the sun (4×10^{26} W), where it is generated by nuclear fusion.
   1. The SI unit for energy is the joule (J).
   2. The SI unit for power is the watt (W).
   3. 1 watt = 1 joule/second.

B. There are two forms of energy: potential (energy stored in the form of position or arrangement) and kinetic (energy of motion).

C. Energy and the Life-Support System.
   1. The energy the Earth receives from the sun is
      a. Reflected back into space (30%).
      b. Converted to heat and warms the planet (~50%).
      c. Used to power the water cycle (23%).
   2. Green plants use sunlight (less than 0.02%) to convert solar energy to chemical energy (photosynthesis).
      a. The sunlight is absorbed by green plant pigments called chlorophylls.
      b. The energy is used to convert carbon dioxide and water to glucose (an energy-rich simple sugar) and oxygen.
      c. Glucose can be stored, or it can be converted to more complex foods and structural materials.

Answers to Self-Assessment Questions

1. b One watt, a unit of power, is equivalent to 1 J/sec.
2. c The industrial sector uses the largest fraction of the nation's energy.
3. c The United States uses about 10% of the world's energy.
4. b An eagle in flight has kinetic energy, the energy associated with motion.
5. d Roughly one-third of Earth's incident solar radiation is reflected back into space (see Table 15.1).

6. a About 0.02% of Earth's incident solar radiation is used in photosynthesis.

15.2 Energy and Chemical Reactions

Learning Objectives: • Classify chemical reactions and physical processes as exothermic or endothermic. • Explain how to calculate the heat of a reaction using bond energies.

A. We usually discuss energy changes in terms of an object or region losing energy and another object or region gaining it.
   1. The system is the part of the universe under consideration; the surroundings are everything else.
   2. Chemical reactions (or physical processes) that result in the release of heat from the system to the surroundings are exothermic. The molecules producing the heat are the "system," and other things near those molecules are the "surroundings." The amount of heat released by the reaction can be shown in the balanced equation as a reaction product.
   3. If energy must be supplied from the surroundings in order for the reaction to occur, the reaction is endothermic. The amount of heat required by the reaction can be shown in the balanced equation as a reactant.

B. Bond Energies.
   1. A chemical reaction occurs when the bonds in the reactant substances break (a process requiring energy), forming molecular fragments. These fragments release energy when they combine to form reaction products. The net amount of energy required (or released) determines whether a reaction is endothermic or exothermic and by how much.
   2. The energy inherent in every possible atom-to-atom bond is called bond energy and is measured in kJ/mol.
   3. Once we have the bond energies, we can calculate the enthalpy change in the reaction (the amount of heat required or released during the process).
      a. Enthalpy change = (sum of energy required to break all the bonds) – (sum of energy released when new bonds are made)

Answers to Self-Assessment Questions

1. a Molecules in the gas state have more energy than those in the liquid state. Therefore, to transition from a gas to a liquid, the molecules must release energy that is an exothermic process. The other processes all require an input of energy, which means they are endothermic.

2. c Heat is released when a candle burns, which indicates that less energy is required to break the bonds in the candle wax than is released when the combustion products form. The process is exothermic.

   When water goes from liquid {$H_2O(l)$} to solid {$H_2O(s)$}, energy is given off, so the process is exothermic (think about the reverse process, heating liquid water to put the molecules into the gas phase—in this case the heat the molecules absorbed is released when they return to the liquid phase).

3. d A substance melts when its components absorb enough energy to allow them to move about with respect to one another. Processes that require the input of energy are endothermic.

4. c The combustion reaction of propane is $C_3H_8 + 5\ O_2 \rightarrow 3\ CO_2 + 4\ H_2O + 2201\ kJ$, which indicates that the heat of combustion of propane is 2201 kJ/mol. The molar mass of C_3H_8 is 44.0 g/mol. We can calculate the amount of heat produced from the combustion of 400 g of propane as follows:

   amount of heat produced, kJ = 400 g C_3H_8(1 mol C_3H_8/44.0 g C_2H_8)(2201 kJ/1 mol C_3H_8) = 20,000 kJ

15.3 Reaction Rates

Learning Objective: • List factors that affect the rates of chemical reactions.

A. Two areas of chemistry that focus on energy are kinetics and thermodynamics.
  1. Chemical kinetics describes the rate at which a chemical reaction occurs, which depends on temperature, concentration of reactants, and the presence of catalysts.
    a. *Temperature*: Generally, the higher the temperature, the faster the reaction. At higher temperatures, molecules move faster and collide more often. Also, more energy is available for breaking chemical bonds.
    b. *Concentration of reactants*: The more molecules there are in a given volume of space, the more likely they are to collide; the more collisions there are, the more reactions are likely to occur.
    c. *Catalysts*: These substances speed up reactions without being used up in the process.
      i. Enzymes are biological catalysts that mediate the reactions in living cells.

Answers to Self-Assessment Questions

1. c Reaction rates decrease (slow down) when the concentrations of reactants are decreased, reducing the likelihood of collisions of reactant species.
2. a Adding water to the contents of a reaction vessel will reduce the concentration of the reactants, which will, in turn, slow the reaction down.
3. c The role of a catalyst is to provide a less demanding pathway that the reaction can follow, increasing the reaction rate. The new pathway is one with a lower activation energy than the pathway for the uncatalyzed reaction.
4. a Manganese dioxide is a catalyst for the decomposition of hydrogen peroxide, enabling the reaction to proceed more rapidly than would otherwise be the case.

15.4 The Laws of Thermodynamics

Learning Objective:• State the first and second laws of thermodynamics, and discuss their implications for energy production and use.

A. The first law of thermodynamics (also called the law of conservation of energy) states that energy is neither created nor destroyed (although it can be changed in form).
B. The second law of thermodynamics states that heat always flows from a hot object to a cooler one and that forms of energy available for *useful* work are continually decreasing because energy spontaneously tends to distribute itself among the objects in the universe.
  1. No engine can operate at 100% efficiency; some energy is converted to heat or friction.
  2. Not all forms of energy are equal; high-grade forms are constantly degraded to low-grade forms.
    a. Mechanical energy (high grade) is eventually changed to heat energy (low grade).
  3. Natural processes tend toward greater entropy (more disorder).
    a. We can reverse the tendency toward disorder, but it costs energy to do it.
    b. The more the energy is spread out, the higher the entropy of the system and the less likely it is that this energy can be harnessed to do useful work.
    c. Spontaneous processes tend toward greater entropy or are exothermic, or both.

Answers to Self-Assessment Questions

1. a The first law of thermodynamics is also known as the law of conservation of energy.
2. c The second law of thermodynamics states that energy goes from more useful to less useful forms.
3. d The second law of thermodynamics states that the total entropy always increases for an isolated system.
4. c Because liquids are more organized than gases, the entropy increases when a substance goes from the liquid to the gaseous state.
5. b Heat is spontaneously transferred from a warm object to a cooler one.

15.5 Power: People, Horses, and Fossils

Learning Objectives: • List the common fossil fuels, and describe how modern society is based on their use.

A. A fuel is a substance that burns readily with the release of significant amounts of energy.
   1. Primitive people obtained their energy (food and fuel) by hunting and gathering.
   2. Domestication of animals increased available energy somewhat; people gained horsepower and oxpower.
   3. Windmills and waterwheels further increased available energy by converting energy to useful work.
   4. Since 1850 and the Industrial Revolution, steam engines and other mechanical devices have provided us with 10,000 times as much energy as was available to primitive people.
   5. Today, more than 85% of the energy used to support our way of life comes from fossil fuels—coal, petroleum, and natural gas that formed during Earth's Carboniferous Period around 300 million years ago.

B. Fuels are reduced forms of matter, and the burning process is an oxidation.

C. Reserves and Consumption Rates of Fossil Fuels.
   1. Fossil fuel reserves are rapidly being depleted.
   2. Of all the fossil fuels that ever existed, we will have used about 90% in 300 years

Answers to Self-Assessment Questions

1. d Coal, petroleum, and natural gas are fossil fuels because they were formed over millennia from the remains of ancient plants and animals.
2. c It is estimated that the supply of coal could last 300 years.
3 d CO_2 is not a fuel. The carbon in CO_2 is in its most oxidized form.
4. d Fuels are reduced forms of matter and release their energy when they are burned.
5. d Forty-six percent of the coal reserves are found in the United States.

15.6 Coal: The Carbon Rock of Ages

Learning Objective: • List the origins, advantages, and disadvantages of coal as a fuel.

A. Coal is a complex combination of organic materials that burn and inorganic materials that produce ash; its energy content is closely related to its carbon content.
   1. Coal was formed millions of years ago from plant material buried under mud.
      a. Cellulose of plants was compressed; it broke down, releasing small hydrogen- and oxygen-rich molecules and leaving behind a material rich in carbon.

B. Coal is an abundant but inconvenient fuel.
   1. Coal is our most plentiful fossil fuel; the United States is estimated to have 25% of the world's reserves.
   2. Electric utilities burn one billion metric tons of coal per year to generate 45% of our electricity.
   3. Coal is inconvenient to use and dangerous to mine.
      a. Most coal is obtained by strip mining.

C. Source of Pollution.
   1. Burning coal produces carbon dioxide, carbon monoxide, sulfur oxides (and sulfuric acid), and particulate matter.
   2. Coal can be cleaned before burning by the flotation method.
      a. Coal has a density of 1.3 g/cm^3. It can be floated away, leaving denser sulfur-containing minerals behind.

D. Coke and Coal Tar.
   1. When coal is heated in the absence of air, volatile material is driven off leaving behind coke, which is mostly carbon and is used in the production of iron and steel.
   2. Coal tar (a source of chemicals) is made by heating coal to drive off volatile matter.
      a. Coal tar and liquid coal oil are good sources of organic chemicals for medical and industrial use.

Answers to Self-Assessment Questions

1. c The coal we know today was formed through the decomposition of ferns, reeds, and grasses that grew around 300 million years ago. Bituminous coal is 90% carbon.
2. b The plants that have been converted into coal were composed of cellulose and other substances formed from carbon, hydrogen, and oxygen.
3. b Bituminous coal is roughly 90% carbon (see Table 15.1).
4. c After burning, the inorganic constituents of coal end up as ash.
5. a Of the fossil fuels, coal contributes the most to SO_x pollution because much of it has a relatively high sulfur content.
6. a The flotation method of cleaning coal takes advantage of the fact that coal is less dense than its major impurities.

15.7 Natural Gas and Petroleum

Learning Objective: • List the characteristics, advantages, and disadvantages of natural gas and petroleum.

A. Composition of Natural Gas in the North American Pipeline.
   1. Methane 83–95%
   2. Ethane 2–6%
   3. Propane 1–2%
   4. Butane and pentane smaller amounts

B. Most natural gas is burned as fuel, but some is separated into fractions.

1. The fractions are cracked to produce ethylene, propylene, and other valuable chemical intermediates.
    a. Ethylene, propylene, and four-carbon compounds are also made by cracking the alkane mixture.
    b. Natural gas is the starting material for many one-carbon compounds.

C. Natural gas is the cleanest fossil fuel in regards to pollution.
    1. Today, 90% of the natural gas located in the United States is obtained by hydrolytic fracturing (fracking).

D. Petroleum is a complex liquid mixture of organic compounds.
    1. Petroleum is thought to be formed mainly from the fats of ocean-dwelling, microscopic animals.
    2. When efficiently burned (when they undergo complete combustion and produce CO_2 and H_2O rather than incomplete combustion and produce CO and H_2O), petroleum products are relatively clean fuels.
    3. Burning petroleum reserves depletes the reserves also used as starting materials for most industrial organic chemicals.

E. Obtaining and Refining Petroleum.
    1. Crude oil is a liquid that can be pumped through pipelines.
    2. To improve its use, crude oil is separated into fractions by boiling in a distillation column.
        a. Lighter hydrocarbon molecules come off the top of the column, and heavier ones come off at the bottom.
    3. Fractions that boil at higher temperatures are often converted to gasoline by cracking or breaking the long molecules into shorter ones, illustrating the way chemists modify nature's materials to meet human needs and desires.

F. Gasoline is a mixture of hydrocarbons.
    1. The gasoline fraction of petroleum as it comes from a distillation column is called straight-run gasoline.

G. Octane ratings of gasoline indicate the extent to which the gasoline will cause knocking.
    1. In 1927, isooctane was assigned an octane rating of 100 and heptane, a straight-chain compound, was assigned an octane rating of zero. Gasoline rated 90 octane would perform the same as a mixture that was 90% isooctane and 10% heptane.
    2. During the 1930s, chemists discovered that the octane rating of gasoline could be improved by heating it in the presence of a catalyst to isomerize some of the unbranched molecules to highly branched molecules with higher octane numbers.
    3. Certain additives also improve the antiknock quality of gasoline.
        a. Tetraethyllead was especially effective, but lead fouls catalytic converters and can lead to learning disabilities in children. Unleaded gasoline became available in the United States in 1974.
        b. Refineries use catalytic reforming to convert low-octane alkanes to high-octane aromatic compounds.
        c. Octane boosters such as ethanol, methanol, *tert*-butyl alcohol, and methyl *tert*-butyl ether (MTBE) have replaced tetraethyllead in gasoline.
        d. MTBE from leaded gasoline storage tanks has polluted groundwater in some areas.

H. Alternative Fuels.
    1. An automobile engine can be made to run on nearly any liquid or gaseous fuel.
    2. Examples are natural gas, propane, diesel fuel, and ethanol.
    3. Fuel cells and batteries are also being used.

I. Energy Return on Energy Invested (EROEI).
    1. Evaluating an energy source involves comparing the amount of energy required to produce the energy source with the amount of energy it releases.

Answers to Self-Assessment Questions

1. a Methane, CH_4, is the principal constituent in natural gas.
2. d Natural gas is the simplest fossil fuel in composition.
3. a U.S. oil reserves peaked in 1972.
4. a Cracking converts larger hydrocarbons into smaller, more useful ones.
5. a The EROEI is greater when equipment is already in place to recover the fuel.
6. a An alkylation unit is used to convert smaller molecules into larger ones.
7. d Isomerization improves the octane rating by converting straight-chain alkanes to branched-chain alkanes.
8. c Catalytic reforming converts low-octane straight-chain molecules into high-octane aromatic ones.
9. d The octane rating for gasoline is a measure of a fuel's tendency to cause knocking or premature firing in an engine.
10. b The mixture containing 10% ethanol and 90% gasoline is common.
11. d Fracking is the process that releases natural gas trapped in the rock beneath the ground.

15.8 Convenient Energy

Learning Objective: • Explain why gaseous and liquid fuels are more convenient to use than solid fuels.

A. Electricity is perhaps the most convenient form of energy.
   1. It flows through wires and can be converted into light, heat, or mechanical energy.
   2. Electricity can be generated by any fuel that can be burned to boil water so the steam produced can turn a turbine.
   3. ~45% of U.S. electric energy comes from coal-burning plants, any one of which is only ~40% efficient; the rest of the energy is generally wasted as heat (thermal pollution).

B. Coal Gasification and Liquefaction.
   1. Coal can be converted to gas or oil, which are easy to transport and freer of sulfur and minerals than the coal from which they came.
   2. The hydrogen is produced by passing steam over hot charcoal to produce synthesis gas, a mixture of hydrogen and carbon monoxide.
   3. Coal liquefaction and gasification require large amounts of energy.
      a. Up to one-third of the energy is wasted in the conversion process.
      b. Liquid fuels made from coal contain sulfur, nitrogen, and arsenic compounds, which contribute to air pollution.
      c. Coal conversions also require large amounts of water and are messy, often resulting in air and water pollution.
   4. The Bergius and Fischer-Tropsch methods are also used to make liquid fuel from coal.

Answers to Self-Assessment Questions

1. c About 45% of the electricity produced in the United States comes from coal-burning plants.
2. b Natural gas is transported mainly by pipelines.
3. b Coal-burning power plants are only about 40% efficient. The other 60% of the energy in the fossil fuels is wasted as heat, although this heat can be used to heat buildings.
4. c Methane, CH_4, is the main fuel produced by coal gasification.
5. d The convenience of a fuel depends on its physical state, which relates to its ease of transportation and use.

15.9 Nuclear Energy

Learning Objectives: • List the advantages and disadvantages of nuclear energy. • Describe how a nuclear power plant generates electricity.

A. Energy released during fission is used to produce steam, which turns a turbine, generating electricity.
   1. In the United States, nuclear power plants produce ~20% of the nation's electricity.
B. Nuclear Power Plants.
   1. Pressurized water reactors are common in the United States, although boiling water reactors have been used in the past.
   2. Nuclear power plants use the same fission reactions as employed in nuclear bombs, but the uranium used in power plants is enriched to only 3–4% U-235 while a bomb requires ~90% U-235.
   3. A *moderator*, such as water or graphite, is used to slow down the fission neutrons so that they can be absorbed by the U-235 nuclei. Water often serves as a moderator.
   4. *Control rods*, made of boron steel or cadmium, are inserted to absorb neutrons and slow the fission reaction. Partial removal of the rods starts the chain reaction.
C. The Nuclear Advantage: Minimal Air Pollution.
   1. Advantages: No soot, fly ash, sulfur dioxide, or other chemical air pollutants are emitted.
      a. Nuclear power does not contribute to global warming, air pollution, or acid rain.
D. Problems with Nuclear Power.
   1. Elaborate safety precautions are required.
   2. Runaway nuclear reactions are unlikely but possible.
   3. Nuclear wastes are highly radioactive and must be isolated for centuries.
      a. Tailings from uranium mines are mildly radioactive and contaminate wide areas.
      b. Slightly more thermal pollution is generated from nuclear power plants than from fossil fuel-burning plants.
E. Nuclear Accidents.
   1. Three incidents heightened public fear of nuclear power: the 1979 accident at the Three Mile Island nuclear power plant near Harrisburg, Pennsylvania; the 1986 accident at Chernobyl, Ukraine; and the Fukushima accident resulting from the earthquake and tsunami in Japan in March 2011.
F. Breeder Reactors: Making More Fuel Than They Burn.
   1. Less than 1% of natural uranium is the fissionable uranium-235 isotope; the rest is nonfissionable uranium-238.
   2. Breeder reactors have a core of fissionable plutonium surrounded by uranium-238. Neutrons from the core convert the uranium-238 to fissionable plutonium-239. The process breeds more fuel than it consumes.
      a. Advantage: There is enough uranium-238 to last a few centuries.
      b. Disadvantages: Plutonium melts at 640 °C; plant operation is inefficient because it is limited to rather cool operating temperatures.
         i. Molten sodium metal, which reacts violently with both air and water, is used as a coolant.
         ii. Core meltdown is more likely than with uranium-235 fuel.
         iii. Plutonium is toxic.
         iv. Reactor-grade plutonium can be converted to nuclear bombs.

3. Breeder reactors can also convert nonfissionable thorium-232 to fissionable uranium-233.
    a. Uranium-233 emits biologically damaging alpha particles.
4. No breeder reactors are operating in the United States.

G. Nuclear Fusion: The Sun in a Magnetic Bottle
1. Thermonuclear reactions power the sun and hydrogen bombs, but controlled, sustainable fusion reactions are yet to be achieved.
2. Controlled fusion would have several advantages.
    a. The principal fuel, deuterium (^2H), is plentiful and is obtained from fractional electrolysis (splitting apart by means of electricity) of water.
    b. The product, helium, is biologically inert; radioactive wastes are minimized.
3. Some Possible Disadvantages.
    a. Radioactive tritium (^3H) might be released and incorporated into living organisms.
    b. Temperatures of $1 \times 10^6 - 2 \times 10^6$ °C are required.
    c. The plasma, which is a mixture of electrons and nuclei, must be contained by a magnetic field or other nonmaterial device.
4. It is unlikely that electricity from fusion will be available for decades.

Answers to Self-Assessment Questions

1. c In a fission reactor, uranium-235 nuclei absorb a neutron and split into two smaller nuclei, releasing large amounts of energy.
2. b The control rods in a nuclear reactor absorb neutrons, preventing them from causing fission and thus controlling the rate of the fission reaction.
3. c Nuclear power plants produce about 20% of the electricity in the United States.
4. a Breeder reactors produce more nuclear fuel than they consume.
5. b There are about 100 nuclear power plants in operation in the United States (see Table 15.7).
6. d Nuclear power plants create more thermal pollution than coal-burning plants.
7. c $^{2}_{1}\text{H} + {}^{3}_{1}H \rightarrow {}^{1}_{0}\text{n} + {}^{4}_{2}He$
8. b Fusion is not currently used in nuclear power plants because it requires temperatures of millions of degrees, which makes the reaction difficult to contain.
9. a France has the largest number of nuclear power plants (see Table 15.7).

15.10 Renewable Energy Sources

Learning Objective: • List important characteristics, advantages, and disadvantages of renewable energy sources.

A. Renewable energy sources account for 8–11% of the energy production in the United States.

B. Solar heating: Solar collectors can be used to heat homes and water for bathing, laundry, and so on.

C. Solar Cells: Electricity from Sunlight.
1. Photovoltaic cells (solar cells) can convert sunlight directly to electricity.
2. Solar cells based on silicon have
    a. Donor crystals doped with arsenic to provide extra electrons (silicon has four valence electrons, arsenic, five).
        i. The material has extra electrons, is negatively charged, and is called an *n-type* (negative) semiconductor.
    b. Acceptor crystals doped with boron (three valence electrons) create positive holes.
        i. The material has a shortage of electrons, leaving a positive hole in the crystal so these materials are called *p-type* semiconductors.

c. Joining the *n*-type and *p*-type crystals forms a photovoltaic cell in which electrons flow from the *n*-type region to the *p*-type region.
d. Sunlight dislodges electrons from donor crystals, creating a current flow from donor cell to receptor crystals.
3. Solar cells have low efficiency (about 12–18%).
4. Solar energy is not available at night or on cloudy days, but it can be stored as heat.

D. Biomass: Photosynthesis for Fuel.
1. Biomass (plants grown for fuel) has several advantages.
   a. It is a renewable resource.
   b. Energy for biomass production comes from the sun.
2. Disadvantages.
   a. Most of the land available for farming biomass is needed for food production.
   b. Plants must be harvested and hauled to where the energy is needed, often over long distances.
3. Biomass can be burned directly as fuel or converted to other fuels.
   a. Plants high in starches and sugars can be used to produce ethanol.
   b. Wood can be used to produce methanol.
   c. Bacteria can convert plant material to methane.
4. Biomass and biofuels currently provide about half of the renewable energy used in the United States, or about 5% of the country's overall energy needs.

E. Green Chemistry: Where will we get the energy?
1. Besides the availability of fossil fuels, we need to consider the impacts of waste generation, toxic substances, transportation costs, feedstocks, and overall energy efficiency as we evaluate energy sources.
   a. Fracking (hydraulic fracturing) to obtain natural gas requires adding chemicals to water which contaminate the aquifers in the area of the activity.
   b. Methanol that enters the atmosphere from the wells is 25 times more effective as a greenhouse gas than CO_2.
   c. The biggest hurdle in developing new technologies is cost.

F. Hydrogen: Light and Powerful.
1. Gram for gram, H_2 yields more energy than any other chemical fuel.
2. Hydrogen is clean, yielding only water as a chemical product.

G. Fuel Cells.
1. A fuel cell is a device in which fuel is oxidized in an electrochemical cell and produces energy directly.
   a. They differ from electrochemical cells in two ways.
      i. Fuel and oxygen are fed in continuously.
      ii. The electrodes are an inert material such as platinum that does not enter the reaction.
   b. In a hydrogen fuel cell, hydrogen is oxidized at the anode, and oxygen is reduced at the cathode.
   c. Fuel cells are used on spacecraft, but they contribute little to the production of electricity on Earth because cheap catalysts have not yet been developed.
2. Hydrogen can be used as a fuel for cars.
   a. Advantages: Exhaust is almost entirely water vapor; H_2 can be produced from renewable sources; using H_2 would decrease our dependence on foreign oil sources.
   b. Disadvantages: Hydrogen-powered cars are expensive; there is no widespread H_2 distribution system in the United States; because of its physical properties, storage, transportation, and dispensing the fuel are problematic.

H. Other Renewable Energy Sources.
   1. Wind power and water power could help solve the energy crisis.
      a. The sun causes the wind to blow and water to evaporate and condense, to fall later as rain.
      b. Blowing wind and flowing water can be used as energy sources.
   2. About 7% of our electricity comes from hydroelectric plants, but nearly all the good dam sites are in use.
   3. Windmills can be used to provide mechanical energy or to generate electricity.
      a. Windmills could meet up to 10% of our electricity needs but provide an almost negligible amount at present.
   4. Wind and water provide relatively clean energy, but suffer from some disadvantages.
      a. Dams flood valuable land. Reservoirs silt up. Sometimes dams break, causing devastating floods.
      b. The wind doesn't always blow; energy storage or an alternative source is needed. Windmills limit the use of the land on which they are placed.

I. Geothermal Energy.
   1. The interior of Earth is heated by immense gravitational forces and by natural radioactivity. The heat comes to the surface in some areas through geysers and volcanoes.
   2. One drawback of geothermal energy is that the wastewater is quite salty and its disposal could be a problem.

J. Oceans of Energy.
   1. Ocean thermal energy takes advantage of the 20 °C difference in temperature between the surface and the depths of the oceans, which is enough to evaporate a liquid and use the vapor to drive a turbine.
   2. The liquid is then condensed by the cold from the ocean depths, and the cycle repeats.
   3. Tides possess great energy. At high tide, water fills a reservoir or bay; at low tide, the water escapes through a turbine to generate electricity.

K. Energy: How much is too much?
   1. The Energy Information Administration of the U.S. Department of Energy projected that world energy consumption will increase by 50% from 2007 to 2035.
   2. Over the past three decades, we have made significant progress in energy conservation: Overall, U.S. industry has reduced its energy consumption per product by almost 30%.

Answers to Self-Assessment Questions

1. c Silicon doped with arsenic is an *n*-type semiconductor.
2. c *p*-Type semiconductors feature crystal sites with positive holes.
3. d Petroleum is a fossil fuel and therefore nonrenewable.
4. a Biodiesel cannot be made from corn starch. Biodiesel is made from the reaction of fats or oils with methyl alcohol. Starch is a polymer of glucose, a sugar.
5. c Water is the major product from burning hydrogen gas.
6. b Hydrogen has the lowest EROEI (Energy Return on Energy Invested).
7. c Nuclear power does not originate in the sun.
8. b Hydrogen is used in almost all present-day fuel cells.
9. c Although wind turbines are expensive to install, they pay for themselves over time.
10. b Hydroelectric is the renewable energy source that provides the most electrical energy (~7%) in the United States today.
11. b Geothermal energy comes from the internal heat of the Earth, which comes to the surface as geysers or volcanoes.
12. d Electric vehicles that run at highway speeds currently exist. The other answer options do represent challenges.

LEARNING OBJECTIVES

You should be able to . . .

1. Perform power and energy calculations. (15.1)
2. Classify chemical reactions and physical processes as exothermic or endothermic. (15.2)
3. Explain how to calculate the heat of reaction using bond energies. (15.2)
4. List factors that affect the rates of chemical reactions. (15.3)
5. State the first and second laws of thermodynamics, and discuss their implications for energy production and use. (15.4)
6. List the common fossil fuels, and describe how modern society is based on their use. (15.5)
7. List the origins, advantages and disadvantages of coal as a fuel. (15.6)
8. List the characteristics, advantages and disadvantages of natural gas and petroleum. (15.7)
9. Explain why gaseous and liquid fuels are more convenient to use than solid fuels. (15.8)
10. List the advantages and disadvantages of nuclear energy. (15.9)
11. Describe how a nuclear power plant generates electricity. (15.9)
12. List important characteristics, advantages, and disadvantages of renewable energy sources. (15.10)
13. Explain how applying the green chemistry principles can help identify improved methods for energy generation.

EXAMPLE PROBLEMS

1. Complete combustion of 16.0 g of methane yields 192 kcal of energy. How much energy is obtained by the combustion of 96.0 g of methane?

   Energy released, kcal = (96.0 g methane)(192 kcal/16.0 g methane) = 1150 kcal

2. Splitting 36.0 g of water into hydrogen and oxygen requires the input of 137 kcal of energy. How much energy is required to split 180 g of water?

   Energy required, kcal = (180 g water)(137 kcal/36.0 g water) = 685 kcal

ADDITIONAL PROBLEMS

1. Burning 4.00 g of hydrogen in sufficient oxygen produces 137 kcal of heat. How much heat is released by the combustion of 20.0 g of hydrogen?

2. What mass of carbon dioxide is formed by the complete combustion of 78.0 g of carbon? The equation is $C + O_2 \rightarrow CO_2$.

3. Complete combustion of 16.0 g of methane (CH_4) yields 192 kcal of energy. How many grams of methane are needed to produce 1850 kcal of energy?

4. Most coal contains 3.0% sulfur. How much sulfur is there in a metric ton (1000 kg) of coal? How much sulfur dioxide is formed by burning this coal? The equation is $S + O_2 \rightarrow SO_2$.

5. A large power plant will burn 25 million kg of coal per day. If the coal is 8% inorganic ash, how much ash is produced per day? Per year? Per decade?

6. If the coal in Problem 5 contains 0.2 ppm of mercury (i.e., 0.2 kg of mercury, Hg, in 1,000,000 kg of coal), what mass of mercury is released into the environment each day? Each year?

SOLUTIONS TO ADDITIONAL PROBLEMS

1. Amount of heat, kcal = (20.0 g H_2)(137 kcal/4.00 g H_2) = 685 kcal

2. Mass of CO_2, g = (78.0 g C)(1 mol C/12.0 g C)(1 mol CO_2/1 mol C)(44.0 g CO_2/mol CO_2)
   = 286 g CO_2

3. Mass of CH_4 = (1850 kcal)(16.0 g CH_4/192 kcal) = 154 g CH_4

4. Mass of S in 1000 kg coal = (1000 kg coal)(1000 g/1 kg)(3.0 g S/100 g coal) = 30,000 g S
   mass of SO_2 formed = (30,000 g S)(1 mol S/32.1 g S)(1 mol SO_2/1 mol S)(64.1 g SO_2/mol SO_2)
   = 60000 g SO_2

5. a mass of inorganic ash/day = (2.5×10^7 kg coal/day)(8 kg inorganic ash/100 kg coal)
   = 2.0×10^6 kg inorganic ash/day
   b. mass of ash/year = (2.0×10^6 kg ash/day)(365 days/year) = 7.3×10^8 kg ash/year
   c. mass of ash/decade = (7.3×10^8 kg ash/year)(10 years/decade) = 7.3×10^9 kg ash/decade

6. a. mass of Hg released/day = (2.5×10^7 kg coal/day)(0.2 kg Hg/1×10^6 kg coal)
   = 5.0 kg Hg/day
   b. mass of Hg released/year = (5.0 kg Hg/day)(365 days/year) = 1800 kg Hg/year

DISCUSSION

Energy, the ability to do work, ultimately comes from the sun. All the food for life comes from photosynthesis, which is powered by the sun. Rates of chemical reactions are increased with increases in temperature, increases in concentration, and in the presence of catalysts. Almost all chemical reactions either require energy (are endothermic) or give off energy (are exothermic). The first law of thermodynamics states that energy is conserved. The second law of thermodynamics states that energy flows from regions of high energy to regions of low energy and that disorder increases. Since the 1850s, humanity has experienced a tremendous increase in the amount of energy available to them, so that now people have access to 10,000 times as much total energy as their ancestors had access to. Ninety percent of our energy comes from fossil fuels. The fossil fuels are coal, petroleum, and natural gas, which are composed of organic compounds that we burn to produce energy. Each fossil fuel has its advantages and disadvantages. We convert a great deal of energy to electricity, a convenient secondary energy source. Although nuclear power is cleaner, radioactive material and wastes have serious handling and disposal issues. There are a number of alternative energy sources, each with its pros and cons: solar energy, biomass energy, wind and water power, geothermal energy, oil shale, and tar sands. Oil can be made from seeds, solid coal can be transformed into more convenient physical states by gasification and liquefaction, hydrogen can be used as a secondary energy source, alcohols can be made from corn, and fuel cells can be a clean secondary energy source. Because the production of any form of energy has its negative consequences, wise stewardship of energy is necessary.

ANSWERS TO ODD-NUMBERED REVIEW QUESTIONS AND SOLUTIONS FOR ODD-NUMBERED END-OF-CHAPTER PROBLEMS

Review Questions

1. The energy on the sun comes from the nuclear fusion of hydrogen to form helium.

3. Coal was the principal fuel used in the United States between 1850 and 1950. Petroleum has been the principal fuel used since 1950.

Problems

5. a. combustion of propane (C_3H_8): $C_3H_8 + 5\ O_2(g) \rightarrow 3\ CO_2 + 4\ H_2O(l)$ + energy
   b. combustion of butane, (C_4H_{10}): $2\ C_4H_{10}(g) + 13\ O_2(g) \rightarrow 8\ CO_2 + 10\ H_2O$ + energy

7. Combustion of tridecane ($C_{13}H_{28}$): $C_{13}H_{28} + 20\ O_2(g) \rightarrow 13\ CO_2 + 14\ H_2O$ + energy

9. In the complete combustion of a hydrocarbon, the hydrocarbon reacts with oxygen to produce carbon dioxide and water.

11. $CH_4(g) + 2\ O_2(g) \rightarrow CO_2(g) + 2\ H_2O(l)$

13. Reactions almost always proceed more rapidly at higher temperatures because the collisions between reactants are more likely to produce reactant products.

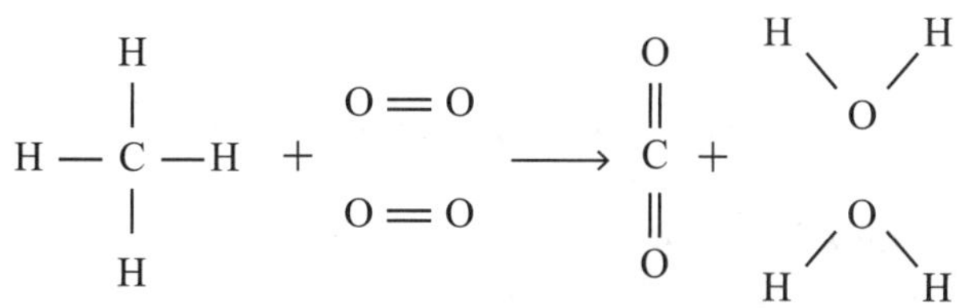

15. a.
   b. energy required to break 4 mol C-H bonds = (4 mol C-H bonds)(413 kJ/mol C-H bonds) = 1652 kJ
   c. energy required to break 2 mol O=O bonds = (2 mol O=O bonds)(498 kJ/mol O=O bonds) = 996 kJ
   d. energy released when 1 mol CO_2 forms = (1 mol CO_2)(2 C=O bonds/mol CO_2)(798 kJ/mol C=O) = 1596 kJ

e. energy released when 2 mol H_2O form = (2 mol H_2O)(2 mol O-H bonds/mol H_2O)
(467 kJ/mol O-H)
= 1868 kJ

f. total energy required = 1652 kJ + 996 kJ = 2648 kJ
total energy released = 1596 kJ + 1868 kJ = 3464 kJ
net energy outcome = 2648 kJ – 3464 kJ = –816 kJ
The negative sign indicates that more energy was released than was required, so the reaction is exothermic.

17. Amount of energy released = (24.5 mol CH_4)(803 kJ/mol CH_4) = 1.97×10^4 kJ

19. 1 calorie = 4.184 joules
Number of kJ released/1.00 g gasoline = (1060 cal/g gasoline)(4.184 joules/cal)(1 kJ/1000 J)
= 4.44 kJ.

21. The amount of energy required to reverse a reaction is the same as the amount of energy released by the reaction, 803 kJ.

23. The first law of thermodynamics states that energy is conserved, which means that the amount of energy in the universe is a constant.

25. Entropy is the measure of the degree of distribution of energy (disorder) in a system. Entropy increases when a fossil fuel is burned because larger molecules (the fuel) are converted into a larger number of smaller molecules.

27. In both cases, the potassium (K) exists in a solid state. In metallic K, the entropy is very low because the K atoms are not in motion and are surrounded by other K atoms. In the case of the KCl in the fertilizer, the entropy is higher because the K^+ ions are surrounded by Cl^- ions and the KCl is mixed with the other components of the fertilizer.

29. The second law of thermodynamics states that heat always flows from a hot object to a cooler one. Therefore, in the case of the ice cream in a bowl, the room is warmer than the ice cream, so heat moves from the warm room to the ice cream, causing it to melt.

31. Natural gas and petroleum are cleaner-burning fuels because their combustion products are primarily CO_2 and H_2O. When coal burns, there is ash produced and the sulfur content is converted into SO_2 which contributes to acid rain. Coal also contains mercury which enters the air and is highly toxic. In addition, the advent of automobiles made the use of liquid energy sources (petroleum) much more convenient than solid energy sources (coal). Transporting the fuel is easier too.

33 a .time remaining, years = (33.4×10^9 barrels of oil)(1 year/6.7×10^9 barrels of oil) = 5.0 years
b. time remaining, years = 10.4×10^9 barrels of oil)(1 year/6.7×10^9 barrels of oil/year) = 1.6 years

35. time remaining, years = 323×10^{12} ft^3)(1 year/24.5×10^{12} ft^3) = 13.2 years

37. The advantages of petroleum as a fuel source are that it is a relatively abundant liquid, which makes it accessible and more convenient to transport than coal or another solid fuel. Its combustion products are generally limited to CO_2 and H_2O, which means that petroleum is generally cleaner burning than coal. Its disadvantage is that petroleum is a complex mixture of hydrocarbons that must be separated into fractions before it can be used. Petroleum sources are not as widely spread around the globe as coal beds are, so political issues can be associated with its availability.

39. Octane rating of gasoline describes the extent to which a gasoline mixture ignites before the spark plug fires, creating a knocking that can damage engines. Some types of hydrocarbons, especially those with branched chains, tend to burn more evenly and were less likely to cause knocking than other hydrocarbons.

41. Asphalt is the residue of the processing of petroleum. Using it in paving roads and highways provides a use for an otherwise "waste" product.

43. About 20% of U.S. electricity is produced by nuclear power plants.

45. The same fission reaction is used in both nuclear power plants and nuclear bombs, but the uranium used in power plants is enriched to only 3–4% U-235, whereas the uranium used in a nuclear bomb is about 90% U-235. Therefore, a power plant uses nuclear material far below the required U-235 concentration needed to create a nuclear explosion so a power plant will not explode.

47. No. The U-235 concentration required for an atomic bomb is much higher (90% vs. 3–4%) than the concentration required for the fission process in a nuclear bomb.

49. Uranium-238 can be converted to fissionable plutonium-239 by bombardment with neutrons. If a reactor is built with a core of fissionable plutonium surrounded by uranium-238, neutrons from the fission of plutonium convert the uranium-238 shield to more plutonium, effectively producing more fuel than it consumes. Even though plutonium is created from uranium-238, the process still requires energy, so the law of conservation of energy is not violated.

51. The main fuel in fusion reactors is deuterium, which is plentiful and can be obtained from fractional electrolysis of water. Also, the primary waste product is helium, which is stable and biologically inert. Plasma is a hot, gaseous mixture of charged particles (nuclei and electrons) made from atoms that have been stripped of their electrons,

53. $^{232}_{90}\text{Th} + ^{1}_{0}\text{n} \rightarrow ^{233}_{90}\text{Th} \rightarrow ^{0}_{-1}e + ^{233}_{91}\text{Pa}$

$^{233}_{91}\text{Pa} \rightarrow ^{233}_{92}\text{U} + ^{0}_{-1}e$

55. Number of mol of CH_4 = (1.8 x 10^{10} kJ)(1 mol CH_4/803 kJ) = 2.2 x 10^7 mol CH_4
Mass of CH_4, t = (2.2 x 10^7 mol CH_4)(16.0 g CH_4/mol CH_4)(1 kg/1000 g)(1 t/1000 kg) = 350 t

57. A photovoltaic cell converts sunlight directly into electricity.

59. 1.3 kW = 1200 W.
Area of solar cell, m^2 = (1300 W)(1 m^2/100 W) = 13 m^2.
Solar cells are not very efficient and need a very large surface area. There also needs to be a means of energy storage for nighttime and very cloudy days.

61. Biomass is plant material used as fuel. Biomass is a renewable resource whose primary combustion products are CO_2 and H_2O (plus ash). Biomass requires a large investment in land and the energy required to plant and harvest the crop. Often the crops are not located close to the areas where the energy is needed, so they must be transported. Plant material is not as dense an energy source as are fossil fuels, for example, because the plant molecules contain a relatively large amount of oxygen, greatly reducing the amount of energy released per gram of compound upon combustion.

63. a. Wind power: Advantages: Wind is clean, free, and abundant. Disadvantages: Rotating blades kill birds; upfront costs of wind turbines are high, but they pay off in the long run. Limitations: Wind does not blow constantly, so some means of energy storage or an alternate energy source must be available.
    b. Geothermal power: Advantages: Works in summer and winter. Disadvantages: Expensive to install (but pays for itself over time). Limitations: Generally limited to areas where steam or hot water are near the surface of the Earth.
    c. Biodiesel: Advantages: Can be synthesized from waste cooking oil and agricultural wastes. Disadvantages: It takes energy to create biodiesel, so we can get more energy from burning biomass than from converting it to biodiesel; Large quantities of corrosive substances (sodium hydroxide and potassium hydroxide) are used to prepare biodiesel. Limitations: Technology isn't as well developed to maximize the energy gain as it will be.
    d. Hydroelectric power: Advantages: Clean. Disadvantages: Reservoirs silt up and dams break, land is lost when reservoirs are formed. Limitations: Most good dam sights in the United States are already in use.
    e. Geothermal home heating system: Advantages: Same as with geothermal power. Disadvantages: Expensive to install but pays off in the long run; Limitations: None.

65. a. $CO(g) + 2\ H_2(g) \rightarrow CH_3OH$
    b. $2\ H_2(g) + O_2(g) \rightarrow 2\ H_2O(l)$

67. This process is exothermic because the reactants release heat to the surroundings when they are mixed.

69. a. number of watts = (2100 kcal)(4.184 kJ/kcal)(1000 J/kJ)(1 W/1 J/s)(1/ 8.64 x 10^4 s) = 102 W
    b. (102 W)(1 hp/745 W) = 0.14. Domesticated horses increased the availability of energy somewhat; 1 hp = 7.45 humanpower.

71. a. Power (W) is energy (J) per second (s). In the case of the banana,
    energy = (170 kcal)(1000 cal/kcal)(4.184 J/cal) = 7.11 x 10^5 J
    Time, in seconds = (2 hr)(60 min/hr)(60 s/min) = 7200 s
    Power = (7.11 x 10^5 J/7200 s) = 99 W
    b. power = (7.11 x 10^5 J/0.0012 s) = 5.9 x 10^8 W.
    Although both the banana and the hand grenade provide the same amount of energy, the hand grenade is 6 x 10^6 times more powerful than the banana because of the amount of time over which the same energy is dispelled.

73. The equation for the combustion of methane is: $CH_4 + 2\ O_2 \rightarrow CO_2 + 2\ H_2O$. The molar mass of CH_4 is 16.0 g/mol, and the molar mass of CO_2 is 44.0 g/mol.
    mass of CH_4 required, t = (19.0 t CO_2/year)(1000 kg/t)(1000 g/kg)(1 mol CO_2/44.0 g CO_2)
    (1 mol CH_4/1 mol CO_2)(16.0 g CH_4/mol CH_4)(1 kg/1000 g)(1 t/1000 kg)
    = 6.91 t
    Note: In the calculation above we converted metric tons to kg and then to g to use the molar mass in units of g/mol. Then we converted the answer in grams back to metric tons as asked for by the problem.

75. mass SO_2 = (2600 tons coal)(0.65 tons S/100 tons coal)(2000 lb/t)(454 g/lb)(64.1 g SO_2/32.1 g S)
= 3.1×10^7 g SO_2
volume = (45000 m)(60000 m)(400 m) = 1.1×10^{12} m^3
concentration of SO_2, $\mu g/m^3$ = (3.1×10^7 g SO_2)(1×10^6 µg/g)/1.1×10^{12} m^3 = 28 µg SO_2/m^3
The SO_2 concentration does not exceed the primary national air quality standard.

77. a. number of years = (9.92×10^6 TWh)(1 yr/1.32×10^5 TWh) = 75.2 yr
b. Americans represent roughly 4.5% of the world population. If the per capita consumption for the rest of the world increased to the per capita level in the United States, the population would be consuming roughly (4)(1.32×10^5 TWh) = 5.28×10^5 TWh. If this were the case, the available energy would last for (9.92×10^6 TWh)(1 yr/5.28×10^5 TWh) = less than 19 years.
In either case, we are not using the energy in a sustainable way, and we need to find ways of replacing nonrenewable energy resources with renewable energy resources, as well as reducing our reliance on energy.
c. The rate of energy use will likely grow instead of remaining the same, so reserves might not last the 75 years calculated in part (a). Worldwide use is unlikely to reach the U.S. level anytime soon, if ever, and so the estimate in part (b) is not at all realistic.

79. The answer is d. All the aspects of a process listed in answers a, b, and c are looked at when green chemistry principles are used to evaluate energy-generation processes.

CHAPTER

16

Biochemistry

A Molecular View of Life

CHAPTER SUMMARY

16.1 Energy and the Living Cell

Learning Objectives: • List the major parts of a cell, and describe the function of each part.
• Name the primary source of energy for plants and three classes of substances that are the sources of energy for animals.

A. Biochemistry is the chemistry of living things and life processes.
   1. The structural unit of all living things is the cell.
      a. Cells are enclosed in lipid membranes; plant cells also have cellulose cell walls.
      b. The cell gains nutrients and gets rid of wastes through the cell membrane.
      c. The cell nucleus contains nucleic acids that control heredity.
      d. Ribosomes are the location of protein synthesis.
      e. Energy is produced in the mitochondria.
      f. Plant cells (but not animal cells) contain chloroplasts in which energy from the sun is converted into chemical energy, which is stored in the plant in the form of carbohydrates.
B. Energy in Biological Systems.
   1. Living organisms can use only certain forms of energy.
   2. Plants' chloroplasts convert radiant energy into chemical energy stored as carbohydrates.
   3. Plant cells can also convert carbohydrates to fats and, with proper inorganic nutrients, to proteins.
   4. Animals cannot use sunlight directly. Animals obtain energy from carbohydrates, fats, and proteins.
   5. Metabolism—the series of coordinated chemical reactions that keep cells alive.
      a. Catabolism is the degradation of molecules to provide energy.
      b. Anabolism is the synthesis of biomolecules.

Answers to Self-Assessment Questions

1. d We obtain energy from carbohydrates, fats, and proteins.
2. a Anabolism is the process of chemically building molecules in living systems.
3. b Metabolism is the entire series of coordinated chemical reactions that keep cells alive.

16.2 Carbohydrates: A Storehouse of Energy

Learning Objective: • Compare and contrast starch, glycogen, and cellulose.

A. Carbohydrates are polyhydroxy aldehydes or ketones or compounds that can be hydrolyzed (split by water) to form polyhydroxy aldehydes or polyhydroxy ketones.
   1. Carbohydrates are composed of carbon, hydrogen, and oxygen.
   2. Usually, the C, H, and O atoms are present in a ratio expressed by the formula $C_x(H_2O)_y$.

B. Some Simple Sugars.
   1. Sugars can be classified in terms of their functional group: Aldoses are polyhydroxy aldehydes, and ketoses are polyhydroxy ketones.
   2. Monosaccharides cannot be hydrolyzed (split apart by water).
      a. Some examples are glucose (also called dextrose, an aldose), fructose (fruit sugar, a ketose), and galactose (a component of lactose, milk sugar, and an aldose).
   3. Disaccharides can be hydrolyzed into two monosaccharides.
      a. Examples:
         i. Sucrose + H_2O $\rightarrow$ glucose + fructose
         ii. Lactose + H_2O $\rightarrow$ glucose + galactose

C. Polysaccharides: Starch and Cellulose.
   1. Polysaccharides are large molecules that yield many monosaccharide units upon hydrolysis.
      a. Examples: starch and cellulose.
         i. Both starch and cellulose are polymers of glucose.
   2. In starch, glucose molecules are connected through an alpha linkage in which the oxygen atom joining the glucose molecules is pointing downward.
      a. Two kinds of plant starch are:
         i. Amylose, in which glucose units are joined in a continuous chain.
         ii. Amylopectin, in which glucose units are joined in a branched chain.
      b. Animal starch is called glycogen.
         i. In glycogen, glucose units are joined in a branched chain.
   3. In cellulose, glucose molecules are connected through a beta linkage in which the oxygen atom joining the glucose molecules is pointing upward.

Answers to Self-Assessment Questions

1. c A ketose is a monosaccharide with an ketone functional group.
2. d An aldose is a polyhydroxy aldehyde.
3. c When they dissolve in water, most monosaccharides form ring structures.
4. d Lactose, or milk sugar, is a disaccharide composed of glucose and galactose.
5. d Amylopectin and glycogen are polysaccharide molecules with branched chains of glucose units.
6. d Carbohydrates are stored in the liver and muscle tissue in the form of glycogen.

16.3 Fats and Other Lipids

Learning Objectives: • Describe the fundamental structure of a fatty acid and of a fat.
• Classify fats as saturated, monounsaturated, or polyunsaturated.

A. Lipids are defined by their solubility, not their structures.

1. Lipids are cellular components that are not soluble in water but are soluble in organic solvents.
2. Some examples of lipids are fats, fatty acids (long-chain carboxylic acids), steroids such as cholesterol, sex hormones, and fat-soluble vitamins.

B. Fats are esters of fatty acids and glycerol (a trihydroxy alcohol).
1. Fats are classified according to the number of fatty acid chains attached to glycerol: if one fatty acid chain is attached, the compound is a monoglyceride; if two fatty acid chains are attached, the compound is a diglyceride; and if three fatty acid chains are attached, the compound is a triglyceride.
   a. Naturally occurring fatty acids almost always have an even number of carbon atoms.
   b. Saturated fatty acids do not contain any carbon-to-carbon double bonds (C=C).
   c. Monounsaturated fatty acids contain one C=C bond in the chain.
   d. Polyunsaturated fatty acids contain two or more C=C bonds.
2. Animal fats are usually solid at room temperature because they are composed of a higher proportion of saturated fatty acids.
3. Vegetable fats (oils) are liquid at room temperature and have a higher proportion of mono- and polyunsaturated fatty acids.
   a. Molecules of polyunsaturated fats include mainly polyunsaturated fatty acids.
4. The degree of unsaturation of a fat is measured by the *iodine number*.
   a. The iodine number is the number of grams of iodine that can add to the double bonds of 100 g of fat.
   b. The higher the iodine number, the more unsaturated the fat.
5. Fats and oils are insoluble in water, and because they are less dense than water, they float on water.

Answers to Self-Assessment Questions

1. d Fats are triglycerides that are solid at room temperature. Oils are also triglycerides but are liquid at room temperature. The degree of unsaturation in the fatty acids determines the melting point of the triglyceride. When there is very little unsaturation, the substance is solid at room temperature; as the degree of unsaturation increases, the melting point of the triglyceride decreases.
2. a S and T are saturated fatty acids, fatty acids with no C=C bonds.
3. c U is a monosaturated (has one C=C bond) fatty acid.
4. d V is a polyunsaturated (has multiple C=C bonds) fatty acid.
5. a Saturated fats often have a high proportion of S and T.
6. d Polyunsaturated fats (oils) often have a high proportion of V.
7. c The iodine number measures the degree of unsaturation in a compound.
8. d In general, vegetable oil molecules have a larger number of C=C double bonds than animal fats.

16.4 Proteins: Polymers of Amino Acids

Learning Objective: • Draw the fundamental structure of an amino acid, and show how amino acids combine to make proteins.

A. All living parts of humans and other organisms contain protein.
1. Proteins serve as the structural material of animals.
   a. Proteins are composed of carbon, hydrogen, oxygen, nitrogen, and often sulfur.

B. Proteins are copolymers of 20 different amino acids (see Table 16.3 in the text).

1. Amino acids have two functional groups, an amino group (—NH_2) located on the carbon next to the carbon of the carboxyl group (—COOH), which is called the *alpha* (α) carbon.

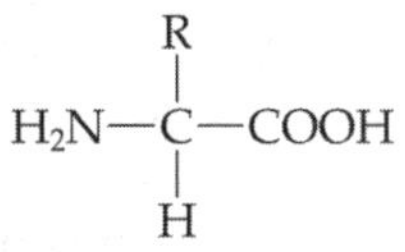

   a. The carboxyl group is acidic and donates an H^+ to the amino group, which is basic and accepts the H^+. This structure is a *zwitterion*.

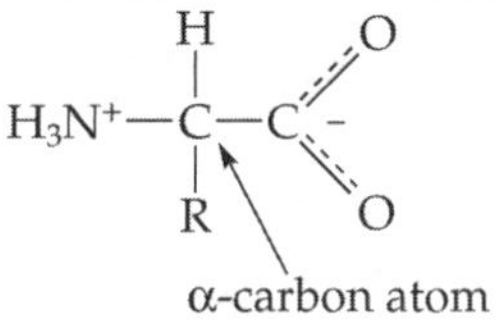

   b. A zwitterion is a molecule containing a positive and negative charge on different parts of the same molecule.
   c. Amino acids differ in the variety of other groups attached to the central carbon atom.

C. Plants synthesize proteins from CO_2, water, and minerals (supplying N and S). Animals must take in proteins as food.
   1. Humans can synthesize some of the amino acids from other molecules. Amino acids that we cannot synthesize must be part of our diet and are called essential amino acids.

D. The Peptide Bond: Peptides and Proteins.
   1. Proteins are polyamides: amino acids linked by many peptide bonds (amide linkages).
      a. A peptide bond (amide linkage) links the $-NH_3^+$ of one amino acid with the $-COO^-$ of another amino acid.

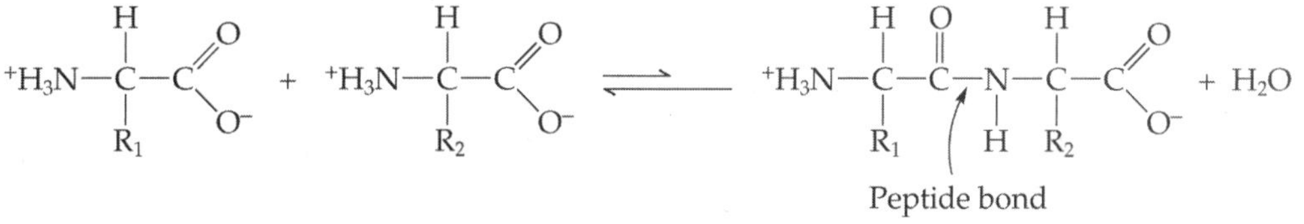

      b. This arrangement leaves a free carboxyl group (—COO^-) at one end of the protein called the C terminal, and a free amino group (—NH_3^+) at the other end, called the N-terminal.
      c. Dipeptides are two linked amino acids, tripeptides are three linked amino acids, and polypeptides are formed when 10 or more amino acids link together.
      d. A protein is a polypeptide that has a molecular weight of more than 10,000 amu.
   2. The sequence of amino acids.
      a. The sequence in which the amino acids are connected in peptides is of critical importance.
         i. Sequences are written using three-letter abbreviations for the amino acids and arranging them with the N-terminal to the left of the sequence and the C-terminal to the right.
         ii. Note that there are two ways of linking just two amino acids. For example, Ala and Ser can link Ala–Ser where the NH_3^+ terminal is part of Ala and the COO^- terminal is part of Ser, or they can link Ser–Ala where the terminals are reversed. These are two distinctly different linkages.
         iii. A minor change in the amino acid sequence of a protein may have disastrous effects for the organism.

iv. *Example*: Sickle-cell anemia is caused by one incorrect amino acid in a 300-unit sequence.

Answers to Self-Assessment Questions

1. a Proteins are polymers of amino acids.
2. d The 20 amino acids making up proteins differ mainly in their side chains, groups attached to the α-C atom.
3. d $H_2NCH_2CH_2COOH$ is not an alpha amino acid because the NH_2 group is not attached to the α-C atom.
4. d $^+H_3NCH_2COO^-$ is a zwitterion or a structure with positive and negative charges in different parts of the same molecule.
5. a Two amino acids are joined through an amide linkage to form a dipeptide.
6. c The N-terminal amino acid and the C-terminal amino acid of the peptide Met-Ile-Val-Glu-Cys-Tyr-Gln-Trp-Ile are the amino acids on the left end (Met) and on the right end (Ile), respectively.
7. c Six different tripeptides can be formed from alanine, valine, and lysine when each appears only once in each product (Ala-Val-Lys, Ala-Lys-Val, Val-Ala-Lys, Val-Lys-Ala, Lys-Val-Ala, and Lys-Ala-Val).
8. d The tripeptides represented as Ala-Val-Lys and Lys-Val-Ala are two different peptides.

16.5 Structure and Function of Proteins

Learning Objectives: • Describe the four levels of protein structure, and give an example of each.
• Describe how enzymes work as catalysts.

A. The structures of proteins have four organizational levels.
   1. The *primary structure* of a protein is its amino acid sequence, which is specified from the amino end (N-terminal) to the carboxyl end (C-terminal).
   2. *Secondary structure:* Polypeptide chains can fold along an axis into regular structures such as the alpha helix and the beta-pleated sheet.
      a. In a *beta-pleated-sheet structure,* molecules are stacked in extended arrays with hydrogen bonds holding adjacent chains together. (An example is silk.)
      b. In an *alpha-helix structure*, a right-handed helix forms when the amino group (NH_3^+) of one amino acid in one turn of the chain forms hydrogen bonds with the carboxyl (COO^-) group of another amino acid. This arrangement allows the protein to be stretched and then regain its shape, like a spring. (An example is wool.)
   3. *Tertiary structure:* Protein folding creates spatial relationships between amino acid units that are relatively far apart in the protein chain.
      a. An example is globular proteins.
   4. *Quaternary structure:* Two or more polypeptide chains can assemble into a multiunit structure. An example is hemoglobin.

B. Four Ways to Link Protein Chains.
   1. Peptide (covalent) bonds fix the primary protein structure.
   2. Four other forces hold the protein in a structural arrangement.
      a. In *hydrogen bonding*, the carbonyl oxygen of one peptide may form a hydrogen bond to an amide hydrogen (N—H).
         i. *Examples:* alpha helix structure of wool protein and pleated-sheet structure of silk protein.

   b. A *salt bridge* (ionic bond) is formed when a proton (H^+) transfer between the acidic side chain of one amino acid and the basic side chain of another results in opposite charges, which then attract each other.
   c. A *disulfide linkage* is formed when two cysteine units (SH) are oxidized. The resulting disulfide bond between the two cysteine units is a covalent bond that is much stronger than a hydrogen bond.
   d. *Hydrophobic interactions* (dispersion forces) are the weak attractive forces between nonpolar side chains.
      i. Significant in the absence of other forces.
      ii. Nonpolar side chains cluster together on the inside folds of proteins, forming several hydrophobic interactions in a given region of the protein.

C. Enzymes: Exquisite Precision Machines.
   1. Enzymes are specialized proteins that act as highly specific biological catalysts.
      a. Enzymes enable reactions to occur at convenient rates and at lower temperatures by changing the reaction path.
      b. The reactive compound, called the substrate, attaches to the enzyme at the active site to form an enzyme–substrate complex that then decomposes into the reaction product and the liberated enzyme.
         i. The substrate need not fit the active site precisely for the reaction to occur; this fit occurs when the flexible configuration of the enzyme accommodates the substrate (the induced-fit model) much like a glove. accommodates the structure of the hand of the person wearing it.
         ii. The substrate and enzyme are held together by bonds between complementary charged groups on the two.
         iii. The formation of new bonds weakens the old bonds to the substrate and facilitates the breaking of old bonds and the formation of new products.
   2. An enzyme can be made ineffective or its rate of catalysis can be slowed down when inhibitor molecules bond to the enzyme at positions remote from the active site, thus changing the three-dimensional shape of the enzyme and preventing further bonding with the substrate.
   3. Inorganic ions serve as *cofactors* necessary for proper functioning of some enzymes.
      a. The protein part of an enzyme is called the *apoenzyme.*
   4. Organic nonprotein cofactors are called coenzymes.
      a. Many coenzymes are vitamins or substances derived from vitamin molecules.
   5. Enzymes are essential to the function of every living cell.

D. Applications of Enzymes.
   1. Enzymes are used in medicine in test strips for diabetes and as a way of breaking up blood clots after a heart attack.
   2. Clinical analysis for enzymes in body fluids or tissues is a common diagnostic technique in medicine.
   3. Enzymes are widely used in a variety of industries, from baby foods to beer, and intact microorganisms are used to make bread, beer, wine, yogurt, and cheese.
   4. Enzymes are components in some detergents where they attack fats, grease, and proteins responsible for stains.

Answers to Self-Assessment Questions

1. c The alpha-helix or beta-pleated sheet produced from hydrogen bonds between N-H and C=O groups are examples of secondary protein structures.

2. d Disulfide bridges are S-S covalent bonds that can form between various parts of a single chain or between two different chains. These bridges represent one of the kinds of intereactions that establish the tertiary structure of a protein.
3. b The arrangement of protein subunits into a specific pattern such as in hemoglobin is an example of quaternary structure in proteins.
4. b Disulfide links are the covalent cross-links between cysteine units in protein chains of fibrous proteins such as hair.
5. d Salt bridges are formed when an acidic side chain of an amino acid on a protein chain reacts with a basic side chain of an amino acid on another chain, forming ions that are attracted to one another.
6. a Dispersion forces exist between a nonpolar side chain of one amino acid on a protein chain and a nonpolar side chain of an amino acid some distance away on the same chain or on another chain.
7. d The substrate is the molecule upon which an enzyme acts; it is the reactant in the reaction catalyzed by the enzyme.
8. d The induced-fit model of enzyme action holds that shapes of active sites can change somewhat to fit a substance.
9. a Cofactors are inorganic metal ions such as Ca^{2+}, Mg^{2+}, and K^{+} required for some enzyme molecules to function.
10. d Catalysts speed up chemical reactions by changing their reaction pathways to reduce the temperature and/or energy requirement that the uncatalyzed reaction would require.
11. a An apoenzyme is the purely protein part of an enzyme.

16.6 Nucleic Acids: Parts, Structure, and Function

Learning Objectives:
- Name the two types of nucleic acids, and describe the function of each type.
- Explain complementary base pairing, and describe how a copy of DNA is synthesized.

A. Nucleic acids are the information and control centers of the cell.

B. There are two kinds of nucleic acids: deoxyribonucleic acid (DNA), which provides a mechanism for heredity and serves as the blueprint for all proteins in an organism, and ribonucleic acid (RNA), which carries out protein assembly.

  1. DNA is found in the cell nucleus and RNA is found in all parts of the cell.
  2. Nucleic acids are chains of repeating nucleotides.
     a. Nucleotides contain a pentose (sugar), a heterocyclic amine base, and a phosphate unit.
     b. The sugar is either ribose (found in RNA) or deoxyribose (found in DNA).
        i. The two sugars differ in the presence of an oxygen atom on the second carbon. Ribose has an oxygen in this position; deoxyribose does not.
  3. The bases are either purines (two fused rings), including adenine and guanine, or pyrimidines (one ring), including cytosine, thymine, and uracil.
  4. The phosphate groups (P_i) are attached to the fifth carbon of the sugar.
  5. The phosphate groups of nucleotides form ester linkages to the hydroxyl groups of sugars of adjoining nucleotides to form nucleic acid chains.
  6. The chain is then composed of a phosphate-sugar backbone with branching heterocyclic bases.
  7. In DNA, the sugar is deoxyribose, and the bases are adenine, guanine, cytosine, and thymine.
  8. In RNA, the sugar is ribose, and the bases are adenine, guanine, cytosine, and uracil.

9. The sequence of bases, or the primary structure of the nucleic acid strand, stores all the information needed to build living organisms.

C. The Double Helix.
   1. The bases in DNA are paired (adenine to thymine) and (guanine to cytosine), through hydrogen bonding.
      a. Watson and Crick determined that DNA was composed of two helixes wound around each other and held in place by base pairing.
   2. In base pairing, a pyrimidine base is paired with a purine base.
      a. In the pyrimidine–purine pair guanine and cytosine, three hydrogen bonds can form. No other such pairing provides such extensive interaction.

D. Structure of RNA.
   1. RNA is a single strand of nucleic acid with some internal base pairing in which the molecule folds back on itself.

E. DNA: Self-Replication.
   1. Chromosomes, found in cell nuclei, contain the hereditary material.
      a. The number of chromosomes varies with species. Human cells have 46 chromosomes, with egg and sperm cells providing half of these chromosomes each.
      b. Chromosomes are composed of DNA and proteins.
   2. Genes are sections of the DNA molecule.
      a. The complete set of genes of an organism is called its genome.
   3. During cell division, each chromosome replicates (produces an exact duplicate of) itself.
   4. When cell division occurs, the DNA is replicated.
      a. Two chains of double helix are pulled apart, and each chain directs synthesis of a new DNA chain.
      b. The chains run antiparallel to one another.
      c. Synthesis begins with a nucleotide from the surrounding cellular fluid pairing with its complementary base on the DNA strand.
      d. As nucleotides align, the enzymes connect them to form the sugar–phosphate backbone of the new chain.
   5. When the cell divides, each daughter cell receives one of the original DNA strands.
   6. The sequence of bases along the DNA strand encodes all the information for building a new organism.
      a. These four bases in the genetic code can form many different combinations.

Answers to Self-Assessment Questions

1. d Nucleotides are the monomer units of nucleic acids.
2. a An RNA nucleotide could contain A, P_i, and ribose.
3. b A possible base pair in DNA is A-T.
4. c A possible base pair in RNA is A-U (uracil replaces thymine in RNA).
5. d Nucleic acid base pairs are joined by hydrogen bonds between the nitrogenous bases.
6. a Replication of a DNA molecule involves breaking the hydrogen bonds between base pairs in the double helix.
7. d The second step of DNA replication involves the pairing up and hydrogen bonding of nucleotides in the proper sequence.

16.7 RNA: Protein Synthesis and the Genetic Code

Learning Objectives: • Explain how mRNA is synthesized from DNA and how a protein is synthesized from mRNA. • List important characteristics of the genetic code.

A. RNA carries the information from DNA to other parts of the cell.
   1. *Transcription* involves the transfer of DNA information to mRNA (messenger RNA) through base pairing.
      a. In RNA, uracil takes the place of thymine, so the allowed base pairing between DNA and mRNA is T-A, C-G, A-U.
   2. *Translation* of the code in mRNA into a protein structure takes place in the ribosomes. The mRNA becomes attached to a ribosome, and the genetic code is deciphered.

B. Transfer RNA (tRNA), located in the cytoplasm, translates the base sequence of mRNA into the amino acid sequence of a protein.
   1. tRNA has an anticodon base triplet that determines which amino acid will attach to the end of tRNA. This amino acid can then be moved into position when the tRNA base pairs with the appropriate bases on mRNA.
   2. When amino acids have been moved into position by tRNAs, these amino acids then bond to each other, forming a peptide chain. In this way, a protein is built up.
   3. The base triplet on the mRNA is called a *codon*; it pairs with its complementary base triplet on tRNA called an *anticodon*.
   4. Some amino acids are specified by several different codons: Others are specified by only one codon.
      a. Three codons are stop signals, calling for the termination of a proton chain.

Answers to Self-Assessment Questions

1. d The conversion of the message on mRNA to a sequence of amino acids in a protein is called translation (the nitrogenous base code is translated into a sequence of amino acids).
2. d The anticodon base triplet is found on the tRNA and determines which amino acid will attach to the end of that tRNA.
3. b There are 20 amino acids and 64 possible triplets or codons.
4. b The fact that each amino acid except tryptophan and methionine is coded for by more than one codon is an example of redundancy.
5. c The codon UGA only codes for stopping or terminating the chain (see Table 16.5).

16.8 The Human Genome

Learning Objective: • Describe recombinant DNA technology, and explain how it is used.

A. DNA is sequenced by using enzymes to cleave it into segments of a few to several hundred nucleotides each.
   1. The pieces are duplicated and amplified using the polymerase chain reaction (PCR), which employs enzymes called DNA polymerases to make millions of copies of the fragment.
   2. An electric current is used to sort the DNA segments by length and the base at the end of each fragment is identified.
   3. Computers are used to compile the short sequences into longer segments that recreate the original sequence of bases in the fragment.

B. Recombinant DNA: Using Organisms as Chemical Factories.
   1. A gene from one organism can be substituted for a defective or missing gene in another organism.
      a. The gene is identified, isolated using the RFLP process, and placed in a separate piece of DNA.

b. The recombined DNA is then transferred into a bacteria DNA (plasmid). The bacteria are cloned, and large amounts of protein coded by the genes are produced.
   i. Human growth hormone, epidermal growth factor, and insulin are produced using recombinant DNA.

C. Gene Therapy.
   1. A functioning gene is introduced into a person's cells to correct the action of a defective gene.

Answers to Self-Assessment Questions

1. a Recombinant DNA cloning involves combining DNA from two different organisms to make a single new organism.
2. c In recombinant DNA cloning, the DNA is cut with restriction enzymes that can catalyze the DNA cleavage only at very specific sites.
3. b Gene therapy involves replacing a nonfunctioning gene with one that is functional.
4. b Transferring some photosynthesis genes from an efficient crop to a less efficient one, thus producing a new plant variety with greater productivity, would involve genetic engineering.

Green Chemistry: Green Chemistry and Biochemistry

A. Finding Natural Renewable Energy Sources.
   1. Chemists are devising ways of converting existing biochemical pathways for fat and carbohydrate syntheses into substitutes for nonrenewable petroleum fuels.
   2. Using biochemistry to make molecules has several advantages:
      a. Reactions normally occur in water, a nontoxic and nonflammable solvent.
      b. When the reaction products are nonpolar, they can often be separated easily.
      c. Most biochemical reactions are rapid and can be performed at mild temperatures and atmospheric pressure.
      d. The atom economy of biochemical reactions is high.
   3. Most of the biochemical reactions are enzyme catalyzed, which poses the greatest challenge to developing biochemical technology because of the specificity of the catalyzed reactions. Adapting biochemical reactions to new substances will require imagination.

LEARNING OBJECTIVES

You should be able to . . .

1. List the major parts of a cell, and describe the function of each part. (16.1)
2. Name the primary source of energy for plants and three classes of substances that are sources of energy for animals. (16.1)
3. Compare and contrast starch, glycogen, and cellulose. (16.2)
4. Describe the fundamental structure of a fatty acid and of a fat. (16.3)
5. Classify fats as saturated, monounsaturated, or polyunsaturated. (16.3)
6. Draw the fundamental structure of an amino acid, and show how amino acids combine to make proteins. (16.4)

7. Describe the four levels of protein structure, and give an example of each. (16.5)

8. Describe how enzymes work as catalysts. (16.5)

9. Name the two types of nucleic acids, and describe the function of each type. (16.6)

10. Explain complementary base pairing, and describe how a copy of DNA is synthesized. (16.6)

11. Explain how mRNA is synthesized from DNA and how a protein is synthesized from mRNA. (16.7)

12. List important characteristics of the genetic code. (16.7)

13. Describe recombinant DNA technology, and explain how it is used. (16.8)

14. Name some advantages of using biochemistry to create useful molecules.

15. Give examples of the use of biochemistry for energy production and other applications.

DISCUSSION

The cell is the structural unit of life. Biochemistry, the chemistry of life, is organized into classes of compounds. Carbohydrates (sugars and starches) are the products of photosynthesis and are the source of energy for living things. Fats and lipids are water-insoluble compounds with important members such as steroids (cholesterol and sex hormones). Proteins, polymers of amino acids, are the structural unit of animals. There are about 20 amino acids found in life on Earth. They are joined with peptide links to form polymers. Proteins play many roles as well as serving as enzymes or biochemical catalysts. There are three levels of structure for each protein: primary (the sequence of amino acids), secondary (the regular patterns of folding in short portions of the chain), and tertiary (the overall folding pattern of the entire chain). Sometimes there is a fourth structural level called the quaternary structure (details the relationship between two or more chains and/or cofactors). Nucleic acids (RNA, DNA) determine the heredity of life. RNA and DNA consist of repeating units of nucleotides that are made up of a sugar, a heterocyclic base, and a phosphate group. The sugar is either ribose (in RNA) or deoxyribose (in DNA). The sequence of bases in the chain determines the genetic code. The two strands of DNA (RNA) are held together with hydrogen bonding. These are broken and new ones are formed during replication (reproduction). Recent advances in identifying DNA (DNA fingerprinting) and modifying DNA (recombinant DNA) are modern issues.

ANSWERS TO ODD-NUMBERED REVIEW QUESTIONS AND SOLUTIONS FOR ODD-NUMBERED END-OF-CHAPTER PROBLEMS

Review Questions

1. Photosynthesis is the process by which plants are able to capture the sun's energy in the form of carbohydrates, which serve as an energy source for themselves and other species. Plants can convert the carbohydrates (initially glucose) into structural material and fats. If sufficient resources are available, plants can also convert glucose into proteins. Animals require a constant input of energy to survive. The source of this energy is ultimately the chemical compounds created by plants.

3. Proteins are found in every cell. Muscles, skin, hair, and nails are mostly proteins.

5. Proteins are polyamides, polymers formed from the joining of amino acids through peptide (amide) linkages.

7. Both are polymers of amino acids. The difference between a polypeptide and a protein is a matter of length. If the molar mass of a polypeptide exceeds 10,000 g/mol, the polypeptide is called a protein.

9. Hydrogen bonding is the intermolecular force holding the base pairs in DNA together.

11. DNA is a double helix while RNA is a single helix with some loops. The pentose in DNA is deoxyribose, while the pentose in RNA is ribose. The four nitrogenous bases in DNA are thymine, adenine, guanine, and cytosine. The four nitrogenous bases in RNA are uracil, adenine, guanine, and cytosine.

13. The polymerase chain reaction (PCR) is a technique for duplicating and producing millions of copies of any specific DNA sequence.

15. Step 1: After scientists determine the base sequence of the gene that codes for a particular protein, they isolate it and amplify it by PCR.
    Step 2: The gene is spliced into a special kind of bacterial DNA called a plasmid.
    Step 3: The recombined plasmid is inserted into a host organism, the bacteria from which the plasmids came.
    Step 4: The plasmids replicate, making multiple exact copies of themselves, a process called cloning.

Problems

17. Carbohydrates are polyhydroxy aldehydes or polyhydroxy ketones or compounds that can be hydrolyzed into polyhydroxy aldehydes or polyhydroxy ketones. Compound a is a polyhydroxy aldehyde, and compound c is a polyhydroxy ketone. The other two structures (compounds b and d) do not fit the requirement.

19. Glycogen is animal starch. Amylose is a plant starch with glucose units joined in a continuous chain. Amylopectin is a plant starch with branched chains of glucose units.

21. The answer is both c and d. Mannose and glucose are monosaccharides.

23. None of the substances listed in Problem 21 are disaccharides. Sucrose and lactose (Problem 22) are both disaccharides.

25. a. Amylose is hydrolyzed into glucose molecules.
    b. Glycogen is hydrolyzed into glucose molecules.

27. Lactose is a disaccharide composed of glucose and galactose. The open-chain forms of both glucose and galactose include both aldehyde and alcohol (hydroxyl) functional groups. The structures of glucose and galactose are identical except for the orientation of the –OH and –H groups on the fourth carbon in the six-carbon chains.

29. The molarity of a solution reflects the number of moles of solute per liter of solution. If the concentration of the trehalose solution was 0.1 M, then 1 L of the solution contained 0.1 mol of trehalose. After heating, the solution concentration rose to 0.2 M, indicating that each mole of trehalose was separated into two parts. Therefore trehalose is a disaccharide.

31. At room temperature, most animal fats are solids while oils are liquids. Oils are obtained principally from vegetable (plant) sources. Structurally, oils are identical with fats, except that oils incorporate a larger proportion of unsaturated (C to C double bonds) fatty acid units than are present in fats.

33. Palmitic acid (a) and stearic acid (d) are both saturated fatty acids. Oleic acid (c) is monounsaturated, while linoleic acid (b) is polyunsaturated (see Table 16.1).

35. a. Linoleic acid has 18 carbon atoms.
    b. Palmitic acid has 16 carbon atoms.
    c. Stearic acid has 18 carbon atoms.

37. Corn oil (the image on the right) and other plant oils generally have higher iodine numbers than solid animal fats (lard, the image on the left). This is because oils have more sites of unsaturation (C=C bonds) in the fatty acid chains, and iodine adds to C=C bonds.

39. The answer is d. Proteins are polymers composed of amino acids joined through peptide (amide) linkages.

41. Of the four amino acids listed, lysine and methionine are both essential amino acids that must be supplied through diet.

43. Structural formulas
    a. threonine

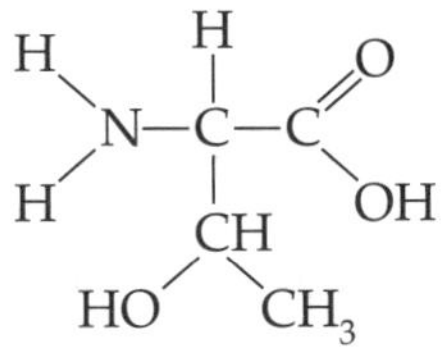

    b. phenylalanine

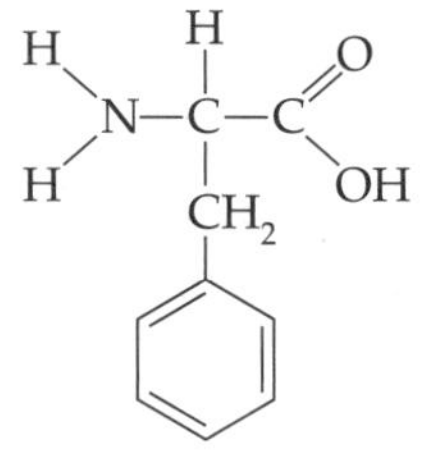

45. Structural formulas
    a. glycylarginine

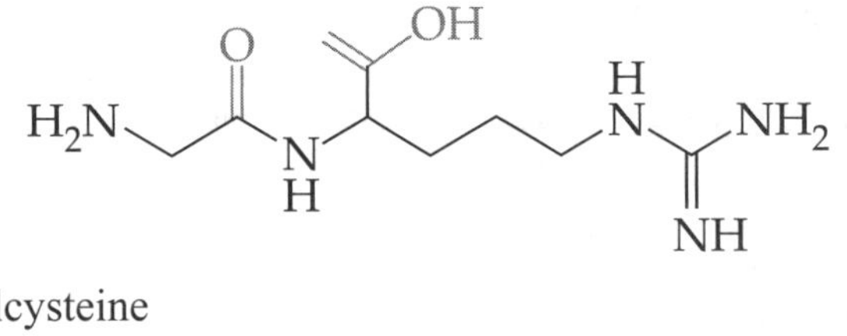

    b. alanylcysteine

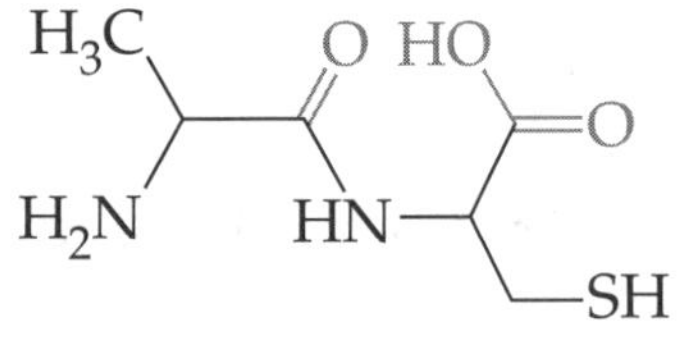

47. The dipeptide is a combination of aspartic acid and phenylalanine: aspartylphenylalanine

49. Proteins are bonded to each other by hydrogen bonds, ionic bonds, disulfide linkages, and dispersion forces.

51. Nucleotide a appears in DNA, while nucleotides b and c appear in RNA. In these cases, the distinguishing feature is the sugar: The sugar in DNA nucleotides is deoxyribose (see choice a), while the sugar in RNA nucleotides is ribose (see choices b and c).

53. The sugar is ribose and the base is uracil.

55. a. Guanine pairs with cytosine in DNA.
    b. Thymine pairs with adenine in DNA.
    c. Cytosine pairs with guanine in DNA.
    d. Adenine pairs with thymine in DNA.

57. Each strand of the parent DNA becomes incorporated in one of the daughter double helices. Therefore, one strand will end up in one daughter's cell nucleus; the other strand will end up in the other daughter cell's nucleus.

59. a. DNA and mRNA are involved in transcription.
    b. mRNA and tRNA are involved in translation.

61. The answer is c. The process in which DNA unzips and makes a strand of mRNA is called transcription.

63. The answer is d. Once it has been synthesized from the DNA in the cell nucleus, the mRNA strand moves to a ribosome in the cytoplasm where the genetic message can be translated into a chain of amino acids.

65. The answer is c. Thymine is found only in DNA. Uracil has a very similar structure and is found in RNA.

67. Transfer RNA (tRNA) molecules carry amino acids to the mRNA, which is the site of protein synthesis in the cell.

69. a. The one-letter abbreviations give a primary protein structure, which is the specific order of amino acids in each of the chains.
    b. The descriptions of the loop and chain linkages give information about the secondary protein structure.

71. a. The base in Problem 53 is uracil, which has a single ring structure and is therefore a pyrimidine .
    b. The base in Problem 54 is also a pyrimidine because it has a single ring structure.

73. a. The base is a purine because it has a fused ring structure.
    b. This compound would be found in RNA because the sugar is ribose.

75. a. The amino acid sequence for ACC-AGC-AUG-GCG is ~Thr-Ser-Met-Ala~.
    b. The amino acid sequence for ACC-AGC-GCG-GCG is ~Thr-Ser-Ala-Ala~.

77. The mRNA codon would have the three-letter base pair of CGU.

79. Water is the solvent typically used in biochemical reactions.

81. The answer is d. The methods listed in answers a., b., and c are all goals designed to produce renewable fuels.

CHAPTER

17

Food

Feasts to Famine

CHAPTER SUMMARY

17.1 Carbohydrates in the Diet

Learning Objective: • Identify dietary carbohydrates, and state their sources and function.

A. Carbohydrates include sugars and starches.
B. Sweet Chemicals: The Sugars.
   1. Monosaccharides.
      a. Glucose (dextrose), often called *blood sugar*, is used by cells to provide energy.
      b. Fructose, called *fruit sugar*, is found in fruit or made from glucose with the help of enzymes.
         i. Fructose is sweeter than glucose.
         ii. High-fructose corn syrup (HFCS) is made by treating corn syrup with enzymes to convert much of the glucose to fructose.
C. Digestion and Metabolism of Carbohydrates.
   1. Glucose and fructose are absorbed directly into the bloodstream from the digestive tract.
   2. Sucrose and lactose are hydrolyzed with the help of enzymes to form simpler sugars.
      Sucrose + $H_2O \rightarrow$ Glucose + Fructose
      Lactose + $H_2O \rightarrow$ Glucose + Galactose
      a. Lactose intolerance is a condition in which a person lacks the correct enzyme to break down lactose.
   3. All monosaccharides are converted to glucose during metabolism.
      a. Galactosemia is a condition due to a deficiency of the enzyme that converts the monosaccharide galactose to glucose.
D. Complex Carbohydrates: Starch and Cellulose.
   1. Starch and cellulose are both polymers of glucose, but the connecting links between the glucose units are different.
      a. The glucose units are connected with alpha linkages in starch and with beta linkages in cellulose.
      b. Because of the spatial differences in linkage, humans can digest starch but not cellulose.

2. Starch is hydrolyzed to glucose during digestion.
   a. More than 50 chemical reactions are needed to produce CO_2, water, and energy from starch. This process is the reverse of photosynthesis.
3. Carbohydrates supply ~4 kcal of energy per gram.
   a. Excess carbohydrates are stored as glycogen and fat.
   b. Carbohydrates are the body's preferred fuel.
4. Cellulose is the most abundant carbohydrate: it is present in all plants, forming their cell walls and other structural features.
   a. Most animals lack the enzymes needed to break the beta linkages in the polymer, but certain bacteria can produce these enzymes.
   b. Cellulose plays an important role in human digestion, providing dietary fiber that absorbs water and helps move food through the digestive tract.

Answers to Self-Assessment Questions

1. b About 1 in 7 people in the world are hungry today (see introduction).
2. c Cellulose is a polymer of beta-glucose (β-glucose). Starch is a polymer of alpha-glucose (α-glucose). Humans lack the enzyme necessary to catalyze the hydrolysis of the beta-linkages.
3. c Glucose and fructose are both monosaccharides. Each of the other possible combinations includes at least one disaccharide.
4. a People who have lactose intolerance are deficient in the enzyme that catalyzes the hydrolysis of lactose to galactose and glucose.
5. b Ordinary corn syrup is primarily glucose (which explains why it is not as sweet as other syrups).
6. d Small quantities of carbohydrates can be stored in liver and muscle tissue as glycogen.
7. c Humans cannot digest cellulose because they lack enzymes for the hydrolysis of beta-glucose linkages.

17.2 Fats and Cholesterol

Learning Objective: • Identify dietary lipids and state their function.

A. Digestion and Metabolism of Fats.
   1. Dietary fats are mainly triacylglycerols, commonly called triglycerides, esters of glycerol and fatty acids.
   2. Fats are digested with the help of enzymes called lipases.
   3. Once they are absorbed, the products of digestion are reassembled as triglycerides (esters of glycerol and fatty acids) that are attached to proteins for transport through the blood.
   4. Fats are stored in adipose tissue to serve as a protective cushion and as insulation.
   5. When needed for energy, fats are hydrolyzed to glycerol and fatty acids. Glycerol can be "burned" to produce energy.
B. Fats, Cholesterol, and Human Health.
   1. Saturated fats and cholesterol are associated with arteriosclerosis, which can result in heart attack and strokes.
   2. Saturated fats have a large proportion of saturated fatty acids (palmitic and stearic acid).
      a. Unsaturated fats (oils) have a large proportion of unsaturated fatty acids (linoleic and linolenic acids).

3. Cholesterol is transported in blood by water-soluble proteins.
    a. A lipoprotein is a complex of cholesterol or fat with protein and is classified by its density.
    b. Very-low-density lipoproteins (VLDL) transport triglycerides.
    c. Low-density lipoproteins (LDL) transport cholesterol to cells for use.
    d. High-density lipoproteins (HDL) transport cholesterol to the liver for eventual excretion.
    e. High levels of LDL increase risk of heart attack and stroke.
    f. Exercise increases levels of HDL.
    g. Fish oil tends to lower cholesterol and triglyceride levels.
4. Omega-3 fatty acids have carbon-to-carbon double bonds beginning at the third carbon from the end opposite the COOH group, the omega end.
    a. Adding omega-3 fatty acids to the diet leads to lower blood cholesterol and triglyceride levels.
5. Fats and oils containing C=C bonds can be hydrogenated by adding H_2 to the double bond.
    a. The arrangement around C=C bonds in naturally occurring fatty acid is *cis*.
    b. Hydrogenation of vegetable oils to produce semisolid fats is an important process in the food industry.
    c. During hydrogenation, the arrangement around some of the C=C bonds is changed from *cis* to *trans*.
    d. Like saturated fatty acids, *trans* fatty acids tend to raise blood levels of LDL cholesterol, which has negative health impacts.
6. Fats yield ~9 kcal of energy per gram.
7. Fats are used as fuel and help build and maintain parts of cells; they are stored as fuel reserves.
8. Health professionals recommend that no more than 30% of total calories come from fats and that no more than one-third of the fat should be saturated.

Answers to Self-Assessment Questions

1. b Fats are esters of glycerol and fatty acids.
2. a Among lipoproteins, LDLs carry cholesterol to the cells for use and are the ones that deposit cholesterol in arteries, leading to cardiovascular disease.
3. b $CH_3(CH_2)_5CH{=}C(CH_2)_7COOH$ is a monounsaturated fatty acid (it has one C=C bond).
4. d $CH_3(CH_2CH{=}CH)_3(CH_2)_7COOH$ is a polyunsaturated omega-3 fatty acid. Omega-3 fatty acids have a C=C bond that starts at the third carbon from the end opposite the –COOH group.
5. a $CH_3(CH_2)_{16}COOH$ is an unsaturated fatty acid because it has no C=C bonds.
6. d With respect to their shape, *trans* fatty acid molecules resemble saturated fatty acids.
7. c Most health professionals recommend that fat represent no more than 30% of our daily caloric intake and that no more than one-third of the fat intake (10%) should be saturated fat.

17.3 Proteins: Muscle and Much More

Learning Objectives: • List the essential amino acids, and explain why we need proteins.
• Describe some protein deficiency diseases and their causes.

A. Proteins are polymers of amino acids. They are needed to make muscles, hair, and enzymes.
B. Protein Metabolism: Essential Amino Acids.

1. Proteins are broken down in the digestive tract into their component amino acids, which are used to synthesize proteins for growth and repair of tissues.
    a. Excess protein can be used as an energy source, supplying ~4 kcal/g.
2. The human body can synthesize 11 of the 20 amino acids. The other 9 (called essential amino acids) must be supplied by the diet.
3. People whose diets fail to provide all the essential amino acids in the needed quantities may suffer from malnutrition even though adequate calories are provided in the diet.
4. Most protein from animal sources contains all the essential amino acids. Gelatin is an example of an inadequate animal protein because it contains almost no tryptophan and has only small amounts of threonine, methionine, and isoleucine.

C. Protein Deficiency in Young and Old.
1. Our requirement for protein is 0.8 g/kg of body weight.
    a. Kwashiorkor is a protein deficiency disease common in parts of Africa and in the United States, mainly among elderly people in nursing homes.

D. Vegetarian Diets.
1. Vegetarian diets are easier on the world's energy needs since it takes less energy to produce plant protein than animal protein.
2. Many ethnic foods provide good protein by combining a legume (deficient in methionine) with a cereal grain (deficient in tryptophan and lysine).
3. A strict vegetarian diet is also likely to be low in vitamin B_{12}, calcium, iron, and riboflavin. Milk, eggs, cheese, fish, and supplements can correct this deficiency.

Answers to Self-Assessment Questions

1. d The numbers of kcal/g provided by carbohydrate, fat, and protein, respectively, are 4, 9, and 4, so proteins supply about the same number of kcal/g as carbohydrates.
2. a Proteins are co-polymers of amino acids.
3. d An essential amino acid must be included in the diet because we are unable to produce them from other starting materials.
4. c Quinoa is the only nonanimal protein that is complete. Combinations of plant proteins such as brown rice and beans provide a complete protein, as do eggs. Peanuts are not animal proteins and are not complete.
5. b The daily protein requirement for an individual is ~0.8 g/kg weight.
   mass of protein = (65 kg body weight)(0.8 g protein/kg body weight) = 52 g protein
6. c One can get complete protein by combining whole wheat bread and peanut butter, which is the combination of a grain and a legume.
7. c Strict vegetarian diets are often deficient in vitamin B_{12} and iron.

17.4 Minerals, Vitamins, and Other Essentials

Learning Objective: • Identify the vitamins and the bulk dietary minerals, and state their function.

A. A variety of inorganic compounds and dietary minerals are necessary for proper growth and repair of body tissues.

B. Dietary Minerals.
1. Inorganic substances, called dietary minerals, represent ~4% of the weight of a human body.

2. Macrominerals: sodium, potassium, calcium, magnesium, chlorine, phosphorus, and sulfur.
   Ca: bones, teeth, blood clotting, milk formation
   P: bones, teeth, nucleic acids, cell membranes
   K: intracellular cation, muscle contraction
   S: amino acids methionine and cysteine
   Na: extracellular cation, fluid pressure
   Mg: enzyme cofactor, nerve impulses
3. Trace elements (iron, copper, and zinc) are needed in smaller amounts but are equally important.
4. Ultratrace elements: manganese, molybdenum, chromium, cobalt, vanadium, nickel, cadmium, tin, lead, lithium, fluorine, iodine, selenium, silicon, arsenic, boron.
   a. Fe: hemoglobin for O_2 transport.
   b. Iodine: proper functioning of thyroid.

C. The Vitamins: Vital, but not all are amines.
1. Vitamins are organic substances that our bodies need for good health but cannot synthesize; they must be included in the diet.
2. Some vitamins were discovered early because of vitamin-deficiency diseases such as scurvy (vitamin C deficiency) and beriberi (thiamine or vitamin B_1 deficiency). The first such compounds characterized were amines, hence the name "vitamin."
3. Vitamins are divided into two broad categories.
   a. Fat-soluble vitamins: A, D, E, and K.
   b. Water-soluble vitamins: B complex and C.
4. Large doses of fat-soluble vitamins can be toxic; excess water-soluble vitamins are excreted in the urine.
   a. Vitamins E and K are fat soluble but metabolized (not stored) and excreted.
5. The body can store fat-soluble vitamins for future use; water-soluble vitamins are needed almost daily.

D. Other Essentials: Fiber and Water.
1. Dietary fiber may be soluble or insoluble.
   a. Insoluble fiber is usually cellulose.
   b. Soluble fiber generally consists of sticky materials called *gums* and *pectins*.
2. Water.
   a. We need between 1.0 L and 1.5 L of water every day, in addition to the water we get from food.

Answers to Self-Assessment Questions

1. b About 4% of our body weight is minerals.
2. c Vitamins are organic molecules that are required in our diet to prevent specific diseases.
3. a Eating seafood is one way to increase your iodine intake.
4. b Iron in red blood cells is found complexed with the hemoglobin.
5. d Because much of it is located in the hemoglobin, which is part of the blood, iron is swiftly depleted from the body by blood loss.
6. b Most of the calcium and phosphorus in the body is located in the bones and teeth.
7. c People with high blood pressure are often advised to reduce their Na^+ intake.
8. d The fat-soluble vitamins are vitamins A, D, E, and K.
9. c Eating foods high in fiber helps fill you up without adding to caloric intake because your body isn't able to digest and metabolize the fiber.

17.5 Starvation, Fasting, and Malnutrition

Learning Objective: • Describe the effects of starvation, fasting, and malnutrition.

A. A body totally deprived of food soon uses up its glycogen reserves and needs to convert to fat metabolism.
   1. Increased dependence on stored fats for energy can result in ketosis, a condition characterized by the appearance of ketone bodies (acetoacetic acid, ß-hydroxybutyric acid, and acetone) in the blood and urine.
      a. Ketosis rapidly develops into acidosis; the blood pH drops, and oxygen transport is hindered, leading to depression and lethargy.
      b. Acidosis is also associated with diabetes.
   2. During the early stages of a total fast, the body will also break down its own structural proteins to try to meet its metabolic needs and to provide glucose to the brain.

B. Processed Food: Less Nutrition.
   1. Making white flour from wheat removes protein, minerals, vitamins, and fiber (bran).
   2. Fruit peels are rich in vitamins and fiber.
   3. Some vitamins are (partially) destroyed by heat; water-soluble vitamins are leached out and discarded in the cooking water.
   4. Over 90% of the food budget of an average U.S. family goes to buy processed foods.

Answers to Self-Assessment Questions

1. b Glycogen stores are depleted in about 1 day.
2. b In the intermediate stages of starvation (1 to 4 days), the body draws on its reserves of fat.
3. d In prolonged starvation (a few weeks), the body uses structural proteins.
4. a The average family spends 90% of its food budget on processed foods, leaving only 10% to spend on fresh fruits and vegetables, meats, and dairy products.

17.6 Flavorings: Spicy and Sweet

Learning Objectives: • List commonly used flavorings in food. • Describe the historic impact of our desire for a variety of flavors and aromas, and of our search for sugar substitutes.

A. Spices, Herbs, and Flavorings.
   1. Spice cake, soft drinks, gingerbread, sausage, and many other foods depend on spices and other additives for most of their flavor.
   2. Natural spices come from plant seeds, roots, or bark and from herbs; natural flavors can be extracted from fruits and plant materials.
   3. Major components of natural and synthetic flavors are often identical, although natural flavorings may be more complicated because they usually contain a wide variety of compounds.

B Other Natural and Artificial Sweeteners.
   1. Stevia is a South American herb that has been used as a sweetener for many years.
      a. Stevia is 30 times sweeter than sucrose.
   2. Cyclamates were banned in the United States in 1970 because they were shown to cause cancer.
   3. In 1977, saccharin was shown to cause cancer in laboratory animals. The FDA moved to ban it, but the move was blocked by the U.S. Congress.

4. Aspartame, approved in 1981, a dipeptide (the methyl ester of aspartylphenylalanine), is another low-calorie sweetener.
5. Acesulfame K and sucralose (Splenda) are low-calorie sweeteners that can survive the high temperatures of cooking.
6. Glycerol and propylene glycol are sweet; they are used as humectants (moistening agents).
7. Sorbitol and xylitol, both polyhydroxy alcohols, are sweet. They have the same caloric content as sugar but don't cause tooth decay.

C. Flavor Enhancers.
1. Some substances, not particularly flavorful themselves, enhance other flavors.
   a. Sodium chloride (table salt) is an example.
2. Monosodium glutamate (MSG) imparts a meaty flavor to foods that contain only small amounts of meat.
   a. Overindulgence in foods high in MSG may cause neurological problems, and it may be teratogenic (cause birth defects when eaten in large amounts by pregnant women).

Answers to Self-Assessment Questions

1. d Monosodium glutamate (MSG) stimulates umami, a protein-like taste.
2. a Natural and artificial vanilla contain the same compound, vanillin, that is responsible for giving vanilla its flavor.
3. c Organic molecules with ester functional groups are responsible for many fruit and other tastes and odors.
4. b Stevia is a natural product, not an artificial one.
5. b Monosodium glutamate is the sodium salt of glutamic acid, an amino acid.

17.7 Other Food Additives: Beneficial or Dangerous?

Learning Objectives: • Identify the beneficial additives in foods, and those that are somewhat controversial. • Define the GRAS list and food additives allowed under its provisions.

A. Food additives are substances other than basic foodstuffs that are intentionally put in products for various reasons related to production, processing, packaging, or storage.
1. In the United States, food additives are regulated by the Food and Drug Administration (FDA).

B. The GRAS List.
1. Under the 1958 food additives amendment, the FDA has established a list of long-used additives "generally recognized as safe" (GRAS).
2. Improved instruments and better experimental design have led to the banning of some GRAS substances on reexamination

C. Additives That Improve Nutrition.
1. Potassium iodide was the first nutrient supplement approved.
   a. Iodine is lacking in foods in some land regions; small amounts of potassium iodide (KI) are added to table salt to prevent goiter (enlargement of the thyroid gland).
2. Vitamin B_1 (thiamine) is added to polished rice to prevent beriberi.
3. Iron, in the form of ferrous carbonate, is added to flour for enrichment.
4. Vitamin C is added to fruit drinks to match the vitamin content found in real fruit juices.
5. Vitamin D is added to milk to prevent rickets.

6. Vitamin A is added to margarine to match the quantity of vitamin A found in butter.
7. Processed foods, even those with added nutrients, seldom match the nutritional value of fresh foods because only some of the nutrients lost in processing are replaced.

D. Additives That Retard Spoilage.
1. Propionic acid, benzoic acid, sorbic acid, and salts of these acids retard spoilage by inhibiting the growth of molds.
2. Sodium nitrite inhibits the growth of bacteria, including those that cause botulism, and is used to maintain the pink color of smoked hams and bologna.
    a. Only 10% of the amount used as a color enhancer is needed to inhibit the growth of botulism.
    b. Nitrites react with HCl in stomach acid and with amines to form carcinogenic nitrosamines, but this reaction is inhibited by ascorbic acid (vitamin C).
3. Sulfur dioxide and sulfite salts are used as disinfectants and preservatives, as bleaching agents, and as a means to prevent foods from turning brown on standing. These substances cause severe allergic reactions in some people.

E. Antioxidants: BHA and BHT.
1. Antioxidants prevent foods containing fats and oils from becoming rancid.
2. BHA and BHT are free-radical reaction inhibitors.
    a. Rancidity involves the reactions of fats and oils with oxygen to form free radicals.
    b. These radicals react with other fat molecules to form new free radicals in a chain reaction.
    c. BHT and BHA react with the radicals, halting the chain process.
3. BHT and BHA cause allergic reactions in some people and fetal abnormalities in rats. Overall, though, BHT has been shown to increase the life span of rats.
4. Vitamin E is a natural antioxidant. Lack of vitamin E causes sterility in rats, but human diets almost always have sufficient vitamin E.
    a. The action of vitamin E as an antioxidant is similar to that of BHT.

F. Color Additives.
1. Some foods are naturally colored.
    a. β-Carotene from carrots (provitamin A)—yellow
    b. Beet juice—red
    c. Grape-hull extract—blue

Answers to Self-Assessment Questions

1. c A manufacturer must provide the FDA with evidence that the ingredient is safe for its intended use.
2. d Sodium nitrite is added to food to inhibit formation of the botulin toxin.
3. c Calcium propionate is added to bread to inhibit the growth of mold.
4. d Antioxidants trap free radicals to inhibit the spoilage of foods in the presence of oxygen.
5. b BHT is an example of an antioxidant and is added to foods containing fats to protect them from becoming rancid.
6. b A total of seven colors have been found to be safe to use in foods, only six of which are in common use today.
7. b Vitamin C, vitamin E, BHA, BHT, and propyl gallate are all antioxidants that are added to prevent fat-containing foods from becoming rancid.

17.8 Problems with Our Food

Learning Objective: • Identify and describe some of the main problems with harmful substances in our food.

A. Many plants and animals are poisonous, including some mushrooms, rhubarb leaves (oxalic acid), and puffer fish.
   1. One of the most poisonous substances known is the botulin toxin, produced by anaerobic bacteria in improperly canned food.

B. Carcinogens.
   1. Carcinogens occur naturally in food.
      a. Charcoal-broiled foods contain 3, 4-benzopyrene, a carcinogen.
      b. Cinnamon and nutmeg contain safrole, a carcinogen.
      c. Aflatoxins are produced by molds that grow on peanuts and grains.

C. Food Contaminants.
   1. Food contaminants come from many sources and can occur at any step between producer and consumer.
      a. For example, antibiotics in animal feed often show up in the meat we eat.
      b. Today's food contamination problems are usually biological, not chemical.
      c. Bacteria from fecal material sometimes found in raw meat or dairy products and, occasionally, in raw fruits and vegetables are responsible for food poisoning.
      d. Viruses such as hepatitis A cause about 50% of all outbreaks of foodborne gastroenteritis in the United States and can be spread by infected food workers.
      e. Parasites are rare in developed countries but are common in developing countries and can be transmitted by eating raw or undercooked pork and fish.

D. The Future of Food Waste—A Green Chemistry Perspective
   1. A study by the United Nations Food and Agriculture Organization discovered that, globally, we generate 1.3 billion tons of food waste every year—one-third of all the food we grow.
   2. In most cases, these food supply chain wastes (FSCW) currently go to landfills, where their decomposition affects soil quality, pollutes local water supplies, and releases the potent greenhouse gas methane.
   3. FSCW is a great example of an economically viable renewable material; instead of paying landfill taxes and fees to dispose of it as waste, the material can be used as a cheap feedstock for chemical processes.
      a. One way to gain value from FSCW is to turn it into a fuel using anaerobic digestion, a technique that converts mixed biomass waste into a fuel and fertilizer through microbial processes.
      b. Alternatively, chemists have used waste straw from agricultural processes to make bioboard using silicate binders that can be used as a replacement for woodchip boards in furniture.

E. A World Without Food Additives.
   1. Food additives seem to be a necessary part of modern society.
      a. Use of some food additives poses potential hazards, but the major problem with our food supply is still contamination by harmful microorganisms, rodents, and insects.

Answers to Self-Assessment Questions

1. b Aflatoxins are carcinogens, compounds produced by molds that can cause cancer.
2. c According to the CDC, about 48 million illnesses and 3000 deaths in the United States result each year from food poisoning caused by bacteria, viruses, and parasites.
3. c Microorganisms pose the major source of food contamination.

LEARNING OBJECTIVES

You should be able to . . .

1. Identify dietary carbohydrates, and state their sources and function. (17.1)
2. Identify dietary lipids, and state their function. (17.2)
3. List the essential amino acids, and explain why we need proteins. (17.3)
4. Describe some protein deficiency diseases and their causes. (17.3)
5. Identify the vitamins and the bulk dietary minerals, and state their functions. (17.4)
6. Describe the effects of starvation, fasting, and malnutrition. (17.5)
7. List commonly used flavorings in food. (17.6)
8. Describe the historic impact of our desire for a variety of flavors and aromas, and of our search for sugar substitutes . (17.6)
9. Identify the beneficial additives in foods and those that are somewhat controversial (17.7)
10. Define the GRAS list and food additives allowed under its provisions. (17.7)
11. Identify and describe some of the main problems with harmful substances in our food. (17.8)
12. Explain how food supply chain waste can be used as a resource in the chemical industry
13. Describe how green chemistry principles are advantageous in recovering essential oils and bioactive food components from waste.

DISCUSSION

We are now ready to apply our knowledge of bonding and of molecules to the chemicals that we eat—to food. Keep in mind that a chemical compound—regardless of where it came from or how it was made—has a constant composition, structure, and properties. With these chemical principles in mind, you can go a long way toward seeing through advertising claims, fad diets, and other food-related phenomena.

Food additives serve a variety of functions. Having a knowledge of chemistry can help you to understand what the additives are and how they work. Chemistry alone cannot determine whether the benefit obtained is worth the risk involved. Whether or not an additive should be used involves a value judgment. Knowledge of chemistry may help you, however, to make a more rational judgment.

ANSWERS TO ODD-NUMBERED REVIEW QUESTIONS AND SOLUTIONS FOR ODD-NUMBERED END-OF-CHAPTER PROBLEMS

Review Questions

1. Carbohydrates are used primarily as an energy source. They also provide thermal insulation and protect organs.

3. Vitamins are organic; minerals are inorganic.

5. Taking an excess of a fat-soluble vitamin is more dangerous than taking an excess of a water-soluble vitamin. The unused portion of the fat-soluble vitamin will be stored and accumulated in adipose tissue, while the unused portion of the water-soluble vitamin will be excreted in the urine.

7. Food additives are substances that are present in food as a result of some aspect of production, processing, packaging, or storage.

9. MSG is monosodium glutamate, a sodium salt of glutamic acid, and is used as a flavor enhancer.

11. Vitamin E (α-tocopherol) is a fat-soluble antioxidant. Vitamin C, which is water soluble, is also an antioxidant.

13. Aspartame is the methyl ester of the dipeptide aspartylphenylalanine.

15. Some food additives are added to increase the nutritional value of the substance; others are added to enhance color and flavor; and yet others to retard spoilage, to ripen or prevent ripening, or to control moisture levels.

Problems

17. The answer is a. A molecule of water is removed between the C-OH group of one monosaccharide and the C-OH group of a second monosaccharide, producing a C-O-C linkage to connect the two portions to form a disaccharide.

19. Starch is a polymer of glucose. The glucose monomers are released from the chain when starch is hydrolyzed.

21. When fat is digested, the fat is hydrolyzed, producing a combination of fatty acids, glycerol, and mono- and diglycerides.

23. Most animal fats are solids. Most vegetable oils are liquids. The fatty acids making up animal fats generally have fewer C=C bonds than the fatty acids making up vegetable oils.

25. An adequate (or complete) protein is one that supplies all of the essential amino acids in quantities needed for the growth and repair of body tissues. Proteins from animal sources (meat, eggs, and milk) are almost all complete proteins (gelatin is not a complete protein).

27. Corn is likely to be lacking in lysine and tryptophan, while beans are likely to be deficient in methionine.

29. Yes, the combination of wheat, barley, beans, lentils, and millet will provide all the essential amino acids because it represents a combination of grains and beans.

31. a. Iodine is necessary for the thyroid gland.
    b. Iron is necessary for the proper functioning of the oxygen-transporting compound, hemoglobin.
    c. Calcium is necessary for proper development of bones and teeth, coagulation of the blood, and maintenance of the heartbeat rhythm.
    d. Sodium is critical in controlling the exchange of fluids between cells and plasma.

33. Pantothenic acid and biotin are water soluble, and phylloquinone is fat soluble.

35. Ascorbic acid is another name for Vitamin C; calciferol is another name for Vitamin D_2; cyanocobalamin is known as Vitamin B_{12}; retinol is known as Vitamin A; and tocopherol is known as Vitamin E.

37. a. Pellagra is caused by the deficiency of niacin (vitamin B_3).
    b. Beriberi is caused by the deficiency of thiamin (vitamin B_1).
    c. Pernicious anemia is caused by the deficiency of cyanocobalamin (vitamin B_{12}).

39. a. Vitamin A (calciferol) is fat soluble.
    b. Niacin is Vitamin B_3 and, like all the B vitamins, is water soluble.
    c. Riboflavin, otherwise known as Vitamin B_2, is water soluble.
    d. Tocopherol is Vitamin E and is fat soluble.

41. The answer is d. Both BHA and BHT are antioxidants that prevent fats from becoming rancid.

43. a. BHA is an antioxidant.
    b. FD&C Blue No. 2 is a colorant.
    c. Saccharin is an artificial sweetener.

45. The answer is d. None of the statements are true about a dietary supplement labeled "natural."

47. Antioxidants are reducing agents, usually free-radical scavengers added to foods to prevent fats and oils from becoming rancid. An example is when a fat molecule reacts with oxygen to form a free radical, then the free radical reacts with another fat molecule, and the reaction repeats indefinitely. Antioxidants stop the chain reaction by reacting with the free radical. Vitamins C and E are naturally occurring antioxidants. BHA and BHT are two examples of synthetic antioxidants.

49. Number of kcal from carbohydrates = (14 g carbohydrates)(4 kcal/g carbohydrate) = 56 kcal
    Percent of calories from carbohydrates = (56 kcal/110 kcal)(100%) = 51%
    Number of kcal from fat = (2.5 g fat)(9 kcal/g fat) = 23 kcal
    Percent of calories from fats = (23 kcal/110 kcal)(100%) = 21%
    Number of kcal from protein = (7 g protein) (4 kcal/g protein) = 28 kcal
    Percent of calories from proteins = (28 kcal/110 kcal)(100%) = 25%
    Note: The fact that these percentages add up to slightly less than 100% (51% + 21% + 25% = 97%) indicates that the kcal equivalents per gram of carbohydrate, fat, and protein are approximations rather than exact numbers.

51. kcal from fat = (5 g fat)(9 kcal/g fat) = 45 kcal
kcal from carbohydrate = (6 g carbohydrate)(4 kcal/g carbohydrate) = 24 kcal
kcal from protein = (3 g protein)(4 kcal/g protein) = 12 kcal
Total kcal/serving = 45 kcal + 24 kcal + 12 kcal = 81 kcal
Percent of kcal/serving from fat = (45 kcal/81 kcal)(100%) = 56%

53. 25% of 1200 kcal = 300 kcal
Mass of fat = (300 kcal)(1 g fat/9 kcal) = 33 g of combined saturated and unsaturated fat
Mass of saturated fat = (33 g fat)(30 g saturated fat/100 g fat) = 11 g saturated fat

55. Omega-6 fatty acids are c. linoleic acid, and e. arachidonic acid.

57. According to the label, one serving provides 160 Calories, 70 of which come from fat. The percent of calories provided by fat is (70 cal/160 cal)(100%) = 44%.
Number of kcal fat in whole milk = (8 g fat)(9 kcal/g fat) = 72 kcal
Percent of energy from fat = (72 kcal/160 kcal)(100%) = 45%

59. Nutritional guidelines indicate that no more than 10% of one's daily diet should come from saturated fats. This means that for someone following a 2200-cal/day diet, a maximum of 220 cal should come from saturated fat. According to the label, one serving (1 cup) of whole milk contains 5 g of saturated fat. This supplies (5 g)(9 cal/g) = 45 cal of energy which is the equivalent of (45 cal/220 cal)(100%) = 20% of the person's daily allotment of saturated fat.

61. a. (13 g sugar)(4 kcal/g sugar) = 52 kcal. (52 kcal/90 kcal)(100%) = 58% kcal from sugar
b. (9 g protein)(4 kcal/g protein) = 36 kcal. (36 kcal/90 kcal)(100%) = 40% kcal from protein. The sugar in milk is lactose.

63. Water-soluble vitamins are readily excreted and so are not stored in the body for long. Because riboflavin, niacin, and cyanocobalamin are water soluble, they are most likely to be needed on a daily basis.

65. Carbon–carbon double bonds are more reactive than carbon–carbon single bonds. Because their fatty acids have more C=C bonds than is the case for fats, oils are much more likely to spoil (become rancid) faster than solids, especially with a lack of effective refrigeration.

67. Adding beta-carotene to foods serves two purposes. First, it adds yellow color to the food, and second, beta-carotene is a precursor to vitamin A, so adding it increases the vitamin content of the food as well.

69. a. The answer is 4. Lycopene is a red pigment that can be found in tomatoes.
b. The answer is 3. Ascorbic acid (Vitamin C) can be found in lemons and grapefruit.
c. The answer is 2. Vitamin A can be found in carrots and squash.
d. The answer is 1. Chlorophyll can be found in algae and green vegetables.

71. Food waste can be converted to fuels through anaerobic digestion by microorganisms.

CHAPTER

18

Drugs

Chemical Cures, Comforts, and Cautions

CHAPTER SUMMARY

18.1 Scientific Drug Design

Learning Objectives: • Identify common drugs as natural, semisynthetic, or synthetic. • Define *chemotherapy*, and explain its origin.

A. Chemists make semisynthetic drugs by modifying molecules from natural sources to improve their properties.
   1. Examples include conversion of salicylic acid from willow bark to acetylsalicylic acid (aspirin), conversion of morphine from opium poppies to heroin, and conversion of lysergic acid from ergot fungus to LSD.

B. There are completely synthetic drugs on the market.

C. Humans in all cultures have used drugs since prehistoric times.

D. Paul Ehrlich (1854–1915) realized that certain chemicals were more toxic to disease organisms than to human cells and could therefore be used to control or cure infectious diseases.
   1. Ehrlich coined the term *chemotherapy* ("chemical therapy")
   2. Ehrlich was awarded the Nobel Prize in Physiology or Medicine in 1908.

Answers to Self-Assessment Questions

1. a Quinine was extracted from Cinchona, so it is a natural product.
2. a Cyclosporins are natural substances extracted from Streptomyces organisms.
3. b Aspirin is a semisynthetic drug prepared by a modification of the salicylic acid extracted from willow bark.
4. d The term *chemotherapy* was first used by Paul Ehrlich to mean "chemical therapy."

18.2 Pain Relievers: From Aspirin to Oxycodone

Learning Objectives: • List common over-the-counter analgesics, antipyretics, and anti-inflammatory drugs, and describe how each works. • Name several common narcotics, describe how each functions, and state its potential for addiction.

A. Aspirin (acetylsalicylic acid) was introduced in 1893 as one of the first successful synthetic pain relievers and has become the largest-selling drug in the world.
   1. The substance originally isolated from willow bark was an effective analgesic (pain reliever), antipyretic (fever reduced), and anti-inflammatory, but it also caused considerable stomach distress and bleeding.
   2. Acetylsalicylic acid is a modified version of salicylic acid that has the desired properties but is gentler on the stomach.

B. Nonsteroidal Anti-inflammatory Drugs (NSAIDs).
   1. The NSAID designation distinguishes these drugs from the more potent steroidal anti-inflammatory drugs such as cortisone and prednisone.
   2. In addition to aspirin, other NSAIDs include ibuprofen (Advil®, Motrin®), naproxen (Aleve®), and ketoprofen (Orudis®).
      a. Acetaminophen, like aspirin, relieves minor aches and reduces fever but is not anti-inflammatory and is not an NSAID.
   3. NSAIDs relieve pain and reduce inflammation by inhibiting the production of prostaglandins, hormone-like lipids derived from a fatty acid.
      a. NSAIDs do not cure the source of the pain but only dull the pain.
      b. Inflammation is caused by an overproduction of prostaglandin derivatives, so inhibition of their synthesis reduces the inflammatory process.
      c. NSAIDs produce side effects such as stomach problems in some individuals.
      d. NSAIDs act as anticoagulants.
      e. NSAIDs reduce body temperature and fevers induced by pyrogens, compounds produced by and released from leukocytes (white blood cells) and others circulating.

C. How NSAIDs Work.
   1. Prostaglandins are produced from the cell membrane component arachidonic acid.
      a. The reaction is catalyzed by cyclooxygenase (COX) enzymes.
   2. Aspirin works by inhibiting the production of prostaglandins.
      a. Prostaglandins are responsible for sending pain messages to the brain.
      b. Inflammation results from overproduction of prostaglandin derivatives.
      c. The anti-inflammatory action of aspirin results from inhibition of prostaglandin synthesis.

D. Acetaminophen: A COX-3 Inhibitor.
   1. Acetaminophen (in Tylenol® among other products)
      a. It is an analgesic; it reduces pain.
      b. It is an antipyretic; it reduces fevers.
      c. It is not an anti-inflammatory, so it is of little use to arthritis sufferers.
      d. It does not promote bleeding, so it can be used by surgical patients.
      e. Overuse is linked to liver and kidney damage.

E. Chemistry, Allergies, and the Common Cold.
   1. Many cold medicines contain antihistamines to relieve symptoms of allergies, sneezing, itchy eyes, and runny nose.
   2. Allergens bind to surfaces of certain cells, triggering the release of histamine which causes allergic symptoms.

F. Combination Pain Relievers.
   1. Anacin®, Excedrin®, and other products are available in various formulations, some of which include caffeine and substances for pain relief, allergy symptoms, and cold and flu treatment.

G. Narcotics.
   1. Narcotics are used to treat pain that does not respond to OTC analgesics.
      a. Narcotics produce narcosis (stupor or general anesthesia) and analgesia (pain relief).
      b. In the United States, only those drugs that produce narcotic effects and are addictive are called narcotics.
   2. Several narcotics are products of opium, which comes from the opium poppy.
      a. Morphine was first isolated from opium in 1805 and was used as treatment of dysentery in the Civil War, leading to many instances of addiction.
      b. Morphine and other narcotics were placed under control of the federal government by the Harrison Act of 1914.
   3. Structural modifications of morphine produce narcotics with various physiological properties.
      a. Codeine, small amounts of which are naturally present in opium, is usually synthesized by methylating morphine.
         i. Codeine is less potent than morphine and less addictive.
      b. Heroin is produced by converting both OH groups of morphine to acetate esters.
         i. Heroin is more addictive, and the addiction is harder to cure than that of morphine.
   4. Deaths from heroin and other narcotics are usually attributed to overdoses, but they can also be caused by the quality of the drugs bought on the street.
   5. Oxycodone is a semisynthetic opioid related to codeine used for relief of moderate to severe pain.
   6. Hydrocodone, a synthetic narcotic, is combined with acetaminophen or another medication.
   7. Methadone, a synthetic narcotic, is used to treat heroin addiction.

H. Morphine Agonists and Antagonists.
   1. A molecule with morphine-like action is called a morphine agonist.
   2. A morphine antagonist inhibits the action of morphine by blocking hormone receptors.

I. Endorphins are naturally produced morphine-like substances.
   1. Endorphins are thought to be released during strenuous exercise and in response to pain.
   2. Capsaicin, the active compound in chili peppers, can also stimulate endorphin release.

Answers to Self-Assessment Questions

1. d An antipyretic is a fever reducer.
2. d NSAIDs act by inhibiting the production of prostaglandins.
3. a Acetaminophen does not act as an anticoagulant.
4. b Allergens trigger the release of histamines.
5. c Currently, there is no cure for the common cold.
6. b Small amounts of codeine occur naturally in opium. Although codeine resembles morphine in its action, it is less potent, has less tendency to induce sleep, and is less addictive.
7. a Heroin is a semisynthetic opiate produced by forming acetate esters from the phenolic OH groups in morphine.
8. b Methadone satisfies an addict's drug craving while, at the same time, allowing him or her to function in society.
9. a Endorphins are naturally produced substances whose effects resemble the effects of opiates.

18.3 Drugs and Infectious Diseases

Learning Objectives: • List the common antibacterial drugs, and describe the action of each. • Name the common categories of antiviral drugs, and describe the action of each.

A. Antibacterial Drugs.
1. A century ago, infectious diseases were the principal cause of death.
2. Antibacterial drugs have dramatically altered that situation.
3 Gerhard Domagk discovered sulfa drugs (the first antibacterial drugs) in 1935.
   a. Used extensively in World War II to prevent wound infections.
   b. Sulfa compounds inhibit the growth of bacteria by mimicking *para*-aminobenzoic acid (PABA), a nutrient needed by bacteria for proper growth.
   c. Bacteria mistake the sulfa drug (sulfanilamide) for PABA and produce molecules that cannot perform growth-enhancing functions.
4. Penicillin, the first antibiotic (a substance derived from molds or bacteria that inhibits the growth of other microorganisms), was discovered by Alexander Fleming in 1928.
   a. Florey and Chain purified penicillin for use in medicine.
   b. There are several penicillins that vary in structure and properties.
   c. Penicillin inhibits the synthesis of bacterial cell walls; without the walls, the cells collapse and die. (Human cells don't have cell walls.)
   d. Disadvantages of penicillin.
      i. Many people are allergic to it.
      ii. Many kinds of bacteria have developed strains that are resistant to penicillins.
5. Penicillins have been partially replaced by related compounds called cephalosporins.
   a. Keflex® is an example of a cephalosporin.
   b. Some bacterial strains are resistant to cephalosporins.
6. Tetracyclines: Broad-spectrum antibiotics effective against a wide variety of bacteria.
   a. The tetracycline antibiotics are characterized by four rings joined side to side.

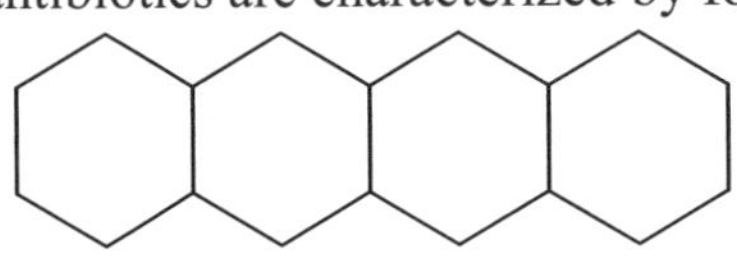

      i. Aureomycin® (chlorotetracycline) was isolated in 1948.
      ii. Terramycin® was isolated in 1950.
      iii. Tetracycline® (the parent compound) was isolated in 1953.
   b. Tetracyclines bind to bacterial ribosomes, inhibiting bacterial protein synthesis and blocking bacterial growth.
   c. Disadvantage: They can cause discoloration of teeth in children.
7. Fluroquinolone antibiotics were first introduced in 1986 and now represent about 33% of the global antibiotic market.
   a. Fluroquinolones have broad-spectrum activity and little in the way of side effects.
   b. Fluorquinolones are effective against bacteria with penicillin resistance.

B. Viruses and Antiviral Drugs.
1. Viral diseases (colds, flu, herpes, AIDS) cannot be cured with antibiotics.
2. Viral diseases are best dealt with by prevention. Vaccination prevents mumps, measles, and other dread viral diseases.

C. DNA Viruses and RNA Viruses.
1. Viruses are composed of nucleic acids and proteins.
   a. The genetic material may be DNA or RNA.

2. DNA viruses replicate in host cells and direct production of viral proteins, which together with the viral DNA assemble into new viruses.
    a. These viruses then invade other cells.
3. RNA virus replication is similar to DNA virus replication.
    a. Some RNA viruses, called retroviruses, synthesize DNA in host cells.
        i. The AIDS virus HIV is a retrovirus that destroys T cells, which protect the body from infections.

D. Antiviral Drugs.
1. There are three important kinds of antiviral drugs.
    a. Nucleoside analogs, which are nucleoside reverse transcriptase inhibitors (NRTIs).
    b. Nonnucleoside reverse transcriptase inhibitors (NNRTIs), which stop the reverse transcriptase from working properly to make more of the retrovirus.
    c. Protease inhibitors, which block the enzyme protease so that new copies of the retrovirus cannot infect new cells.
2. A few modestly effective antiviral drugs have been found.
    a. Acyclovir® helps control (but does not cure) chickenpox, shingles, cold sores, and genital herpes infections.
3. Antiretroviral drugs prevent the reproduction of retroviruses such as HIV and are used against AIDS.
    a. Azidothymidine® (AZT) slows the onslaught of AIDS.

E. Basic Research and Drug Development.
1. Gertrude Elion, George Hitchings, and James Black helped design antiviral drugs that block receptors in infected cells.

F. Prevention of Viral Diseases with Vaccination
1. For many viral diseases there are no effective drugs, but these diseases can often be prevented by vaccination, which causes the body to develop immunity to the virus microorganism.
    a. Polio outbreaks have been ended in the United States and throughout most of the world through use of polio vaccines.
    b. Smallpox vaccinations were given until 1972 when the tiny statistical risk from taking the vaccine was considered greater than the risk of acquiring the disease.
    c. The MMR (measles, mumps, rubella) vaccine was developed in the 1970s and is ordinarily given to children before they enter kindergarten.

Answers to Self-Assessment Questions

1. a Sulfa compounds inhibit the growth of bacteria by mimicking *para*-aminobenzoic acid (PABA), a nutrient needed by bacteria for proper growth.
2. a Pencillins inhibit the synthesis of bacteria cell walls.
3. a Antibiotics are substances derived from molds or bacteria that inhibit the growth of other microorganisms.
4. b Taking antibiotics for a stomach virus will increase the chance of developing drug-resistant bacteria.
5. b Viruses consist of a protein coat surrounding a core of nucleic acid.
6. d Most RNA viruses replicate in host cells by replicating RNA strands and synthesizing viral proteins.
7. c Retroviruses replicate in host cells by synthesizing DNA and forming new viruses.
8. c Sulfa drugs are antibacterial drugs, not antiretroviral drugs.
9. d The best treatment for viral diseases such as diphtheria, mumps, and smallpox is to prevent them through vaccinations.

18.4 Chemicals Against Cancer

Learning Objective: • Describe the action of the common types of anticancer drugs.

A. Antimetabolites: Inhibition of Nucleic Acid Synthesis.
   1. Anticancer metabolites block DNA synthesis, which blocks an increase in the number of cancer cells.
      a. Cancer cells that are dividing rapidly need large quantities of DNA and, therefore, are greatly affected by the DNA shortage.
         i. 6-Mercaptopurine substitutes for adenine and guanine and thus inhibits the synthesis of nucleotides incorporating adenine and guanine, slowing DNA synthesis.
         ii. 5-Fluorouracil and 5-fluorodeoxyuridine inhibit the formation of a thymine-containing nucleotide required for DNA synthesis.
         iii. Methotrexate is a folic acid antagonist that interferes with cellular reproduction and is used to help control leukemia.

B. Alkylating Agents: Turning Old Weapons into Anticancer Drugs.
   1. Alkylating agents can transfer alkyl groups to important biological compounds.
   2. Some alkylating agents are used against cancer.
      a. Mustard "gas" is a sulfur-containing blister agent used in chemical warfare in World War I.
      b. Nitrogen mustards were developed in about 1935 as chemical warfare agents.
      c. Nitrogen mustards such as cyclophosphamide are effective anticancer drugs.

   1. Cisplatin, an alkylating agent, binds to DNA and blocks its replication, inhibiting cell division.

C. Other Anticancer Agents.
   1. Alkaloids from vinca plants are effective against leukemia and Hodgkin's disease.
   2. Actinomycin from molds is used against Hodgkin's disease and other cancers.
      a. It binds to the double helix of DNA, blocking the formation of RNA on the DNA template. Protein synthesis is inhibited.

Answers to Self-Assessment Questions

1. a 6-Mercaptopurine mimics the base adenine.
2. b 6-Mercaptopurine mimics the base adenine, forming a false nucleotide and thus slowing DNA synthesis.
3. a Methotrexate acts as a folic acid antagonist, slowing cell division.
4. b Cisplatin inhibits cell division by binding to DNA and blocking its replication.
5. b Cyclophosphamide is an alkylating agent often used to treat leukemia and other cancers.
6. b Cyclophosphamide acts by cross-linking DNA chains and blocking replication.

18.5 Hormones: The Regulators

Learning Objectives: • Define the terms *hormone*, *prostaglandin*, and *steroid*, and explain the function of each. • List the three types of sex hormones, and explain how each acts and how birth control drugs work.

A. Hormones are chemical messengers produced in the endocrine glands.
   1. Hormones cause profound changes in parts of the body often far removed from the gland that secretes the substance.
   2. Table 18.2 in the text lists a variety of hormones and their physiological effects.

B. Prostaglandins: Hormone Mediators.
   1. Prostaglandins are hormone-like lipids synthesized from unsaturated fatty acids that contain 20 carbon atoms (such as arachidonic acid).
      a. Each prostaglandin has a five-membered ring in its structure.
   2. Prostaglandins act together with hormones to regulate smooth-muscle activity and blood flow.
   3. Medically, prostaglandins and derivatives are used to
      a. Induce labor.
      b. Lower or raise blood pressure.
      c. Inhibit stomach secretions.
      d. Relieve nasal congestion.
      e. Relieve asthma.
      f. Inhibit the formation of blood clots.
      g. Synchronize breeding in cattle.

C. Diabetes.
   1. The pancreas produces the hormone insulin, which allows the body to increase cellular usage of glucose.
   2. Diabetes occurs when the pancreas does not produce enough insulin (Type 1) or when the insulin is not properly used by the body (Type 2).

D. Steroids.
   1. All steroids have the same skeletal four-ring structure.

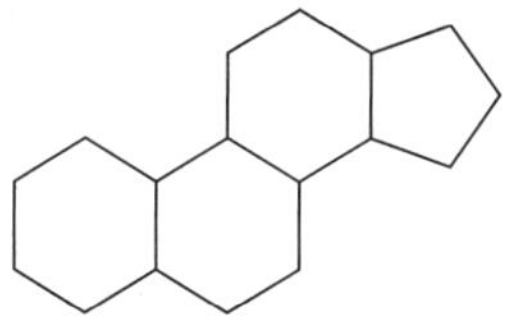

   2. Not all steroids have hormonal activity.
      a. Cholesterol is a steroid but not a hormone.
      b. Cortisol is a steroid hormone produced by the adrenal glands.
   3. Many natural and semisynthetic drugs are based on the steroidal ring system.
      a. Medically, cortisol and related compounds such as prednisone are anti-inflammatory drugs.

E. Sex Hormones.
   1. Sex hormones are steroids.
      a. Androgens are compounds that stimulate the development or control the maintenance of masculine characteristics and are secreted by the testes.
      b. Estrogens are compounds that control female sexual functions, such as the menstrual cycle and the development of breasts.
      c. Two important groups of female hormones are the estrogens and progesterones.
         i. Estrogens, produced in the ovaries, are female hormones.
         ii. Two important estrogens are estradiol and estrone.

2. Sex hormones are used therapeutically.

F. Chemistry and Social Revolution: The Pill.
   1. Progesterone is an effective birth control drug when injected.
   2. Synthetic analogs of progesterone, called progestins, are effective birth control drugs.
   3. Oral progestins incorporate an ethynyl group (−C≡CH).
   4. Oral birth control pills usually combine an estrogen (to regulate the menstrual cycle) with a progestin that signals a state of false pregnancy so that ovulation does not occur.

G. Emergency Contraceptives.
   1. Several products, called emergency contraceptives or morning-after pills, can be used to prevent pregnancy after unprotected intercourse.
   2. There are two kinds of emergency contraception pills (ECPs): combination pills with both estrogen and progestin (synthetic analogs of the natural substances), and pills containing only progestin.
   3. An interuterine device (IUD) is also used for emergency contraception.
   4. ECPs are not the same as the so-called medical abortion pills such as mifepristone (or RU-486), which works after a woman becomes pregnant and the fertilized egg has attached to the uterine wall.

H. Risks of Taking Birth Control Pills.
   1. Side effects are mostly minor.
   2. FDA advises all women who smoke, especially those over 40, to use some other method of contraception.

I. The Minipill.
   1. Minipills are available that contain only small amounts of progestin and no estrogen.
   2. Minipills are not quite as effective as the combination pills but have fewer side effects.

J. A Pill for Males?
   1. Women usually bear the responsibility for contraception because:
      a. It is easier to interfere with a once-a-month event (ovulation) than a continuous process (sperm production).
      b. In males, the pituitary hormones FSH and LG are required for continued production of sperm and of the male hormone testosterone.
      c. Several research groups have developed what may be a safe, effective, and reversible contraceptive for males, a pill containing progestin and testosterone.

Answers to Self-Assessment Questions

1. a Hormones are produced in the endocrine glands and act throughout the body.
2. a Prostaglandins are synthesized in the body from arachidonic acid.
3. a To date, there are no male contraceptives available to the general public.
4. b The gonads are stimulated to produce sex hormones and eggs or sperm by FSH and LH.
5. c The "pregnancy hormone" is progesterone.
6. c The group that makes a birth control drug effective orally is the ethynyl group (HC≡C–).

18.6 Drugs for the Heart

Learning Objective: • Describe the action of four types of drugs used to treat heart disease.

A. Major diseases of the heart and blood vessels include
   1. Ischemic ("lacking oxygen") coronary artery disease.
   2. Heart arrhythmias (abnormal heartbeat).
   3. Hypertension (high blood pressure).

4. Congestive heart failure.

B. Lowering Blood Pressure.
   1. Hypertension, or high blood pressure, is defined as pressure, in mmHg above 140/90; normal blood pressure is defined as less than 120/80.
   2. Hypertension is the most common cardiovascular disease, affecting one in four adults in the United States.
   3. There are four major categories of drugs for lowering blood pressure.
      a. Diuretics, which cause the kidneys to excrete more water, thus lowering the blood volume.
      b. Beta blockers, which slow the heart rate and reduce the force of the heartbeat.
      c. Calcium channel blockers, which are vasodilators, inducing muscles around the blood vessels to relax.
      d. Angiotensin-converting enzyme (ACE) inhibitors, which inhibit the action of an enzyme that causes blood vessels to contract.

C. Normalizing Heart Rhythm.
   1. Tachycardia (too rapid heartbeat) and abnormal heart rhythms are quite common.
   2. Atrial fibrillation, a type of irregular heartbeat, is so common that defibrillator devices are now available on many airplanes and in various other public places.

D. Treating Coronary Artery Disease.
   1. Angina pectoris, a common symptom of coronary artery disease, is caused by an insufficient supply of oxygen to the heart, usually due to partial blockage of the coronary arteries by lipid-containing plaque (arteriosclerosis).
   2. Medical treatment usually involves dilation of the blood vessels to increase blood flow and slowing the heart rate to decrease its demand for oxygen.
   3. Some drugs used to treat high blood pressure are effective, as are some organic nitro compounds, especially amyl nitrite and nitroglycerin.
      a. These compounds act by releasing nitric oxide (NO), which relaxes constricted vessels.
   4. The foxglove plant was used by ancient Egyptians and Romans to treat heart failure.
      a. Foxglove contains a mixture of glycosides, one of which is digoxin which produces the steroid digitoxigenin, which affects the rhythm and strength of the heart muscles' contractions.
      b. Digoxin is still used to treat patients with heart failure.

Answers to Self-Assessment Questions

1. a Amphetamines are central nervous system stimulants, not drugs to control blood pressure. See Section 18.7.
2. d Increasing lipid plaque deposition in blood vessel would be a cause of heart problems, not a cure for those problems.
3. c Nitric oxide is released by amyl nitrite and nitroglycerin to relax the smooth muscles in blood vessels.

18.7 Drugs and the Mind

Learning Objectives: • Explain how the brain amines norepinephrine and serotonin affect the mind and how various drugs change their action. • Identify some stimulant drugs, depressant drugs, and psychotropic drugs, and describe how they affect the mind.

A. Psychotropic drugs affect the mind; there are three classes of these drugs:

1. Stimulants increase alertness, speed mental processes, and elevate the mood.
    a. Amphetamines, caffeine, and cocaine are stimulants.
2. Depressants reduce the level of consciousness and the intensity of reactions to environmental stimuli.
    a. Ethanol, barbiturates, opiates, and tranquilizers are depressants.
3. Hallucinogenic (psychotomimetic, psychedelic) drugs alter qualitatively a person's perception of the surroundings.
    a. LSD, mescaline, and marijuana are hallucinogens.

B. Chemistry of the Nervous System.
1. Nerve cells (neurons) carry messages between the brain and other parts of the body.
    a. Axons of a nerve cell can be very long; however, the nerve impulse must be transmitted to the next dendrite across short fluid-filled gaps (synapses) via chemical messengers called *neurotransmitters*.
    b. Neurotransmitters determine to a large degree how you think, feel, and move about.
    c. Each type of neurotransmitter binds to a specific type of receptor site to complete the intended action.
2. Many drugs (and poisons) act either by blocking or mimicking the action of these natural neurotransmitters.
3. Many neurotransmitters are amines.

C. Biochemical Theories of Brain Diseases.
1. Epinephrine, often called adrenaline, is secreted by the adrenal glands when a person is under stress or is frightened.
2. Norepinephrine (NE) is a relative of epinephrine and is synthesized from the amino acid tyrosine.
    a. NE causes euphoria or, in large excess, a manic state.
3. Serotonin, a brain amine, is involved in sleep, appetite, memory, learning, sensory perception, mood, sexual behavior, and regulation of body temperature.
4. NE and related compounds fall into several general categories:
    a. NE agonists, or drugs that enhance or mimic the action of NE, are stimulants.
    b. NE antagonists, or drugs that block the action of NE, slow down various processes.
    c. Serotonin agonists are used to treat depression, anxiety, and obsessive-compulsive disorder; serotonin antagonists are used to treat migraine headaches and to relieve the nausea caused by cancer chemotherapy.

D. Brain Amines and Diet: You feel what you eat.
1. Serotonin is synthesized from the amino acid tryptophan.
    a. A high-carbohydrate meal allows maximum tryptophan to reach the brain (where it is converted to serotonin).
    b. Protein-rich diets lower the level of serotonin in the brain.
2. Norepinephrine is synthesized from the amino acid tyrosine.

E. Right-Handed and Left-Handed Molecules.
1. Stereoisomers are isomers having the same structural formula but differ in the arrangement of atoms or groups of atoms in three-dimensional space.
2. Enantiomers are nonsuperimposable mirror images.
    a. Enantiomers have a chiral carbon, a carbon atom to which four different groups are attached.
    b. Enantiomers fit enzymes differently, and they have different effects.

F. Anesthetics.
1. Anesthetics are substances that cause lack of feeling or awareness.
    a. General anesthetics act on the brain to produce unconsciousness and general insensitivity to pain.
    b. Local anesthetics cause loss of feeling in a part of the body.

2. Diethyl ether (ether) was the first general anesthetic (1846).
   a. Ether is relatively safe but causes nausea.
   b. Ether is highly flammable.
3. Nitrous oxide is a quick-acting anesthetic. It is administered with oxygen to prevent brain damage.
4. Chloroform was once used as an anesthetic.
   a. It has a narrow safety margin; the effective dose is close to the lethal dose.
   b. It causes liver damage.
   c. It can react with oxygen to form deadly phosgene gas.
   d. It is nonflammable.
5. Modern anesthetics include fluorine-containing compounds such as sevoflurane, desflurane, and isoflurane.
   a. These compounds are nonflammable and are relatively safe for the patient.
   b. Women who work in operating rooms where halothane is used have higher rates of miscarriage than the general population.
6. Modern surgical practice often combines the following:
   a. A tranquilizer to decrease anxiety.
   b. An intravenous anesthetic to produce unconsciousness quickly.
   c. A narcotic pain medication to block pain.
   d. An inhalant anesthetic to provide insensitivity to pain and keep the patient unconscious, often combined with oxygen and nitrous oxide to support life.
   e. A relaxant to relax the muscles and make it easier to insert a breathing tube.

G. Local Anesthetics.
1. Local anesthetics render one part of the body insensitive to pain but leave the patient conscious.
   a. Cocaine was the first local anesthetic.
2. Many local anesthetics are ester derivatives of, or related to, *p*-aminobenzoic acid.
   a. Procaine (Novocain) was introduced in 1905.
   b. Lidocaine and mepivicaine are widely used today.

H. Dissociative Anesthetics: Ketamine and PCP.
1. Ketamine is a dissociative anesthetic because it disconnects a person's perceptions from his or her sensations.
   a. Ketamine is widely used in veterinary medicine.
2. PCP (phencyclidine) is a dangerous drug but has found use as an animal tranquilizer.
   a. PCP is fat soluble. Stored in body fat, it is mobilized when fat is metabolized, causing flashbacks.

I. Depressant Drugs.
1. Ethyl alcohol, obtained from the fermentation of the sugars in fruits and grains, has been used by humans for centuries.
   a. Although people sometimes think of alcohol as a stimulant, it is actually a depressant, slowing down physical and mental activity.
   b. The mechanism of intoxication is not well known, but researchers have found that ethanol disrupts receptors for two neurotransmitters—one that inhibits impulsiveness and the other that excites certain nerve cells.

J. Barbiturates.
1. Barbiturates are cyclic amides that act on GABA receptors and can be used to produce mild sedation, deep sleep, or even death.
2. While more than 2500 barbiturates have been synthesized, only a few have found widespread use in medicine.

a. Pentobarbital (Nembutal®) is a short-acting hypnotic drug used to calm anxiety.
b. Phenobarbital (Luminal®) is a long-acting barbiturate used as an anticonvulsant for people suffering from seizure disorders such as epilepsy.
c. Thiopental (Pentothal®) is used as an anesthetic.
3. Large doses of barbiturates serve as sleeping pills.
4. Barbiturates are especially dangerous when ingested along with ethyl alcohol.
a. The combination produces an effect 10 times greater than the sum of the effects of two depressants, called synergism, or a synergistic effect.
5. Barbiturates, like ethanol, are intoxicating and strongly addictive.
6. Barbiturates are cyclic amides that resemble thymine.

K. Antianxiety Agents.
1. Antidepressant drugs are now the most commonly prescribed drugs.
2. One class of antianxiety drugs (anxiolytics) is the *benzodiazepines*, compounds that feature seven-member heterocyclic rings.
a. Diazepam (Valium®) is a classic antianxiety agent.
b. Clonazepam is an anticonvulsant as well as an antianxiety drug.
c. Lorazepam is used to treat insomnia.
3. The first antipsychotic drugs, sometimes called *major tranquilizers*, were compounds called phenothiazines.
a. Phenothiazines act in part as dopamine antagonists, blocking postsynaptic receptors for dopamine, a neurotransmitter important in the control of detailed motion, memory and emotions, and exciting brain cells
4. A second generation of antipsychotics, called *atypical antipsychotics*, includes drugs used to treat schizophrenia, acute manic episodes of bipolar disorder, and depression.
a. These drugs have served to greatly reduce the number of patients confined to mental hospitals.
5. The oldest class of antidepressants is the tricyclic antidepressants that block reabsorption of neurotransmitters such as norepinephrine and serotonin.
a. These drugs have serious side effects and have been largely replaced by newer drugs.
6. Antidepressants commonly prescribed today include *selective serotonin-reuptake inhibitors* (SSRIs), which are prescribed for anxiety syndromes such as panic disorder, obsessive-compulsive disorders, and premenstrual syndrome.

L. Stimulant Drugs.
1. Amphetamines, a variety of synthetic amines related to β-phenylethylamine, are similar in structure to epinephrine and norepinephrine.
a. Amphetamine and methamphetamine are inexpensive and have been widely abused.
b. Methamphetamine has a more pronounced psychological effect than amphetamine and is readily prepared from an antihistamine and household chemicals.
c. Amphetamine exists as enantiomers.
i. Benzedrine® is a mixture of the two isomers.
ii. Dexedrine® is the pure dextro isomer.

M. Cocaine, Caffeine, and Nicotine.
1. Cocaine is obtained from the leaves of a shrub that grows almost exclusively on the eastern slopes of the Andes Mountains.
a. Cocaine arrives illegally in this country as broken lumps of the free base (crack cocaine).
b. Cocaine acts by preventing the neurotransmitter dopamine from being taken back up from the synapse after it is released by nerve cells and so the cells fire wildly, depleting the dopamine supply.
c. Cocaine is a stimulant that increases stamina and reduces fatigue, but the stimulant effects are short-lived, followed by depression.

2. Coffee, tea, and cola soft drinks naturally contain caffeine, a mild stimulant.
    a. An effective dose of caffeine is about 200 mg.
3. Nicotine is a drug found in smoking and chewing tobacco.
    a. The lethal dose (when injected) for humans is estimated at about 50 mg.
    b. Nicotine is powerfully addictive.

N. Hallucinogenic Drugs.
1. Hallucinogenic drugs are consciousness-altering substances that induce changes in sensory perception, qualitatively altering the way users perceive things.
    a. LSD is a semisynthetic drug that is a powerful hallucinogen whose powers were discovered by Hofmann in 1943.
2. LSD is a hallucinogen related to lysergic acid and other ergot fungus alkaloids.
3. As little as 10 μg of this powerful drug can cause hallucinations.
4. Psilocybin and mescaline are compounds found in mushrooms and peyote cactus, respectively. The effects of these drugs are shorter-lasting than those of LSD.

O. Marijuana.
1. Marijuana is the leaves, flowers, seeds, and small stems of the *Cannabis sativa* plant.
2. The principal active ingredient is tetrahydrocannabinol (THC).
3. Marijuana reduces pressure in the eyes of glaucoma patients and relieves nausea of cancer patients undergoing radiation and chemotherapy.
4. Marijuana is now legal in several states in the United States.

Answers to Self-Assessment Questions

1. a Neurotransmitters are chemicals released from the tiny vesicles (small fluid-filled sacs) into synapses, not from vesicles in postsynaptic cells. They carry messages across the synapse between an axon and a dendrite.
2. d Norepinephrine and epinephrine are synthesized metabolically from tyrosine, an amino acid.
3. d Depression may be the result of abnormal metabolism of serotonin.
4. a When used as an anesthetic, nitrous oxide can cause brain damage if it is not combined with O_2.
5. c Globally, ethyl alcohol is the most used and abused depressant.
6. b Benzodiazepines are commonly used as antianxiety medications.
7. d Several stimulant drugs are chemical derivatives of β-phenylethylamine.
8. d Dextroamphetamine is composed of molecules that are all the right-handed isomer of the two enantiomers.
9. d LSD acts on the serotonin receptors.
10. a While agonists mimic the action of a drug, antagonists block that action.

18.8 Drugs and Society

Learning Objectives: • Differentiate between drug abuse and drug misuse. • Describe how a new drug is developed and brought to market.

A. Drug Abuse: Using a Drug for Its Intoxicating Effect.

B. Drug Misuse: Inappropriate Use of a Drug to Treat a Specific Illness.

C. Design and Approval of New Drugs
1. Preliminary testing is not done on humans.
    a. Primary testing is done on bacteria, cell cultures, and animal models.
2. If animal tests look promising, then testing is done on humans.

a. Phase I trials are done on a small group of relatively healthy volunteers to look for side effects, dosage amounts, and means of administration.
b. Phase II trials involve a larger group of people and focus on the effectiveness of the drug against the disease for which it is designed.
c. Phase III trials involve a still larger group.
d. In Phase II and III trials, the effects of the drug are compared to other therapies and to placebos (inactive substances given in the form of medication to a patient).

3. Placebo effects.
   a. The placebo effect confers health benefits from a treatment that should have no effect.
4. Orphan drugs are drugs that treat orphan or very uncommon diseases that are often of genetic origin.

D. Green Pharmaceutical Production.
   1. Because drugs must be selective and safe, newer drugs tend to be structurally complex and require lengthy syntheses.
   2. Drug synthesis often uses large amounts of solvents and complex reagents, so drug manufacturers commonly generate more than 100 kg of waste for every kilogram of drug manufacture.
   3. The E-factor (see Chapter 8 essay) is the mass of generated waste for each kilogram of manufactured product.
   4. Syntheses in water, enzymatic processes, catalysts that mimic biological processes, and processes without hazardous reagents are increasingly used to make drug production more efficient and green.
   5. Green chemistry has become more important as drug molecules have become increasingly complex and more difficult to product.
      a. The expense and high E-factors common to drug manufacturing provide incentives to develop green chemistry approaches.
      b. Reducing drug manufacturing expense also improves worldwide access to essential medicines.

Answers to Self-Assessment Questions

1. c Using an antibiotic to treat a cold is an example of drug misuse because antibiotics are not effective in treating colds.
2. b Using OxyContin (oxycodone) for its intoxicating effect is an example of drug abuse.
3. b A placebo is an inactive drug that looks like real medication.
4. d In a double-blind experiment, some patients are given the drug being tested and others are given a placebo.
5. d An orphan drug is a drug used to treat orphan or rarely-occurring diseases.

LEARNING OBJECTIVES

You should be able to . . .

1. Identify common drugs as natural, semisynthetic, or synthetic. (18.1)

2. Define *chemotherapy*, and explain its origin. (18.1)

3. List the common over-the-counter analgesics, antipyretics, and anti-inflammatory drugs, and describe how each works. (18.2)

4. Name several common narcotics, describe how each functions, and state its potential for addiction. (18.2)

5. List the common antibacterial drugs, and describe the action of each. (18.3)

6. Name the common antiviral drugs, and describe the action of each. (18.3)

7. Describe the action of the common types of anticancer drugs. (18.4)

8. Define the terms *hormone*, *prostaglandin*, and *steroid*, and explain the function of each. (18.5)

9. List the three types of sex hormones, and explain how each acts and how birth control drugs work. (18.5)

10. Describe the action of four types of drugs used to treat heart disease. (18.6)

11. Explain how the brain amines norepinephrine and serotonin affect the mind and how various drugs change their action. (18.7)

12. Identify some stimulant drugs, depressant drugs, and psychotropic drugs, and describe how they affect the mind. (18.7)

13. Differentiate between drug abuse and drug misuse. (18.8)

14. Describe how a new drug is developed and brought to market. (18.8)

DISCUSSION

The first half of this chapter concentrates on chemical substances that are used to relieve pain or distress, to cure or alleviate disease, to prevent pregnancy, and for a variety of other purposes. Once again, keep in mind that a chemical substance has a specific set of properties that are invariant. Each compound may have some properties that are desirable and some that are undesirable; drugs may have nasty side effects as well as desired therapeutic properties.

The second half of this chapter is devoted to drugs that affect our mental state. It is important to realize that our moods, our sense of feeling "up" or "down," and our sense of and tolerance of pain are all influenced by the presence or absence of chemical molecules that help regulate our body functions. We now know that many of the drugs that have been used and abused for centuries are capable of eliciting certain effects because they somehow mimic a natural neurotransmitter or body regulator molecule. This mimicking usually involves some structural similarity that allows the drug molecule to bind to a receptor site intended for the normal body regulator molecule.

ANSWERS TO ODD-NUMBERED REVIEW QUESTIONS AND SOLUTIONS FOR ODD-NUMBERED END-OF-CHAPTER PROBLEMS

Review Questions

1. Natural drugs are molecules that come from natural sources, which usually means that they've been extracted from plant or animal sources by physical processes as opposed to materials formed by chemical reactions. Semisynthetic drugs are made by modifying molecules derived from natural sources to improve their therapeutic properties. Synthetic drugs are synthesized in laboratories and drug companies.

3. An antibiotic is a drug that kills or slows the growth of bacteria; originally limited to formulations derived from living organisms.

5. Cortisol is a hormone released in the body during stressed or agitated states. Prednisone is a synthetic hormone similar to cortisone.

7. Prostaglandins act as mediators of hormone action. Prostaglandins act near the site where they were produced, they are rapidly metabolized, and they have different effects on different tissues. Hormones act throughout the body.

9. Two dissociative anesthetics are Ketamine and PCP.

11. A narcotic is a drug that produces stupor and relief of pain.

13. Brain amines are made from dietary amino acids; for example, high carbohydrate diets produce high serotonin levels in the brain.

15. No, tranquilizers are not a cure for schizophrenia. These drugs relieve the symptoms, but the symptoms return once use of the drugs is terminated.

Problems

17. Ethyl alcohol and marijuana are examples of natural drugs (there are many more). Aspirin (acetylsalicylic acid), LSD, and heroin are examples of semi-natural drugs. Hydrocodone is an example of a synthetic narcotic.

19. Chemotherapy is the use of drugs to treat cancer.

21. The chemical name for aspirin is acetylsalicylic acid.

23. Aspirin provides pain relief and reduces fever, but it is also an anticoagulant and has anti-inflammatory effects; acetaminophen provides pain relief and reduces fever, but it is not an anticoagulant (nor an anti-inflammatory drug), so it does not promote bleeding, making healing from surgical procedures proceed more smoothly.

25. The acetaminophen molecule includes a phenolic group (-OH on a benzene ring) and an amide group.

27. The answer is c. As long ago as the late 1800s the Bayer Company in Germany advertised heroine as a sedative for coughs.

29. The answer is a. Acetaminophen relieves aches and reduces fever but is not anti-inflammatory, so it is not among the members of the NSAID group.

31. To create codeine, the phenolic hydroxyl group (–OH) in morphine is replaced by a methoxy group (–OCH_3). Codeine resembles morphine in its action but is less potent, has less tendency to induce sleep, and is less addictive.

33. Sulfanilamide

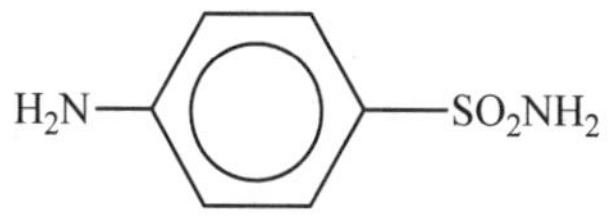

para-aminobenzoic acid

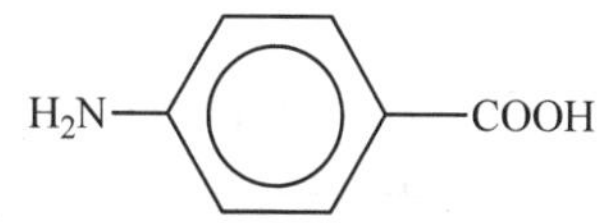

35. Broad-spectrum antibiotics are effective against a wide variety of bacteria.

37. The answer is b. Quinolone antibiotics were introduced in 1986 and are often used against penicillin-resistant bacteria.

39. The three classes of antiretroviral drugs are:
   a. Nucleoside analogs (or nucleoside reverse transcriptase inhibitors), which substitute for nucleosides in viral DNA crippling the retrovirus and slowing down its replication.
   b. Non-nucleoside reverse transcriptase inhibitors, which stop the reverse transcriptase from working properly to make more of the retrovirus.
   c. Protease inhibitors, which block the enzyme protease so that new copies of the retrovirus cannot infect new cells.

41. The answer is b. Cyclophosphamide is used in cancer treatment. The other three drugs are antimetabolites.

43. Yes, some drugs that treat cancer can also cause cancer, as illustrated by some alkylating agents. The nitrogen mustard HN_2 causes lung, mammary, and liver tumors when injected into mice, but it can be used with some success in the management of certain human tumors.

45. The compound is 6-mercaptopurine, which is an antimetabolite.

47. A hormone is a chemical messenger produced in the endocrine glands that is released in one part of the body and signals physiological change in other parts of the body. A prostaglandin is a hormone-like lipid derived from a fatty acid. Prostaglandins function similarly to hormones in that they act on target cells. However, they differ from hormones in that they act near the site where they are produced, can have different effects in different tissues, and are rapidly metabolized.

49. The progesterone structure includes two ketone groups and a C=C double bond, all built into the four fused rings that are characteristic of steroids.

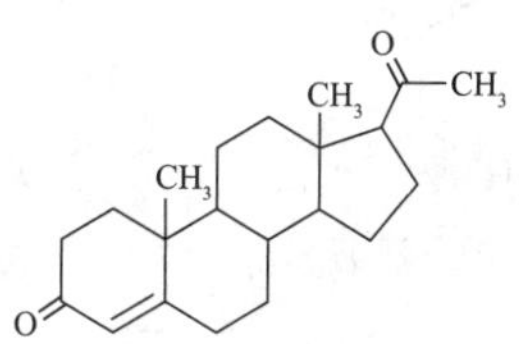

51. Mifepristone (RU-486) causes the uterus to expel an implanted egg and end a pregnancy at an early stage.

53. Emergency contraceptives, also called morning-after pills, can be used to prevent pregnancy after unprotected intercourse. There are two kinds of emergency contraceptives: combination pills with both estrogen and progestin (synthetic analogs of the natural substances) and pills containing only progestin. Given in high doses, emergency contraceptives half development of the uterine lining and inhibit ovulation and fertilization. Intrauterine devices (IUDs) are also used, as is RU-486. IUDs cause inflammation that makes the endometrium unsuitable for implantation, and RU-486 causes the uterus to expel an implanted egg.

55. The answer is a. Diuretics cause the kidneys to excrete more water, thus lowering the blood volume and decreasing pressure.

57. The answer is c. Nitric oxide, NO, is effective in dilating blood vessels, which can both relieve chest pain and alleviate symptoms of erectile dysfunction.

59. A psychotropic drug is one that affects the human mind and the way we perceive things. Psychotropic drugs are generally divided into three classes: stimulants, depressants, and hallucinogens.

61. The answer is a. Antimetabolites are not psychotropic drugs. Instead, they are used to treat cancer.

63. The answer is b. Lidocaine was first introduced in the 1940s and continues to be widely used.

65. The answer is c. Coniine is a neurotoxic alkaloid found in hemlock.

67. a. Nitrous oxide, N_2O, is quick acting but not very potent. Concentrations of 50% or greater must be used for it to be effective. When nitrous oxide is mixed with ordinary air instead of oxygen, not enough oxygen gets into the patient's blood, and permanent brain damage can result.
    b. Halothanes are generally considered safe for patients, but female operating room workers suffer from a higher rate of miscarriages than women in the general population.
    c. Diethyl causes nausea.

69. The answer is b. The FDA is not involved in drug pricing.

71. The answer is b. In a double-blind study, neither the patients nor the doctors know who is receiving the real drug and who is receiving the placebo.

73. 8-methoxy-2,4-dione would be an example.

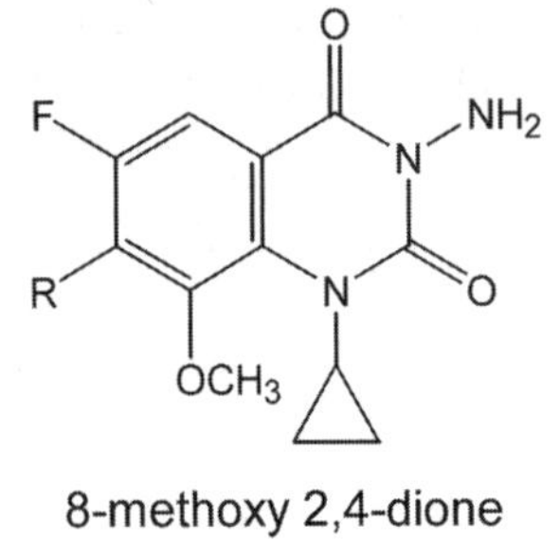

8-methoxy 2,4-dione

75. a. daily dose, g = (140 lb)(0.454 kg/lb)(40 mg ampicillin/kg body weight)(1 g/1000 mg)
= 2.5 g

If the medicine is given three times a day, each dose is 2.5 g/3 doses = 0.83 g/dose.

b. number of capsules/dose = (0.83 g/dose)(1000 mg/g)(1 capsule/500 mg) = 1.7 capsule/dose
847 mg/dose or about 1.7 pills for a pill containing 500 mg of active ingredient.

77. The structure shared by all steroids is four fused rings.

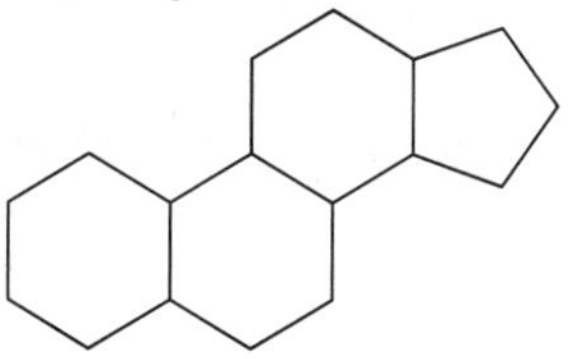

79. A compound with the chemical formula $C_{17}H_{21}NO$ would not be soluble in water because it is nonpolar and water is polar. This compound would be soluble in fat because fat is nonpolar, and the fact that this molecule has only one oxygen atom and one nitrogen atom in a structure containing 17 carbon and 21 hydrogen atoms means that the large majority of bonds in the structure are nonpolar.

81. The structure of cyclohexylamine

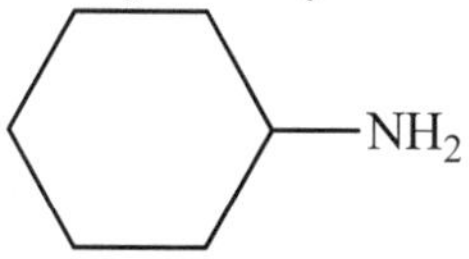

83. Club drugs such as 3,4-methylenedioxymethamphetamine (MDMA or Ecstasy), gamma hydroxybutyrate, Rohypnol, ketamine, and metamphetamine all interact with alcohol, increasing their potency and possibly causing death.

85. The LD_{50} of the timber rattlesnake is lower than the LD_{50} of the Southern copperhead. This means that the toxin from the timber rattlesnake is more toxic than the toxin from the Southern copperhead snake because less of it will cause a fatal reaction.

87. The LD50 of procaine is 50 times greater than that of nicotine, which means that it takes much less nicotine to kill 50% of the test population than it would take of procaine to achieve the same end.
Likely lethal dose of procaine = (50 kg human)(50 mg/kg) = 2500 mg or 2.5 g procaine
Likely lethal dose of nicotine = (50 kg human)(1 mg/kg) = 50 mg nicotine

89. The structure of codeine is

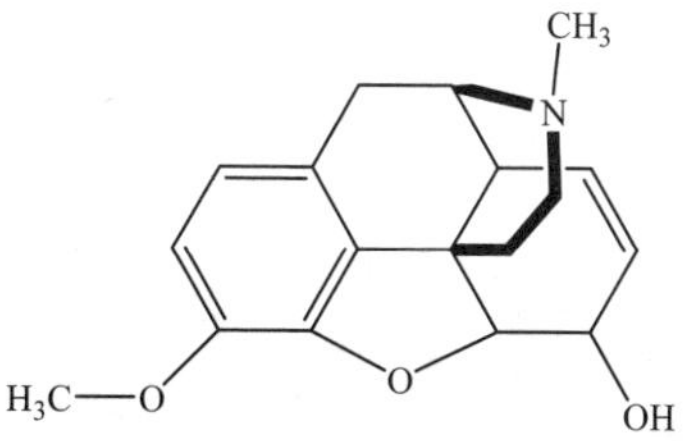

91. The structure of oxycodone is

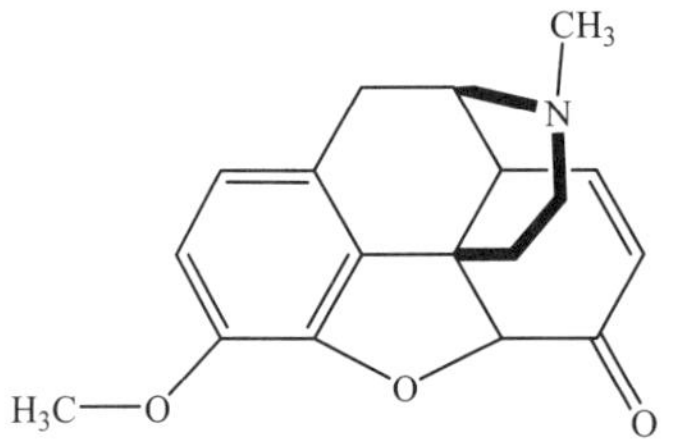

93. *Gamma*-butyrolactone (GBL) contains a cyclic ester. Hydrolyzing the ester will convert GBL to the open-chain structure $HOCH_2CH_2CH_2COOH$.

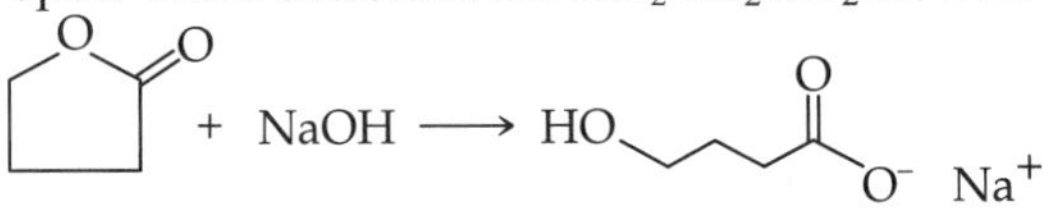

95. The death rate associated with birth control pills is about 3 in 100,000, only one-tenth of the death rate associated with childbirth, so taking birth control pills is safer than childbirth. For women over 40 who smoke 15 cigarettes per day and take oral contraceptives, the risk of death from stroke or heart attack is 1 in 5000, so smoking while taking birth control drugs is more dangerous than childbirth which in turn is more dangerous than taking birth control drugs.

97. NSAIDs relieve pain and reduce inflammation by inhibiting the production of prostaglandins, the compounds involved in sending pain messages to the brain. Inflammation is caused by overproduction of prostaglandin derivatives, and inhibiting their synthesis reduces the inflammatory effect.

99. The structure shown is Keto profen (see Figure 18.1). One of the oxygen-containing groups is a ketone (the one between the two aromatic rings), and the other is a carboxylic acid.

101. Perhaps the most helpful approach to this problem is to count the carbon atoms in the structure (there are 20) and to notice the five-carbon ring. If you compare the general structure here with those in Figure 18.12 you'll see the resemblance to the prostaglandins (and arachidonic acid). There are two alcohol functional groups, one on the ring and the other on the lower of the two hydrocarbon tails; there is a ketone (C=O) on the five-carbon ring; and there is a carboxylic acid group (COOH) at the end of the upper hydrocarbon tail.

CHAPTER

19

Fitness and Health

The Chemistry of Wellness

CHAPTER SUMMARY

19.1 Calories: Quantity and Quality

Learning Objectives: • List the recommendations (sources and percentages) for calories from fats and other sources in the American diet. • Describe the special dietary requirements of athletes.

A. Total calorie intake is important, but the distribution of calories is even more important.
   1. The 2000-calorie-per-day diet is an average, depending on weight, age, and level of activity.
   2. Adults should keep fat intake between 20% and35% of all calories.
      a. Only 7–10% of fat calories should come from saturated fats
      b. A minimum of fat calories should come from *trans*-fatty acids.
      c. Cholesterol intake should be less than 300 mg/day.
B. Nutrition and the Athlete.
   1. Recommended ranges for energy nutrients for most people are 20–35% of calories from fat, 45–65% of calories from carbohydrates, and 10–35% of calories from protein.
   2. Athletes generally need more calories because they expend more energy than the average sedentary individual.
      a. Those calories should come mainly from carbohydrates.
   3. Athletes do not need extra protein in their diets.
      a. Muscles are built through exercise, not from eating protein.
      b. Protein metabolism produces toxic wastes that tax the liver and kidneys.
      c. Athletes need the Dietary Reference Intake (DRI) quantity of 0.8 g protein/kg body weight.
   4. When muscle contracts against resistance, creatine (an amino acid) is released.
      a. Creatine stimulates production of protein (myosin), thus building muscle.
      b. If exercise is ended, the muscle begins to shrink after about two days.

Answers to Self-Assessment Questions

1. c Good nutrition and exercise are needed for good health and fitness.
2. c Undereating has been shown to extend the life span of mice.

3. d Lard is an animal fat containing saturated fatty acids, which are the most detrimental to health.
4. b Maximum fat intake is 35%, so 35% of 2200 kcal is 770 kcal.
   Mass of fat = (770 kcal)(1 g fat/9 kcal) = 86 g fat
5. b Maximum saturated fat intake is 20%, so 20% of 2000 kcal is 200 kcal.
   Mass of saturated fat = (200 kcal)(1 g fat/9 kcal) = 22 g saturated fat
6. b Statins are drugs that can help reduce cholesterol buildup in blood vessels.
7. c Dietary Reference Intakes (DRIs) are nutrient recommendations that are useful primarily for planning and assessing diets.
8. c A 66-kg athlete requires a daily protein intake of about 0.8 g of protein/kg body weight, or 53 g of protein each day.

19.2 Vitamins, Minerals, Fluids, and Electrolytes

Learning Objective: • Describe the special dietary requirements of athletes.

A. Good nutrition requires sufficient carbohydrates, fats, and proteins; proper proportions of essential minerals and vitamins; and adequate water.
   1. Vitamins are organic substances that the human body needs but cannot manufacture in sufficient quantities, so they must be included in the diet.
   2. Minerals are inorganic elements that the body needs, so they must be included in the diet.
   3. The DRI values for vitamins and minerals set by the Food and Nutrition Board are the amounts that will prevent deficiency diseases and are readily supplied by any well-balanced diet.
      a. Vitamin A, a fat-soluble vitamin, is essential for good vision, bone development, and skin maintenance, and may confer resistance to some cancers.
      b. Vitamin B complex includes eight compounds, several of which serve as coenzymes and are important in maintaining the skin and the nervous system.
         i. Vitamin B_6 (pyridoxine) is a coenzyme for more than 100 different enzymes and helps people with arthritis by shrinking the connective tissue membranes that line the joints.
         ii. B_3 (niacin) offers relief from arthritis and lowers cholesterol levels.
         iii. Vitamin B_{12} deficiency can lead to pernicious anemia. Because vitamin B_{12} is not found in plants, this deficiency can be a problem for vegetarians unless they take supplements.
         iv. Folic acid, a B vitamin, is critical to the development of the nervous system of a fetus, but taken in excess of 1000 μg can cause nerve damage.
      c. Vitamin C is essential for efficient functioning of the immune system.
      d. Vitamin D is a steroid hormone that protects children against rickets by promoting adsorption of calcium and phosphorus from foods to produce and maintain healthy bones.
         i. Too much vitamin D can lead to excessive calcium and phosphorus adsorption.
         ii. DRI for vitamin D increases with age; the recommended upper limit is 2000 IU.
      e. Vitamin E is a mixture of tocopherols, phenols with hydrocarbon side chains.
         i. Its antioxidant activity helps maintain the cardiovascular system.
         ii. Low levels are associated with sterility and muscular dystrophy.
         iii. The upper limit for vitamin E is 400–800 IU/day depending on side effects. Higher levels present the risk of excessive bleeding.

B. Body Fluids and Electrolytes.
   1. An electrolyte is a substance that conducts electricity when dissolved in water.

2. In the body, electrolytes are ions required by cells to maintain their internal and external electric charge and thus control the flow of water molecules across the cell membrane.
   a. The main electrolytes are Na^+, K^+, and Cl^-. Others include Ca^{2+}, Mg^{2+}, sulfate (SO_4^{2-}), hydrogen phosphate (HPO_4^{2-}), and bicarbonate (HCO_3^-).
3. Water is an essential nutrient that is best replaced after respiration, sweating, and urination by drinking water rather than other beverages.
   a. Dehydration can be serious or even deadly.

Answers to Self-Assessment Questions

1. d Vitamins cannot be synthesized by the body and must be included in the diet.
2. a Beta-carotene is a precursor of vitamin A.
3. c Large doses of vitamin A are most likely to be harmful because vitamin A is fat soluble and excesses are stored in the adipose (fatty) tissue.
4. a Vitamin A is essential for good bone development.
5. b Vitamin E is a fat-soluble antioxidant.
6. c The major electrolytes in body fluids are Na^+, K^+, and Cl^-.
7. c Plain water is the best replacement for water lost except in prolonged exercise.
8. c As little as 5% of the body weight lost in fluids can cause heat exhaustion (see Section 19.2).

19.3 Weight Loss: Diets and Exercise

Learning Objectives: • Explain how weight is lost through diet and exercise. • Calculate weight loss due to calorie reduction and to exercise.

A. Diets with fewer than 1200 kcal/day are likely to be deficient in necessary nutrients, particularly the B vitamins and iron.
B. Biochemistry of Hunger.
   1. Two peptide hormones, ghrelin and peptide YY, are produced by the digestive tract and are linked to short-term eating behaviors.
      a. Ghrelin is an appetite stimulant produced by the stomach; PYY acts as an appetite suppressant.
   2. The hormone insulin and a substance called leptin, produced by fat cells, determine longer-term weight balance.
      a. Leptin, a protein consisting of 146 amino acid units, is produced by fat cells.
   3. Other substances involved in weight control include cholecystokinin (CCK), a peptide formed in the intestine that signals that we have eaten enough food, and a class of compounds called melanocortins, which act on the brain to regulate food intake.
C. Crash Diets: Quick = Quack.
   1. Any weight-loss program that promises a loss of more than a pound or two a week is likely to be dangerous quackery.
      a. Many use a diuretic to increase urine output. Weight loss is water loss, which is quickly regained when the body is rehydrated.
      b. Other diets, low in carbohydrates, depend on glycogen depletion.
         i. Glycogen molecules have many OH groups that are attached to water molecules through hydrogen bonding. Each pound of glycogen carries about 3 pounds of water.
         Depleting 1 pound of glycogen results in a weight loss of 4 pounds.
         ii. The weight is quickly regained when carbohydrates are returned to the diet.

2. The largest amount of fat an individual can lose in a day, even with a total fast, is about 0.69 pounds.
    a. The body won't burn just fat; if carbohydrates are not supplied in the diet, the body will break down muscle tissue to make glucose.
    b. Any diet that restricts carbohydrate intake results in a loss of muscle mass as well as fat.
    c, Weight loss through dieting includes loss of muscle mass as well as fat; weight regained (without exercise) is pure fat.

D. Exercise for Weight Loss.
1. The most sensible weight-loss program is to follow a balanced low-calorie diet that meets the DRI for essential nutrients and to engage in a reasonable exercise program.

Answers to Self-Assessment Questions

1. c There are 3500 kilocalories of energy stored in 1 pound of adipose tissue.
2. c To lose 1 pound per week, you need to burn 500 kilocalories (3500 kcal/7 days) each day more than you eat.
3. d The hormone PYY is an appetite suppressant.
4. c The hormone insulin lowers blood glucose levels.
5. c Leptin seems to prevent weight loss when an adequate amount of food is not available.
6. b The adult body can store about 1 pound of glycogen.
7. b Crash diets depend on glycogen depletion for quick weight loss.

19.4 Measuring Fitness

Learning Objectives: • Describe several ways to measure fitness. • Calculate BMI values.

A. The male body requires about 3% body fat, while the average female body needs 10–12%.
1. One way to estimate body fat is by measuring a person's density.
2. A simpler way is by measuring the waist and hips and dividing the waist measurement by the hip measurement.
    a. The waist/hip ratio should be ≤ 1 for males and ≤ 0.8 for females.
3. Newer scales provide a measure of body fat content using *bioelectric impedance analysis*, which involves passing a small electric current through the body.
    a. Fat has a greater impedance (resistance to varying current) than does muscle.

B. Body Mass Index.
1. *Body mass index (BMI)* is a commonly used measure of fatness defined as weight (in kg) divided by the square of the height (in meters).
    a. A BMI < 18.5 indicates that a person is underweight.
    b. A BMI from 25–29.9 indicates that a person is overweight.
    c. A BMI > 30 indicates that a person is obese.
2. When measuring in pounds and inches, the equation is:
    $\text{BMI} = [(705)(\text{body weight, lb})]/(\text{height, in})^2$.

C. V_{O2} Max: A Measure of Fitness.
1. V_{O2} max is the maximum amount, in mL/kg body weight, of oxygen that a person can use in 1 minute.
    a. The higher the V_{O2} max, the greater is the fitness.

Answers to Self-Assessment Questions

1. d Body mass index (BMI) is calculated by weight (kg) ÷ [height (m)]^2.

2. c A person with a body mass index (BMI) of 22 is in the ideal weight range.
3. d V_{O2} max is the maximum amount of oxygen used during one minute of exercise.

19.5 Some Muscular Chemistry

Learning Objectives: • Differentiate between aerobic exercise and anaerobic exercise, and describe the chemistry that occurs during each. • Describe how muscles are built and how they work.

A. Energy for Muscle Contraction: ATP.
   1. Some of the energy from glucose or fatty-acid metabolism is stored as adenosine triphosphate (ATP).
      a. Muscle contains the proteins actin and myosin in a loose complex called actomyosin.

B. Aerobic Exercise: Plenty of Oxygen.
   1. When muscle contraction begins, glycogen is converted to pyruvic acid.
   2. If sufficient oxygen is present (as in aerobic exercise), the pyruvic acid is oxidized to carbon dioxide and water.

C. Anaerobic Exercise and Oxygen Debt.
   1. If sufficient oxygen is not available (as in anaerobic exercise), pyruvic acid is reduced to lactic acid.
   2. This lactic acid buildup leads to a pH drop and deactivation of muscle enzymes, described as "muscle fatigue."
   3. The overworked muscles incur an oxygen debt that needs to be repaid after the strenuous exercise is over.
   4. When glycogen stores are depleted, muscle cells can switch to fat metabolism.
      a. Fats are the main source of energy for sustained activity of low to moderate intensity.

D. Muscle Fibers: Which kind do you have?
   1. There are two classes of muscle fibers.
      a. Fast-twitch (Type IIB) for anaerobic activity.
      b. Slow-twitch (Type I) for aerobic activity.
   2. Slow-twitch fibers (Type I): endurance activities.
      a. These are best suited for aerobic work of light or moderate intensity for sustained periods of time.
      b. High respiratory capacity and myoglobin levels of slow-twitch fibers help supply oxygen for sustained exercise, like long-distance running.
      c. Slow-twitch fibers have a low ability to hydrolyze glycogen and low actomyosin catalytic activity.
   3. Fast-twitch fibers (type IIB): allow for bursts of power.
      a. These are best suited for anaerobic short bursts.
      b. Low respiratory capacity and myoglobin levels of fast-twitch fibers are designed for quick bursts of energy like sprints.
      c. Fast-twitch fibers have high capacity for glycogen hydrolysis and high catalytic activity of actomyosin, which facilitate rapid ATP production and ability to hydrolyze ATP quickly.

E. Building Muscles.
   1. Endurance training increases myoglobin levels in muscles (slow-twitch).
   2. Weight training develops fast-twitch muscles. Muscles increase in size.

Answers to Self-Assessment Questions

1. d Glucose is used in cells to energize ATP.
2. d Glycogen is converted to pyruvic acid when muscle contraction begins.
3. a Pyruvic acid is converted to CO_2 and H_2O during aerobic exercise.
4. c Pyruvic acid is converted to lactic acid during anaerobic exercise.
5. c Lactic acid production results in oxygen debt.
6. a Slow-twitch muscle fiber is classified as type I.
7. c The presence of myoglobin causes the reddish color of slow-twitch muscle fibers.

19.6 Drugs, Athletic Performance, and the Brain

Learning Objectives: • Describe the physiological effects of restorative drugs, stimulant drugs, and anabolic steroids. • Explain how endorphins, neurotrophins, and tobacco can affect the brain and body.

A. Restorative drugs are used to remedy the effects of performance—pain, soreness, and injury.
   1. Painkillers include aspirin, acetaminophen, ibuprofen, and methyl salicylate.
   2. Anti-inflammatory drugs include aspirin and ibuprofen and cortisone derivatives.

B. Stimulant Drugs.
   1. Caffeine can conserve glycogen, speed heart rate, and increase metabolism, but the effect is small.
   2. Cocaine, like amphetamines, stimulates the central nervous system and increases alertness and muscle tension. It can also mask fatigue and give a sense of increased stamina.

C. Anabolic Steroids.
   1. Anabolic steroid hormones increase muscle mass.
   2. Side effects in males include testicular atrophy and loss of function, impotence, acne, liver damage (including liver cancer), edema, elevated cholesterol levels, and growth of breasts.
   3. Anabolic steroids (derived from male sex hormones) make women more masculine. In addition to muscles, they develop baldness, extra body and facial hair, a deepened voice, and menstrual irregularities.

D. Drugs, Athletic Performance, and Drug Screening.
   1. Use of drugs to increase athletic performance is illegal and physically dangerous.
      a. Stimulants give a short-lived and false sense of confidence.
      b. Following the high of stimulants is extreme depression.
   2. Blood and urine screening for illegal drugs is becoming standard practice.

E. Exercise and the Brain.
   1. Vigorous exercise causes the body to produce its own painkillers, endorphins.
   2. These substances can produce euphoric highs.
   3. Endorphins are addictive; athletes suffer withdrawal symptoms when they can't exercise.
   4. Exercise also directly affects the brain by increasing its blood supply and by producing neurotrophins, substances that enhance the growth of brain cells.

F. No Smoking.
   1. Cigarette smoking is the chief preventable cause of premature death in the United States.
      a. At every age, the death rate is higher among smokers than among nonsmokers.
      b. More than 7000 different chemicals have been identified in cigarette smoke, about 70 of which are known carcinogens.

2. E-cigarettes, which are essentially vaporizers for nicotine solutions, are safer than cigarettes because they produce fewer gases.
    a. Nicotine itself is poisonous, and the nicotine solutions in e-cigarettes contain potentially lethal concentrations of nicotine.

G. Green Chemistry: Your fitness benefits the planet.
    1. Although the chemical reactions that power your body when you walk or bike produce CO_2, they are based on renewable fuels—the carbohydrates in your food.
    2. The reaction of carbohydrates with oxygen powers your body and leads to CO_2 as a by-product.
    3. The plants that produced the carbohydrates you eat incorporated CO_2 from the atmosphere through photosynthesis; thus, your energy source is carbon-neutral because the CO_2 that you produce is balanced by the CO_2 used to make the fuel your body uses.

Answers to Self-Assessment Questions

1. c Cortisone is used to reduce swelling in joints.
2. d Caffeine is thought to aid endurance athletes by stimulating the release of fatty acids.
3. c Anabolic steroids help build larger muscles.
4. a Endorphins are chemicals produced by the body that are similar to morphine.
5. c Exercise is thought to help relieve depression by increasing the levels of neurotrophins.
6. a CO in cigarette smoke makes breathing less efficient because it blocks oxygen transport by hemoglobin.
7. d Nicotine is an addictive drug that is the active ingredient in the solutions present in e-cigarettes..

LEARNING OBJECTIVES

You should be able to . . .

1. List the recommendations (sources and percentages) for calories from fats and other sources in the American diet. (19.1)
2. Describe the special dietary requirements of athletes. (19.1)
3. Describe the dietary requirements for vitamins, minerals, and water. (19.2)
4. Explain how weight is lost through diet and exercise. (19.3)
5. Calculate weight loss due to calorie reduction and to exercise. (19.3)
6. Describe several ways to measure fitness. (19.4)
7. Calculate BMI values. (19.4)
8. Differentiate between aerobic exercise and anaerobic exercise, and describe the chemistry that occurs during each. (19.5)
9. Describe how muscles are built and how they work. (19.5)

10. Describe the physiological effects of restorative drugs, stimulant drugs, and anabolic steroids. (19.6)

11. Explain how endorphins, neurotrophins, and tobacco can affect the brain and body. (19.6)

12. Explain how green chemistry principles can help us make transportation choices.

13. Describe the relationship between walking or biking and personal health, especially the implications for weight loss.

DISCUSSION

In this chapter, we consider the application of chemistry to healthy living, with emphasis on sports and athletics. Science has transformed many athletic events. We now have a better understanding of the action of muscles and of the effects of diet and exercise. Chemistry has provided a plethora of new materials for clothing and equipment. It also has provided drugs that help (and harm) athletes. Chemical principles can help you to separate fact from fancy and to guard against quackery and harmful practices.

ANSWERS TO ODD-NUMBERED REVIEW QUESTIONS AND SOLUTIONS FOR ODD-NUMBERED END-OF-CHAPTER PROBLEMS

Review Questions

1. Meat and other animal products are the main dietary source of saturated fats.

3. This type of weight loss is due to depletion of glycogen stores and water (dehydration). It is not a useful way to lose weight and keep it off because the weight is quickly regained when carbohydrates and liquids return to the diet.

5. Three ways that chemistry has had an impact on sports are:
    1. The development of a better understanding of muscle physiology.
    2. The better understanding of drugs.
    3. The development of improved sporting equipment.

7. One method involves measuring a person's density (mass to volume ratio). Measuring mass with a scale is relatively easy, but measuring volume with a dunk tank is subject to considerable error and does not take into account the volume of air in the lungs. A second method involves using bioelectric impedance analysis where a person stands on the scale in bare feet and a small electric current is sent through the body. By measuring the impedance of the body, the percentage of body fat can be calculated based on height and weight. The problems with this method involve taking into account many variables such as bone density, water content, and location of fat.

9. A high-fat diet can be problematic because fat metabolism produces ketone bodies that stress the kidneys. If the high-fat diet does not include sufficient carbohydrates, then energy for brain activity is low. If the diet does not include sufficient essential proteins, then muscle development and construction of other critical proteins are limited.

Problems

11. Athletes generally need more calories because they expend more calories than a sedentary individual. Most athletes do not need more protein than nonathletes. Protein metabolism produces more toxic wastes that tax the liver and kidneys. The added energy they require is best obtained through the consumption of additional carbohydrates.

13. The benefit derived from vitamin B_{12} is that, by taking it, the individual can avoid having pernicious anemia.

15. Arthritis is helped by vitamin B_6 (pyridoxine) because this compound shrinks the connective tissue membranes that line the joints. Vitamin B_6 is a cofactor for more than 100 different enzymes.

17. Vitamin D promotes adsorption of calcium and phosphorus. Taking too much vitamin D (more than 2000 IU) causes absorption of excessive amounts of calcium and phosphorus, which leads to formation of calcium deposits in various soft body tissues, including those of the heart.

19. A diuretic promotes water loss through urine production. The antidiuretic hormone ADH keeps water in the body.

21. Thirst is not a good indicator of dehydration because it is a delayed response to water loss and may be masked by symptoms of dehydration such as exhaustion, confusion, headache, and nausea. You are not always thirsty when you are dehydrated.

23. Fad diets often include a diuretic that promotes water loss. The weight is regained as soon as the body is rehydrated. Other quick-weight-loss diets depend on depleting the body's stores of glycogen. No fat is lost, and the weight is quickly regained when the dieter resumes eating carbohydrates.

25. Walking time required = (420 kcal)(1 hr/210 kcal) = 2.0 hr

27. 1 lb = 0.454 kg
    mass of protein, g = (220 lb player)(0.454 kg/lb)(0.8 g protein/kg body weight) = 80 g

29. distance run = (2000 kcal)(1 km/100 kcal) = 20 km

31. a. Leptin is produced by the fat cells. It causes weight loss in mice by decreasing their appetite and increasing their metabolic rates. Letpin's main role seems to be to protect against weight loss in times of scarcity rather than against weight gain in times of plentiful food.
    b. Ghrelin is an appetite stimulant produced by the stomach.

33. BMI = $[(705)(\text{body weight, lb})]/(\text{height, in})^2$

    BMI = $[(705)(186\text{ lb})]/(70\text{ in})^2 = 27$

35. Weightlifting develops fast–twitch muscle fibers that increase in size with repeated anaerobic exercises. The process does not create new (and different) muscles but strengthens those we have.

37. No, eating more protein does not result in larger muscles. Aerobic exercise is the best means of increasing muscle mass.

39. a. Anaerobic metabolism provides energy for intense bursts of vigorous activity.
    b. Aerobic metabolism provides energy for prolonged low levels of activity.

41. High levels of myoglobin are appropriate for muscle tissue geared to aerobic oxidation because they store the oxygen needed for aerobic oxidation.

43. A high capacity for glycogen use and a high catalytic activity of actomyosin allow tissue rich in fast-twitch (Type IIB) fibers to generate ATP rapidly and to hydrolyze this ATP rapidly during intense muscle activity.

45. No, eating extra protein does not build muscles. Engaging in weightlifting and other exercises is the best way to build muscles.

47. Cortisone derivatives reduce swelling (inflammation) in damaged joints and tissues. Side effects include increased fluid retention, hypertension, ulcers, and a disturbance of the sex hormone balance.

49. In runner's high, endorphins that block pain are produced in the body.

51. Exercise directly affects the brain by increasing its blood supply and producing neurotrophins which enhance the growth of brain cells. Exercises involving complex motions lead to an increase in the connections between brain cells.

53. Some of the health problems related to smoking are heart disease, stroke, lung cancer, emphysema, pneumonia, and cancer of the pancreas, bladder, breast, kidney, and cervix.

55. number of kcal/day = (160 lb)(15 kcal/lb) = 2400 kcal/day

57. Mass, g = (90 kg)(1000 g/kg) = 90000 g; volume, mL = (110 L)(1000 mL/L) = 110000 mL
density = mass/volume = 90000 g/110000 mL = 0.82 g/mL.
The person is fat because the person's density more resembles the density of fat tissue than the density of lean tissue.

59.

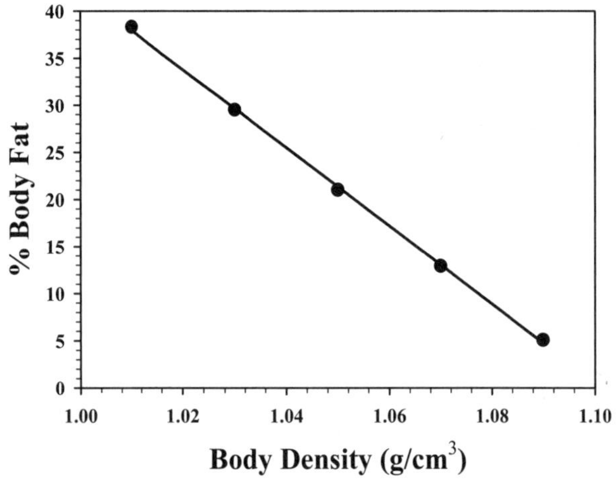

a. From the graph, a body density of 1.037 g/cm^3 is approximately equivalent to 27% body fat.

b. mass of individual, kg = (165 lb)(0.454 kg/lb) = 74.9 kg
mass of water displaced, kg = (14 lb)(0.454 kg/lb) = 6.4 kg
volume of water displaced, L = (74.9 kg – 6.4 kg)(1 L/0.9951 kg) = 68.8 L
density = 74.9 kg/68.8 L = 1.09 g/cm^3
From the graph, the individual has about 5% body fat.

c. Person B is more likely to be an athlete because the density of person A is less than that of person B, so person A's constitution likely contains a higher percentage of fat than that of person B.

d. The 5.30 kg underwater is obtained with 1.57 L of air, which displaces 1.562 kg of water. The mass of the individual when he's underwater is greater than 5.30 kg by 1.562 kg, making a "real" mass underwater of 6.86 kg. Therefore, his density is:
Density = [(89.34)/89.34 – 6.86)] = 1.083 g/cm^3.
The air in the lungs helps keep a person afloat.

61. a. 8 ft 11 in = 107 in (199 kg)(1 lb/0.454 kg) = 438 lb
BMI = [(705)(438 lb)]/[107 in]^2 = 27

b. (486 kg)(1 lb/0.454 kg) = 1070 lb (184 cm)(1 in/2.54 cm) = 72.4 in
BMI = [(705)(1070 lb)]/[72.4]^2 = 144

63. a. volume of O_2 = (100 mL blood)(15 g hemoglobin/100 mL blood)(1.34 mL O_2/1 g hemoglobin) = 20.1 mL O_2

b. volume of O_2 = (6.0 L blood)(1000 mL/L)(15 g hemoglobin/100 mL blood)(1.34 mL O_2/1 g hemoglobin) = 1200 mL O_2

65. pounds lost = (45 kcal/mi)(2 mi/day)(80 days)(1 lb/3500 kcal) = 2.1 lb
The individual will lose 2.1 pounds by riding a bike rather than driving to school.

67. The answer is c. Carbon dioxide is a greenhouse gas that contributes to global warming.

CHAPTER

20

Chemistry Down on the Farm

...and in the Garden and on the Lawn

Introduction

A. Green plants use photosynthesis to make sugars from carbon dioxide and water. The reaction also produces oxygen, replenishing the atmosphere.
B. Structural elements of plants (carbon, hydrogen, oxygen) are derived from air and water. Other nutrients come from the soil. Energy is supplied by the sun.
   1. Using other nutrients, particularly compounds of nitrogen and phosphorus, plants can convert the sugars produced by photosynthesis to proteins, fats, and other chemicals that we use as food.
C. In early agricultural societies, nearly all the energy used came from renewable resources.
   1. Much of the energy obtained from food was used to produce food.
   2. One unit of human work energy might produce 10 units of food energy.
D. Early societies usually recycled nutrients by returning wastes to the soil.
   1. Properly practiced, this kind of agriculture could be continued for centuries without seriously depleting the soil, but farming on this level supports relatively few people.
   2. Making use of fertilizers, pesticides, and energy from fossil fuels, modern high-production farming has greatly expanded yields.

20.1 Farming with Chemicals: Fertilizers

Learning Objectives: • List the three primary plant nutrients as well as several secondary nutrients and micronutrients, and describe the function of each. • Differentiate between organic fertilizers and conventional fertilizers.

A. The three primary plant nutrients are nitrogen, phosphorus, and potassium.
   1. Plants cannot use N_2 directly, but some bacteria can fix nitrogen (converting it to a combined, soluble form).
      a. Lightning also "fixes" nitrogen.
   2. Plants take up nitrogen as nitrate (NO_3^-) or ammonium ions (NH_4^+), which combine with carbon compounds from photosynthesis to form the amino acids that make up proteins.
   3. Population growth in the late nineteenth century led to an increased demand for fertilizers.

B. Nitrogen.
   1. Fritz Haber developed a process for combining nitrogen and hydrogen to make ammonia.
   2. By oxidizing part of the ammonia molecule to nitric acid and reacting the acid with ammonia, the Germans were able to make ammonium nitrate, an explosive.
      a. Ammonium nitrate is a valuable nitrogen fertilizer.
      b. Ammonia can also be combined with carbon dioxide to make urea.
   3. The water in wells close to fields can contain high amounts of nitrate which is hazardous to the development of babies and young children, causing blue baby syndrome.

C. Phosphorus Fertilizers.
   1. The limiting factor in plant growth is often the availability of phosphorus.
      a. Plants incorporate phosphates into DNA and RNA and other compounds essential to plant growth.
      b. In the 1840s, chemists learned how to treat animal bones with sulfuric acid to form soluble superphosphate.
      c. Most modern phosphate fertilizers are produced by treating phosphate rock with phosphoric acid, which produces water-soluble calcium dihydrogen phosphate.
      d. Fluorides are often associated with phosphate ores and can be a contamination problem.
   2. Phosphate reserves are limited and not renewable; phosphates are scattered and lost through use.
   3. Phosphate runoff from farm fields causes algae blooms in a process termed eutrophication which is a big concern of environmentalists.

D. Potassium Fertilizers.
   1. Plants use potassium as the K^+ ion.
   2. The role of potassium ions, along with sodium ions, in plants is related to fluid balance in cells, the formation and transport of carbohydrates, and the assembly of proteins from amino acids.
   3. Uptake of potassium ions from the soil leaves the soil acidic.
      a. Each time a potassium ion enters the root tip, a hydronium ion enters the soil to maintain electrical balance.
   4. The potassium in fertilizers is usually in the form of potassium chloride (KCl), which is mined in many parts of the world.
   5. Potassium salts are a nonrenewable resource; they are scattered and lost through use.

E. Other Essential Elements (Secondary Plant Nutrients).
   1. Three secondary nutrients (magnesium, calcium, and sulfur) are needed in moderate amounts.
      a. Calcium, in the form of lime (CaO), is used primarily to neutralize acidic soils and is also a nutrient.
      b. Magnesium ions (Mg^{2+}) are incorporated into chlorophyll molecules, which are necessary for photosynthesis.
      c. Sulfur is a constituent of several amino acids and is necessary for protein synthesis.
   2. Plants need eight other elements, called micronutrients (boron, copper, iron, manganese, molybdenum, nickel, zinc, and chloride), in small amounts.
   3. There are other elements plants require in very small quantities (boron, chlorine, copper, iron, manganese, molybdenum, nickel, and zinc). These elements are generally present in most soils.

F. Fertilizers: A Mixed Bag.
   1. "Complete fertilizers" contain the three main nutrients (NPK).
      a. The first number on the fertilizer represents the percent nitrogen.
      b. The second is phosphorus, calculated as percent P_2O_5.
      c. The third is potassium, calculated as percent K_2O.
      d. The remainder of the fertilizer is composed of inert ingredients.

2. To be used by plants, nutrients must be soluble in water.

G. Organic Fertilizers.
   1. "Organic" means the fertilizers come from natural animal and plant products.
   2. Organic fertilizers generally must be broken down by soil organisms to release the actual nutrients that the plants use, which can be a slow process.
   3. Organic material has the advantage of providing humus, a substance consisting of partially decayed plant or animal matter.
      a. Humus improves the texture of soils, thus improving water retention, reducing the loss of nutrients through leaching, and helping to resist erosion.
      b. Fresh human manure should not be used to fertilize fruits and vegetables to avoid the possibility of transmitting human pathogens.
      c. Pig, dog, and cat feces should not be used as fertilizers because they can carry internal parasitic worms from those animals to humans.

Answers to Self-Assessment Questions

1. d The three primary nutrients required by plants are nitrogen, potassium, and phosphorus.
2. d Nitrogen is taken up in the form of NO_3^-.
3. d The second number in the fertilizer label 24-5-10 represents 10% P_2O_5.
4. b $(NH_4)_2HPO_4$ furnishes two primary plant nutrients, nitrogen and phosphorus.
5. a Uptake of potassium by plants leaves the soil acidic.
6. c Phosphate rock is treated with H_2SO_4 to make the phosphate more available to plants.
7. b The secondary plant nutrients are Ca, Mg, and S.
8. d Mg^{2+} is essential to the function of chlorophyll.
9. a The first of the three numbers on a fertilizer bag indicates the percent nitrogen in the product. A fertilizer listed as 8-0-24 contains 8% nitrogen.
10. c. Organic fertilizers generally must be broken down by soil organisms; thus, in addition to the three primary nutrients, they leave the soil enriched with humus and other by-products.
11. c. Lime, or calcium oxide (CaO), can be added to soil to reduce its acidity.
12. d. Adding iron to the soil can help prevent the yellowing of plant leaves (see Table 20.2).
13. c. Fourteen elements are important to plant growth and survival: the three primary nutrients, the three secondary nutrients, and eight micronutrients.

20.2 The War against Pests

Learning Objectives: • Name and describe the action of the main kinds of pesticides. • List several biological pest controls, and explain how they work.

A. Since the earliest recorded days, insect pests have destroyed crops and spread disease.
   1. The use of modern chemical pesticides (substances that kill organisms that we consider pests) may be all that stands between us and some insect-borne plagues.
   2. Pesticides prevent consumption of a major portion of our food supply by insects and other pests.

B. Early insecticides included lead and arsenic compounds, pyrethrum, and nicotine sulfate.
   1. Only a few insects are harmful; many are beneficial.
      a. Most pesticides are indiscriminate; they kill both harmful and beneficial insects.
      b. Because pesticides kill living things, they should be classified as biocides (substances that kill living things) rather than insecticides (substances that kill insects).

C. DDT: The Dream Insecticide—Or Nightmare?
   1. DDT, a chlorinated hydrocarbon, was found to be effective against a variety of insects and of low toxicity to humans.
   2. DDT is easily synthesized from cheap, readily available chemicals.
   3. The World Health Organization estimates that chlorinated hydrocarbon pesticides such as DDT have saved 25 million lives and prevented hundreds of millions of illnesses.
   4. DDT has been banned in most developed countries, but it is still used in many developing countries.
   5. DDT is especially toxic to fish and birds. It interferes with calcium metabolism, making eggshells weak and easily broken. This was described by Rachel Carson in *Silent Spring*.
   6. DDT and its metabolites break down very slowly in the environment, a situation known as pesticide persistence.
   7. Chlorinated hydrocarbons, such as DDT and PCBs, are generally unreactive, so they remain on plants for long periods of time.

D. Biological Magnification: Concentration in Fatty Tissues.
   1. Because they are fat soluble, chlorinated hydrocarbons tend to become concentrated in animals higher on the food chain, a process known as biological magnification.
   2. DDT and other chlorinated hydrocarbons such as PCBs are nerve poisons, and DDT interferes with calcium metabolism essential to the formation of healthy bones and teeth.
   3. DDT is still used in some countries where malaria and typhus are great health problems.
      a. It is applied mainly on the inside walls of houses, so the environmental impact is much less than that of the earlier widespread use of DDT in agriculture.
      b. In 2000, more than 120 countries signed a treaty to phase out persistent organic pollutants (POPs), a group of chemicals that includes DDT.

E. Development of Insect Resistance
   1. Insects develop resistance to insecticides by natural selection, and this resistance can spread quickly.

F. Organophosphorus Compounds.
   1. Chlorinated hydrocarbon pesticides have been replaced in part by organic phosphorus compounds such as malathion, parathion, and diazinon.
   2. Organophosphorus compounds are nerve poisons; they interfere with the conduction of nerve signals in insects.
   3. With the exception of malathion, these compounds generally are more toxic to mammals but less persistent in the environment.
   4. Because they break down more quickly, residues are seldom found in food.

G. Carbamates: Targeting the Pest.
   1. Like phosphorus compounds, carbamates are nerve poisons, but they act over a shorter span of time.
   2. Most carbamates, such as carbaryl, carbofuran, and aldicarb, are narrow-spectrum insecticides, directed against only a few insect species.
      a. Chlorinated hydrocarbons and organic phosphorus compounds are broad-spectrum insecticides; they are effective against many species.
         i. Carbaryl is especially harmful to honeybees.
   3. Carbamates break down rapidly in the environment and do not accumulate in fatty tissue.

H. Organic Pesticides.
   1. Organic pesticides are derived from natural sources (rather than made synthetically) and include:
      a. Insecticidal soap that breaks down the insect's outer covering.
      b. Pyrethrins obtained from the perennial plant pyrethrum, which are low in toxicity to most mammals.

c. Rotenone, a natural product obtained from the roots and stems of several tropical and subtropical plants.
d. Boric acid, used mainly to control ants and cockroaches, but moderately toxic to humans.
e. Cryolite, used on food crops and ornamental plants to block caterpillars and other grazing insects from feeding.
f. Diatomaceous earth, the fossilized remains of a type of algae whose shells consist largely of silica.
g. Ryania, a botanical insecticide obtained from a tropical plant that is highly toxic to caterpillars of fruit moths, codling moths, and corn earworms.

2. Just because a product is considered organic doesn't mean it is safe.
   a. Some insecticides are as toxic as, or more so than some synthetic pesticides.

I. Biological Insect Controls.
   1. The use of natural enemies is another way of controlling pests.
      a. These biological controls include predatory insects, mites, and mollusks; parasitic insects; and microbial controls that are insect pathogens such as bacteria, viruses, fungi, and nematodes.

J. Genetically Engineered Insect Resistance.
   1. One highly successful biological approach is the breeding of insect- and fungus-resistant plants.
      a. Genetically modified (GM) cotton is resistant to cotton bollworms; GM corn is resistant to cotton root worm and corn borers.
   2. Development of GM plants is highly controversial.

K. Sterile Insect Technique.
   1. Sterile insect technique (SIT*)* is a method of insect control that involves rearing large numbers of males, sterilizing them with radiation or chemicals or by cross-breeding, and then releasing them in areas of infestation.
      a. Successful SIT programs have been conducted against screwworm, the Mediterranean fruit fly, and the codling moth.

L. Pheromones: The Sex Trap.
   1. A pheromone is a chemical that is secreted externally by an insect to mark a trail, send an alarm, or attract a mate.
      a. Insect sex attractants are usually secreted by females to attract males.
   2. Most sex attractants have complicated structures, and all are secreted in extremely tiny amounts.
   3. Pheromones are usually too expensive to play a huge role in insect control, though the use of recombinant DNA methods to synthesize them holds promise for future work.

M. Juvenile Hormones.
   1. Juvenile hormones control the rate of development of young insects.
   2. Synthesis of juvenile hormones is difficult and expensive, and they can only be used against insects that are pests at the adult stage.

Answers to Self-Assessment Questions

1. b A pesticide is a product that kills other organisms as well as the target.
2. a Genetically modified crops are designed to be resistant to pests without the use of added pesticides.
3. d Through the process of biomagnification, great blue heron will have the highest level of pesticides.
4. d Pesticides that are sprayed on crops may dissolve in rain water and ultimately run off into rivers and lakes.

5. b Compared to DDT, organic phosphorus insecticides are less persistent and more toxic.
6. a Although carbamates are composed of carbon, hydrogen, and oxygen, they are not naturally occurring and are therefore not considered organic pesticides.
7. c The SIT method of insect control works best when large numbers of males can be sterilized and then released in areas of infestation.
8. d A pheromone, or sex attractant, is usually produced by a female to attract a male.
9. b Using ladybugs and praying mantises to control insect pests is a way of allowing natural enemies to control insect populations.

20.3 Herbicides and Defoliants

Learning Objective: • Name and describe the action of the major herbicides and defoliants.

A. A weed is a rapidly growing plant that is hard to get rid of and tends to choke out plants we intend to grow.
   1. Herbicides are used to kill weeds.
   2. Some herbicides are defoliants, which cause leaves to fall off plants.

B. 2,4-D and 2,4,5-T.
   1. It wasn't until the introduction of 2,4-D in 1945 that the use of herbicides became common.
   2. 2,4-D and its derivatives are growth-regulator herbicides, which are especially effective against newly emerged, rapidly growing broad-leaved plants.
      a. One 2,4-D derivative, 2,4,5-T, is especially effective against woody plants and works by defoliation.
   3. Agent Orange, a mixture of 2,4-D and 2,4,5-T, was used in Vietnam to remove enemy cover and destroy crops.
   4. Herbicide contaminants called dioxins are thought to cause birth defects.
   5. Continuing concern about dioxin contamination led the EPA to ban 2,4,5-T in 1985.

C. Atrazine and Glyphosate.
   1. Atrazine kills plants by stopping the electron transfer reactions of photosynthesis.
      a. When atrazine is used in corn fields, the corn plants deactivate atrazine and are not killed. Weeds cannot deactivate atrazine and are killed.
   2. Glyphosate, a derivative of the amino acid glycine, kills all vegetation.
      a. Genetically modified soybeans, corn, alfalfa, and other plants that are resistant to glyphosate have been developed.
      b. Widespread use of glyphosate has resulted in many glyphosate-resistant weeds.

D. Paraquat: A Preemergent Herbicide—and More.
   1. A preemergent herbicide is one used to kill weed plants before crop seedlings emerge.
   2. Paraquat inhibits photosynthesis in all plants but is rapidly broken down in the soil.
   3. Paraquat is one of the most potent toxins used in agriculture. Farm workers exposed to paraquat have an increased chance of developing Parkinson's disease and other health problems.

E. Organic Herbicides.
   1. Herbicides made from natural ingredients do not work as rapidly or as well as synthetic herbicides, and may need to be used in fairly high concentrations, making them expensive for widespread use.

F. Genetically Engineered Herbicide Resistance.
   1. To avoid having crops affected by herbicides, herbicide-resistant genetically modified organisms (GMOs) are being developed.
      a. Herbicide-resistant corn and soybeans have been developed so far.

G. Human Risks.
   1. The EPA estimate that physicians diagnose between 10,000 and 20,000 pesticide poisonings a year among agricultural workers.
   2. The World Health Organization estimates there are 300,000 deaths of agricultural workers each year worldwide and that pesticides adversely affect the health of 2–5 million people annually.

Answers to Self-Assessment Questions

1. d The harmful component in Agent Orange was dioxin found in 2,4,5-T.
2. a Paraquat is one of the most potent toxins used in agriculture.
3. d The herbicide atrazine acts by inhibiting photosynthesis.
4. d Glyphosate is considered a preemergent herbicide because it can be used to treat the soil before a crop is planted and kills weeds before they emerge through the soil.
5. b Almost all selective herbicides contain 2,4-D.

20.4 Sustainable Agriculture

Learning Objective: • Describe sustainable agriculture and organic farming.

A. Sustainable agriculture is the ability of a farm to produce food and fiber indefinitely, without causing irreparable damage to the ecosystem.
   1. Sustainable agriculture has three main goals: a healthy environment, farm profitability, and social and economic equity.
   2. Sustainable agriculture is similar to the practices that organic farmers and gardeners have followed for generations.
   3. Modern agriculture is energy intensive.
   4. Organic farming is carried out without synthetic fertilizers or pesticides.
   5. Organic farms use less energy, but they require more human labor (a renewable resource) than conventional farms.
   6. In addition to organic practices, sustainable agriculture involves buying local products and using local services when possible, reducing transportation costs.

Answers to Self-Assessment Questions

1. d Sustainable agriculture is farming that meets human needs without degrading or depleting the environment.
2. d Organic farming differs from conventional agriculture in that organic farmers use no synthetic pesticides.
3. c Organic farmers use more energy than conventional farmers, often because they use more human labor as opposed to mechanical labor, and human labor is less efficient.

20.5 Looking to the Future: Feeding a Growing, Hungry World

Learning Objective: • Assess the challenges in feeding an increasing human population.

A. In 1830, Thomas Malthus predicted that population would increase faster than the food supply and that unless the birth rate was controlled, poverty and war would have to serve as restrictions on the increase.

1. Malthus said population grows geometrically while food growth is arithmetic.
2. In arithmetic growth, a constant amount is added in each time period.
3. In geometric growth, the increment increases in size for each time period.
4. The doubling time for a population growing geometrically can be approximated using the Rule of 72: Doubling time = 72/percentage of annual growth

B. Can we feed a hungry world?
1. The United Nations Population Division projects that Earth's population will reach 9 billion in 2045.
2. Virtually all the world's arable land is under cultivation now.
3. Even quadrupling current food production would meet our needs for only a few decades.
   a. We lose farmland every day to housing, roads, erosion, encroaching deserts, and increasing salt content of irrigated soils.
   b. Ultimately, the only solution will be to stabilize the population.
4. Despite a growing movement toward sustainable agriculture, we are still greatly dependent on a high-energy form of agriculture that uses synthetic fertilizers, pesticides, and herbicides and depends on machinery that burns fossil fuels.

C. Green Chemistry: Safer Pesticides through Biomimicry and Green Chemistry.
1. Generally, a safer pesticide means higher selectivity toward the pest organism and fewer effects on the surrounding environment and beneficial organisms.
2. In creating more targeted pesticides, scientists focus on structure–function relationships to find specific modes of action that are unique to the targeted pest, in a process called biomimicry.

Answers to Self-Assessment Questions

1. a The production levels of corn and soybeans, irrespective of their food and nonfood use, does not necessarily address the starvation problem if the locations of the corn and soybeans aren't distributed as needed or can't be afforded by those in need.
2. c The use of improved agricultural practices has made it possible to provide food for increasing populations, especially in developed countries.
3. b Use the Rule of 72 to estimate this: doubling time = 72/3.32 = 22 years. Because the calculation gives an estimate, we can see that at the current rate of population growth, Uganda's population will double in about 20 years.
4. b India's growth rate is such that within a decade it might surpass China as the most populous country on Earth.

LEARNING OBJECTIVES

You should be able to . . .

1. List the three primary plant nutrients as well as several secondary nutrients and micronutrients, and describe the function of each. (20.1)

2. Differentiate between organic fertilizers and conventional fertilizers. (20.1)

3. Name and describe the action of the main kinds of pesticides. (20.2)

4. List several biological pest controls, and explain how they work. (20.2)

5. Name and describe the action of the major herbicides and defoliants. (20.3)

6. Describe sustainable agriculture and organic farming. (20.4)

7. Assess the challenges in feeding an increasing human population. (20.5)

8. Identify the benefits of pesticides to humans and the challenges their use presents.

9. Give examples of how pesticides can be designed to reduce hazards to humans and the environment.

DISCUSSION

In this chapter we consider the chemical requirements for healthy plants and describe some of the threats to successful agriculture in the form of pests and weeds. Use of modern of pesticides and herbicides has helped increase crop production but has also presented environmental challenges. Organic farming avoids the use of synthetic pesticides and herbicides, allowing for much more sustainable agricultural practices. According to Malthus, food production grows arithmetically while population grows geometrically, so food production cannot keep up with population growth into the future. Additional scientific and technical advances may help increase food production or control population growth, but, at present, the prospect of food shortages is real. Applying green chemistry principles to the development of pesticides and herbicides has reduced the environmental threats caused by use of those crop-enhancing products.

EXAMPLE PROBLEMS

1. The LD_{50} of the herbicide paraquat is 150 mg/kg orally in rats. If its toxicity in humans is comparable, what is the lethal dose for a 50-kg (110-lb.) person?

   Lethal dose = (50 kg person)(150 mg paraquat/kg person) = 7500 mg paraquat

2. In 1986, it was estimated that world population growth had dropped to an annual rate of 1.7%. At this rate, how long will it be before the world population has doubled? Use the Rule of 72:

$$\frac{72}{1.7} = 42 \text{ years}$$

ADDITIONAL PROBLEMS

1. The U.S. population was estimated in 1986 to be increasing at a rate of 0.8% per year (excluding illegal immigration). At that rate, when will the population have doubled?

2. The LD_{50} of pyrethrins is 1.2 g/kg orally in rats. If its toxicity in humans is similar, what is the lethal dose for a 70-kg person?

ANSWERS TO ADDITIONAL PROBLEMS

1. Doubling time = 72/0.08% = 90 yr

2. Lethal dose = (70 kg)(1.2 g/kg) = 84 g

ANSWERS TO ODD-NUMBERED REVIEW QUESTIONS AND SOLUTIONS FOR ODD-NUMBERED END-OF-CHAPTER PROBLEMS

Review Questions

1. Most of the matter of a growing plant comes from carbon dioxide and water through the photosynthesis process.

3. A preemergent herbicide is one used to kill weed plants before crop seedlings emerge.

5. Malthus predicted that population would increase faster than the food supply. He stated that unless the birthrate was controlled, poverty and wars would serve as population restrictions.

Problems

7. The three structural elements of a plant are carbon, hydrogen, and oxygen.

9. Nitrogen (N_2) can be converted into a soluble form such as ammonia by lightning, bacterial action, and industrial production from hydrogen and nitrogen using the Haber process.

11. Anhydrous ammonia is gaseous ammonia that has been compressed until it liquefies ("anhydrous" means "without water"). Anhydrous ammonia is compressed and stored in tanks from which it is applied directly to the soil as fertilizer.

13. Each time a positive potassium ion (K^+) enters the root tips of a plant, a hydronium ion (H_3O^+) must leave the plant so that the plant can maintain electrical neutrality.

15. DDT is easily synthesized from cheap, readily available chemicals and is very effective against a variety of insect pests. Following its use, DDT-resistant houseflies were reported, as was DDT's toxicity to fish. In addition, DDT causes birds to produce eggs with thin shells, and very low concentrations of DDT interfere with the growth of plankton and the reproduction of crustaceans such as shrimp. Finally, DDT remains toxic long after its initial application.

17. Chlorinated hydrocarbons are soluble in fats and bioaccumulate as one species feeds on those species lower in the food chain.

19. Pheromones are chemicals secreted to mark a trail, send an alarm, or attract a mate. They attract male insects into traps, allowing a determination of which pests are present and the level of the infestation. Workers can then undertake measures to minimize crop damage. Pheromones have been used by grape growers in Germany and Switzerland.

21. The sterile insect technique (SIT) is a method of insect control that involves rearing large numbers of males, sterilizing them through radiation, chemicals, or cross-breeding, and then releasing them in areas of infestation. SIT is not widely used because it is very expensive and time consuming.

23. Mass of DDT that would kill a 70-kg individual = (70 kg)(0.5 g DDT/kg body weight) = 35 g

25. Lethal dose of diquat dibromide = (56 kg)(600 mg diquat dibromide/kg) = 3.4×10^4 mg or 34 g
Lethal dose of paraquat = (56 kg)(150 mg paraquat/kg) = 8.4×10^3 mg or 8.4 g

27. The gerbil population increases geometrically; 24 weeks represents four doubling times, so that at the end of 24 weeks there would be $2^4 = 32$ gerbils.
Arithmetic growth occurs when a constant amount is added in each growth period. For example, adding $10.00 to a savings account each week.

29. Doubling time = 72/0.493% = 146 years. This growth rate may be changed by factors such as the availability of food, the increase or decrease in mortality due to disease or war, or political restrictions on birthrates.
Number of months = ($1000)(1 month/$50) = 20 months. This is an example of arithmetic growth.

31. Of the 7.7 billion acres of ice-free land, 14.9% of it is used for urban areas, roads, and mines.
Number of acres used for urban areas, roads, and mines = (7.7×10^9 acres)(0.149) = 1.1×10^9 acres

33. The photosynthetic reaction: $6\ CO_2 + 6\ H_2O \rightarrow C_6H_{12}O_6 + 6\ O_2$

35.
1. Boric acid is an example of (b)—it is not an organic compound but is used in organic farming.
2. Diatomaceous earth is (b) used in organic farming but is not an organic chemical.
3. Lead arsenate is (d)—neither organic and not used in organic farming.
4. Malathion is (a)—organic in a chemical sense but not used in organic farming.
5. Rotenone is (c)—an organic compound used in organic farming.

37. The chemical formula for potassium nitrate is KNO_3. This compound supplies both potassium ion (K^+) and nitrate ion (NO_3^-), which is a usable form of nitrogen for plants.
a. Ammonium monohydrogen phosphate supplies nitrogen and phosphorus.
b. $2\ NH_3 + H_3PO_4 \rightarrow (NH_4)_2HPO_4$

39. $CuO(s) + H_2SO_4(aq) \rightarrow CuSO_4(aq) + H_2O(l)$

41. Lime is another name for calcium oxide, CaO. Lime is added to neutralize acidic soils.

43. The reactions are $2\ KOH + CO_2 \rightarrow K_2CO_3 + H_2O$
$K_2CO_3 + CO_2 + H_2O \rightarrow 2\ KHCO_3$.
Because these reactions do not occur in nature, potassium bicarbonate made by these reactions is not organic.

45. a. percent population increase = [(37,500,000 – 34,500,000)/34,500,000](100%)
= 8.7%

b. Use the Rule of 72 to estimate the doubling time. Doubling time = 72/8.7 = 8.3 years.
c. Factors likely to change this rate are changes in birthrate or death rate (war, famine, etc.).

47. The answers are a and b. Farmers use pesticides both to increase food production and to reduce the chance of crop failure.

49. The answer is d. All of the benefits listed in choices a, b, and c are gained through the application of green chemistry principles to the design of pesticides.

CHAPTER

21

Household Chemicals

Helps and Hazards

CHAPTER SUMMARY

21.1 Cleaning with Soap

Learning Objectives: • Describe the structure of soap, and explain how it is made and how it works to remove greasy dirt. • Explain the advantages and disadvantages of soap.

A. Cleaning with Soap.
   1. In developing societies, clothes were cleaned by beating them with rocks in a stream.
      a. Sometimes saponins, soapy compounds found in the leaves of soapworts and soapberry plants, were used as the first detergents.
   2. Plant ashes contain potassium carbonate and sodium carbonate, compounds that react with water to form an alkaline solution with detergent properties.
      a. These alkaline ashes were used by Babylonians 4000 years ago.
      b. Europeans used plant ashes to wash their clothes as recently as 100 years ago.
      c. Sodium carbonate is still sold today as washing soda.

B. Soap Making: Fat + Lye → Soap Plus Glycerol.
   1. Sodium hydroxide, also called lye, is produced by heating sodium carbonate (produced by the evaporation of alkaline water) with lime (from limestone or seashells).
      $Na_2CO_3 + Ca(OH)_2 \rightarrow 2\ NaOH + CaCO_3$
   2. Sodium hydroxide was heated with animal fats or vegetable oils to produce soap.
      a. Soaps are the salts of a long-chain carboxylic acid.
      b. In modern commercial soap-making, fats and oils are often hydrolyzed to fatty acids and glycerol (also called glycerin) with superheated steam and the fatty acids are then neutralized to make soaps.
         i. Hand soaps usually contain additives such as dyes, perfumes, creams, and oils.
         ii. Scouring soaps contain abrasives such as silica and pumice.
         iii. Some soaps have air blown in before they solidify to lower their density so they float.
         iv. Potassium soaps are softer than sodium soaps and produce a finer lather, so they are used alone or in combination with sodium soaps in liquid soaps and shaving creams.

3. Soaps used in shampoos and other cosmetics are also made by reacting fatty acids with triethanolamine.

C. How Soap Works.
   1. Oil and greases hold dirt to skin and fabrics.
   2. A soap has an ionic (hydrophilic, water-soluble) end (called a head) and a hydrocarbon (hydrophobic, oil-soluble) end called a tail.
      a. Soap molecules break oils into tiny globules called micelles by sticking their hydrocarbon tails into the oil while the ionic heads remain in the aqueous phase.
      b. The oil droplets don't coalesce owing to the repulsion of the charged groups.
   3. The oil and water form an emulsion, with soap acting as the emulsifying agent.
   4. Any agent that stabilizes the suspension of nonpolar substances, such as oil and grease, in water, including soap, is called a surface-active agent (or surfactant).

D. Disadvantage and Advantages of Soaps.
   1. Disadvantage: "Hard" water contains calcium, magnesium, or iron ions, which form insoluble salts with the fatty acid anions that precipitate as "bathtub ring."
   2. Advantages: Soap is an excellent cleanser in soft water; it is relatively nontoxic; it is derived from renewable resources (animal fats and vegetable oils); and it is biodegradable.

E. Water Softeners.
   1. Water softeners are used to remove calcium, magnesium, and iron ions from water.
   2. Sodium carbonate (washing soda) acts in two ways.
      a. It makes the water basic so that the fatty acids won't precipitate.
      b. The carbonate ions precipitate the hard-water ions and keep them from forming soap scum.
   3. Trisodium phosphate acts similarly to sodium carbonate.
      a. It makes the water basic.
      b. The phosphate ions precipitate calcium and magnesium ions.
   4. Water softening tanks absorb the calcium, magnesium, and iron ions on polymeric resins, exchanging them with sodium ions to soften the water.

Answers to Self-Assessment Questions

1. d Sodium phosphate may act as a water softener but is not a soap. Soaps are salts of fatty acids.
2. c $CH_3(CH_2)_{14}COONa$ is a soap formula. The compound is a sodium salt of a fatty acid.
3. b Sodium chloride does not precipitate the cations that make water hard (Ca^{2+}, Mg^{2+}, and Fe^{2+}).
4. a The hydrocarbon portion of a soap molecule is nonpolar (hydrophobic), so it is attracted to oil, which is also nonpolar or hydrophobic.
5. b The ionic end of a soap molecule is charged, so it is hydrophilic or attracted to water.
6. d When used in cleaning, soap removes oily dirt by forming micelles.
7. a In the cleansing action of soap, the nonpolar hydrocarbon tail intermingles with the nonpolar oily dirt, and the ionic head with the polar water.
8. d Hard water contains Ca^{2+}, Mg^{2+}, and Fe^{2+} ions. These are exchanged for Na^{+} ions (which do not precipitate soaps) in a water softener.
9. d The calcium and magnesium salts of soaps are insoluble in water, so they precipitate out and are not available to clean.

21.2 Synthetic Detergents

Learning Objectives: • List the structures, advantages, and disadvantages of synthetic detergents. • Classify surfactants as amphoteric, anionic, cationic, and nonionic, and describe how they are used in various detergent formulations.

A. ABS Detergents: Nondegradable.
   1. Alkylbenzenesulfonate (ABS) detergents were derived from propylene, benzene, and sulfuric acid, followed by neutralization.
      a. ABS detergents worked well in acidic and hard waters but were not degraded by microorganisms in wastewater treatment plants or in nature, threatening the groundwater supply.
      b. Foaming rivers led to their ban and replacement with biodegradable detergents.

B. LAS Detergents: Biodegradable.
   1. Linear alkylsulfonates (LAS) are derived from ethylene, benzene, and sulfuric acid, followed by neutralization.
      a. LAS detergents are biodegradable.
      b. They work better than soap in acidic solution and in hard water.

C. Laundry Detergent Formulations.
   1. Detergent products used in homes and commercial laundries usually contain a variety of ingredients.
   2. Surfactants are classified according to the ionic charge, if any, on their working part.
      a. Soaps and ABS and LAS detergents are all anionic surfactants; they have a negative charge on the active part.
         i. Anionic surfactants are used in laundry and hand dishwashing detergents, household cleaners, and personal cleansing products.
      b. Nonionic surfactants have no electrical charge.
         i. They are low sudsing and are not affected by water hardness.
         ii. They are typically used in laundry and automatic dishwasher formulations.
      c. Cationic surfactants have a positive charge on the active part.
         i. These surfactants are not particularly good detergents, but they do have germicidal properties.
         ii. They are used as cleansers and disinfectants in the food and dairy industries and as sanitizing agents in some household cleaners.
      d. Amphoteric surfactants carry both a positive and a negative charge.
         i. They react with other acids and bases and are noted for their mildness, sudsing, and stability.
         ii. They are used in personal cleansing products such as shampoos for babies and in household cleaning products such as liquid hand-washing soaps.
   3. Any substance added to a surfactant to increase its detergency is called a builder.
      a. Most builders are water softeners and may include phosphates.
         i. Many locales have banned the use of phosphates because they promote eutrophication of lakes.
      b. Other builders include zeolites, complex aluminosilicates that tie up the hard-water ions.
   4. Almost all detergents include fluorescent dyes called optical brighteners.
      a. Optical brighteners make clothes appear brighter by absorbing ultraviolet light (invisible) and reemitting it as blue light (visible).
      b. Optical brighteners appear to have low toxicity to humans but do cause skin rashes in some people.
   5. Some detergent formulations include lipases (enzymes that work on lipids) to help remove fats, oils, and blood.

D. Dishwashing Detergents.
   1. Liquid detergents for washing dishes by hand generally contain one or more surfactants; few contain builders, but they may contain enzymes to help remove greases and protein stains.
   2. Detergents for automatic dishwashers are different from those for hand dishwashing.
      a. They are often strongly alkaline (some contain sodium hydroxide) and should never be used for hand dishwashing.
      b. They depend mainly on their strong alkalis, heat, and on vigorous agitation of the machine for cleaning.

Answers to Self-Assessment Questions

1. d In order to remove grease, a detergent must have an ionic head that will dissolve in water and a nonpolar tail that will dissolve in grease so that micelles can form.
2. b The hydrocarbon chain of ABS is branched and resistant to biodegradation.
3. b $CH_3(CH_2)_9CH(CH_3)C_6H_4SO_3Na$ is an LAS detergent.
4. a Hard-water ions react with carbonate ions to form products such as $CaCO_3(s)$.
5. a An ion exchange resin retains hardness ions (Ca^{2+}, Mg^{2+}, and Fe^{2+}), exchanging them for Na^+ ions when it softens water.
6. b An optical brightener converts ultraviolet light to blue light.
7. d DEA is neither an acid nor a base (its functional group is an amide), so it is nonionic.
8. a Because it is so alkaline, detergent for automatic dishwashers is the most likely to irritate skin.
9. c Because it is so alkaline (basic), the pH of dishwashing detergent is high.

21.3 Laundry Auxiliaries: Softeners and Bleaches

Learning Objectives: • Identify a fabric softener from its structure and describe its action. • Name two types of laundry bleaches and describe how they work.

A. Fabric Softeners.
   1. The working part of a cationic surfactant is a positive ion.
   2. Most cationic surfactants are quaternary ammonium salts; they have four alkyl groups attached to a nitrogen atom that has a positive charge.
   3. Quaternary ammonium salts with two long hydrocarbon tails (and two smaller groups on nitrogen) act as fabric softeners.
      a. They form a film on fibers, lubricating them for flexibility and softness.

B. Laundry Bleaches: Whiter Whites.
   1. Bleaches are oxidizing agents that remove stains from fabrics.
   2. There are two main types of laundry bleaches: chlorine bleaches and oxygen bleaches.
      a. Liquid laundry bleaches ("chlorine bleaches") are dilute solutions of sodium hypochlorite (NaOCl).
         i. These solutions release chlorine rapidly and can damage fabrics and turn polyester fabrics yellow.
         ii. "Ultra" chlorine bleaches have a higher concentration of NaOCl.
      b. Solid bleaches release chlorine slowly in water to minimize damage to fabrics.
   3. Oxygen-releasing bleaches usually contain sodium percarbonate or sodium perborate.
      a. In hot water they release hydrogen peroxide, which acts as a strong oxidizing agent.
      b. Borates are somewhat toxic.
   4. Bleaches work by acting on certain light-absorbing chemical groups, called chromophores, which cause a substance to be colored.

Answers to Self-Assessment Questions

1. d $CH_3(CH_2)_{16}CH_2N^+(CH_3)_2CH_2(CH_2)_{16}CH_3Cl^-$ could serve as a fabric softener. Note the two long alkyl chains attached to the nitrogen atom.
2. d Quaternary ammonium compounds (compounds with four groups bonded to the nitrogen atom) are the most common fabric softeners.
3. a In hot water, oxygen bleaches work by releasing hydrogen peroxide, which is an oxidizing agent.
4. d NaOCl is the main component of household bleach, often called "chlorine bleach."
5. d Bleaches act by oxidizing chromophores, the substances that give the color we want to remove.
6. b Perborate bleaches do work well on cotton fabrics.

21.4 All-Purpose and Special-Purpose Cleaning Products

Learning Objective: • List the major ingredient(s) (and their purposes) of some all-purpose cleaning products and some special-purpose cleaning products.

A. All-Purpose Cleaners.
   1. All-purpose cleaners for use in water may contain surfactants, sodium carbonate, ammonia, solvent-type grease cutters, disinfectants, bleaches, deodorants, and other ingredients.
   2. Household ammonia has many uses.
      a. It can be used undiluted to loosen baked-on grease, burned-on food, or diluted to clean glass.
      b. Mixed with detergent, ammonia removes wax from linoleum.
   3. Baking soda is a mild abrasive cleanser that absorbs odors from refrigerators and freezers.
   4. Vinegar is a good grease cutter but should not be used on marble because it reacts with marble, causing surface pitting.

B. Hazards of Mixing Cleaners.
   1. Mixing bleach with other household chemicals can be quite dangerous.
      a. Mixing hypochlorite bleach with hydrochloric acid produces poisonous chlorine gas.
         i. Mixing bleach with toilet bowl cleaners that contain HCl is especially dangerous because bathrooms are seldom well ventilated.
      b. Mixing bleach with ammonia is extremely hazardous because two of the gases produced are chloramine (NH_2Cl) and hydrazine (NH_2NH_2), both of which are toxic.

C. Special-Purpose Cleaners.
   1. Toilet bowl cleaners are acids that dissolve the "lime" buildup that forms in toilet bowls.
      a. Solid crystalline cleaners usually contain sodium bisulfate ($NaHSO_4$) and the liquid cleaners include hydrochloric acid, citric acid, or some other acidic material.
   2. Most scouring powder cleansers contain an abrasive such as silica (SiO_2) that scrapes soil from hard surfaces and a surfactant to dissolve grease.
      a. Some scouring powders include bleach.
      b. Scouring powders can scratch surfaces.
   3. Glass cleaners are volatile liquids that evaporate without leaving a residue.
      a. A common glass cleaner is isopropyl alcohol (rubbing alcohol) diluted with water.
      b. Most commercial glass cleaners contain ammonia or vinegar.
   4. Drain cleaners often contain sodium hydroxide, either in solid form or as a concentrated liquid.
      a. Sodium hydroxide reacts with water to generate heat that melts the grease that was clogging the pipes.

b. Some products also contain bits of aluminum metal that react with the sodium hydroxide solution to form hydrogen gas that bubbles out of the clogged area, creating a stirring action.
c. Many liquid drain cleaners contain bleach (NaOCl), as well as concentrated sodium hydroxide.
5. The active ingredient in most oven cleaners is sodium hydroxide.
   a. Sodium hydroxide converts greasy deposits to soap.
6. Hand sanitizers contain a high concentration of alcohol that kills bacteria and evaporates rapidly during use, leaving hands clean and dry.
   a. A surfactant provides additional germicidal properties, and a thickening agent makes the sanitizer easier to apply.
7. Home mechanics use waterless hand cleaners with lanolin, pumice, an oil with a fragrance, and a surfactant.
8. Green cleaners generally have more ingredients derived from plants.
   a. While more of the ingredients in these products come from plants, the lack of regulations makes green cleaners difficult to evaluate.

D. Practicing Green Chemistry at Home.
   1. Buy and use only what you need.
   2. Choose safer products.
   3. Choose products that minimize depletion of resources.
   4. Select reusable, recyclable, or readily degradable materials when possible.
   5. Keep informed about new developments and product safety regulations.
   6. Prevent accidents by reading product labels, using proper protective gear, maximizing ventilation, and choosing the safest products.

Answers to Self-Assessment Questions

1. c Mixing acidic toilet cleaner with bleach produces chlorine gas (Cl_2), which is highly toxic.
2. a Household ammonia cannot be safely used for cleaning aluminum because it erodes the surface.
3. c Vinegar cannot be safely used for polishing marble surfaces because it causes pitting.
4. d Baking soda (sodium bicarbonate) cannot be used safely for neutralizing bleach.
5. b Mixing vinegar (an acid) and ammonia (a base) together forms ammonium acetate, a salt.
6. d Mixing hypochlorite bleach and ammonia forms chloramine and hydrazine, which are both toxic gases.
7. c Lime ($CaCO_3$) deposits in toilet bowls are best removed by treatment with HCl.
8. a The most effective way to keep drains open is to keep grease and hair from going down them.

21.5 Solvents, Paints, and Waxes

Learning Objectives: • Identify compounds used in solvents and paints, and explain their purposes.
• Identify waxes by their structure.

A. Solvents.
   1. Solvents are used to remove paint, varnish, adhesives, waxes, and other materials.
      a. Most organic solvents are volatile and flammable, and many have toxic fumes.
      b. Long-term sniffing of the fumes can cause permanent damage to vital organs, especially the lungs, as well as irreversible brain damage and heart failure.

B. Paints.
  1. Paint is a broad term used to cover lacquers, enamels, varnishes, oil-base coatings, and a number of different water-base finishes.
  2. A paint contains three basic ingredients: a pigment, a binder, and a solvent.
     a. Since 1977, titanium oxide, a brilliant white nontoxic pigment has replaced lead carbonate as the foundational pigment to which small amounts of colored pigments or dyes are added to achieve the desired color.
     b. The binder, or film former, is a substance that binds the pigment particles together and holds them on the painted surface.
        i. In oil paints, the binder is usually tung oil or linseed oil (oil-based paints)
        ii. In water-based paints the binder is a latex-based polymer such as polyvinyl acetate.
     c. The solvents keep the paint fluid until it is applied to a surface and is usually alcohol, a hydrocarbon, an ester, or water.
     d. Paint may also contain additives: a drier (or activator) to make the paint dry faster, a fungicide to act as a preservative, a thickener to increase the paint's viscosity, an anti-skinning agent to keep the paint from forming a skin inside the can, and a surfactant to stabilize the mixture and keep the pigment particles separated.

C. Waxes.
  1. A wax is an ester of long-chain, organic (fatty) acid and a long-chain alcohol.
     a. Waxes are produced by plants and animals mainly as protective coatings.
     b. Beeswax, carnauba wax, spermaceti wax, and lanolin are some examples.

Answers to Self-Assessment Questions

1. d Organic solvents in household and commercial products are very toxic.
2. c The three basic components of a paint are a pigment, a solvent, and a binder.
3. d $CH_3(CH_2)_{14}COOCH_2(CH_2)_{28}CH_3$, an ester of a long-chain organic acid and a long-chain alcohol, is a wax.
4. c Waxes are esters with two long chains (see question 3 for an example).
5. c Because of the toxicity of lead, modern paint pigments substitute titanium dioxide (TiO_2) for lead carbonate ($PbCO_3$).

21.6 Cosmetics: Personal Care Chemicals

Learning Objectives: • Identify the principal ingredients in various cosmetic products, and discuss their purposes. • Describe the chemical nature of skin, hair, and teeth.

A. Cosmetics are defined by the U.S. Food, Drug, and Cosmetic Act of 1938 as "articles intended to be rubbed, poured, sprinkled, or sprayed on, introduced into, or otherwise applied to the human body or any part thereof, for cleansing, beautifying, promoting attractiveness, or altering the appearance."
  1. Soap, antiperspirants, and antidandruff shampoos are excluded from this definition.
  2. The main difference between drugs and cosmetics is that drugs must be proven safe and effective before they are marketed, while cosmetics generally do not have to be tested before being marketed.

B. Skin Creams and Lotions.
  1. Skin is the body's largest organ, with an area of about 18 ft^2.
  2. The outer layer of skin is the epidermis, divided into two layers: dead cells and the corneal layer.

a. The corneal layer of the epidermis is mainly keratin, a tough, fibrous protein with a moisture content of about 10%.
b. Below 10% moisture, skin is dry and flaky.
c. Above 10% moisture, microorganisms flourish.
d. Sebum (skin oils) protects the skin from loss of moisture.

3. Lotions and creams are applied to the dead cells of the corneal layer.
   a. Lotions are emulsions of oil in water that help hold moisture in.
      i. Typical oils include mineral oil, petroleum jelly, and natural fats and oils.
   b. Creams are emulsions of water in oil that help hold moisture in.
   c. Creams and lotions protect the skin by coating and softening it.
      i. Emollients are skin softeners that provide a protective coating on the skin to prevent loss of moisture.
   d. Moisturizers that hold moisture in skin usually contain lanolin or collagen.
   e. Some cosmetics contain humectants that hold water by hydrogen bonding.
   f. Some cosmetics contain exfoliants that cause dead cells to fall off more rapidly than without them.

C. Shaving Creams.
1. Shaving creams consist of a soap or soap-like compound, glycerol, and water, plus fragrances, preservatives, and sometimes a foaming agent.
2. The soap forms a smooth, slippery surface on the skin and hair. Glycerol is attracted to water by hydrogen bonding and acts as a humectant.

D. Sunscreens.
1. Sunscreen lotions offer physical or chemical protection from ultraviolet radiation. The first common ingredient was *para*-aminobenzoic acid, but it is seldom used today and has usually been replaced by octyl methoxycinnamate.
   a. Skin protection factors (SPFs) indicate how many times longer a person can remain in the sun without burning.

E. Lipsticks and Lip Balms.
1. Lipsticks and lip balms are similar to skin creams in composition and function.
   a. They are made of an oil and a wax, with a higher proportion of wax than in creams to make them firmer than creams.

F. Eye Makeup.
1. Various chemicals are used to decorate the eyes.
2. Some people have allergic reactions to ingredients in eye makeup or get eye infections caused by bacterial contamination.
   a. It is recommended that eye makeup be discarded after 3 months.

G. Deodorants and Antiperspirants.
1. Deodorants contain perfume to mask body odor and a germicide to kill odor-causing bacteria.
2. Antiperspirants retard perspiration.
   a. Nearly all antiperspirants have aluminum or zirconium chlorohydrate as the only active ingredient.
   b. Aluminum or zirconium chlorohydrate is an astringent; it constricts the openings of sweat glands.

H. Toothpaste: Soap with Grit and Flavor.
1. Toothpastes have two essential ingredients: a detergent and an abrasive.
   a. Sodium lauryl sulfate is a typical detergent used in toothpastes, but any pharmaceutical-grade soap or detergent is satisfactory.
   b. Other ingredients include flavors, colors, aromas, and sweeteners.

2. Tooth decay is caused by bacteria that convert sugars to plaque, a biofilm composed of thousands of bacteria, small particles, proteins, and mucus deposited on teeth when they are not well cleaned.
3. Fluorides harden tooth enamel, reducing the incidence of decay.
   a. Fluorides convert hydroxyapatite to fluorapatite, a harder material that is more resistant to decay.
4. Hydrogen peroxide and baking soda are used to prevent gum disease, which is the major cause of adult tooth loss.

I. Perfumes, Colognes, and Aftershaves.
1. Perfumes are among the most ancient and widely used cosmetics.
   a. A good perfume may have one hundred or more constituents.
   b. The components are divided into three categories, called notes, based on differences in volatility.
      i. The most volatile fraction is called the top note and is made up of relatively small molecules responsible for the odor when the perfume is first applied.
      ii. The middle note (or heart note) is intermediate in volatility and responsible for the lingering aroma after most of the top-note compounds have vaporized.
      iii. The end note (or base note) is the low-volatility fraction and is made up of compounds with large molecules, often with musky odors.
2. A perfume usually consists of 10–25% fragrant compounds and fixatives dissolved in ethyl alcohol.
3. Colognes are perfumes diluted with ethyl alcohol or an alcohol-water mixture.
   a. Colognes are only about one-tenth as strong as perfumes, usually containing only 1–2% perfume essence.
4. Aftershave lotions are similar to colognes.
   a. Most are 50–70% ethanol, the remainder being water, perfume, and food coloring.
   b. Some contain menthol for a cooling effect on the skin; others contain an emollient to soothe chapped skin.

J. Some Hairy Chemistry.
1. Hair is composed of the fibrous protein keratin.
2. Protein molecules in hair are strongly held together by four types of forces.
   a. Hydrogen bonds—disrupted by water.
   b. Salt bridges—destroyed by changes in pH.
   c. Disulfide linkages—broken and destroyed by permanent wave and hair-straightening treatments.
   d. Dispersion forces.

K. Shampoo.
1. When hair is washed, the keratin absorbs water. The water disrupts hydrogen bonds and some salt bridges.
   a. The hair is softened and made more stretchable.
   b. Acids and bases are particularly disruptive to salt bridges, making pH control important.
      i. The number of salt bridges is maximized at a pH of 4.1.
2. Hair shafts are dead; only the root is alive. The hair is lubricated by sebum.
   a. Dirt adheres to sebum.
   b. Washing hair removes the oil and dirt.
3. Modern shampoos use a synthetic detergent as a cleansing agent.
   a. Shampoo for adults usually has an anionic surfactant, such as sodium dodecyl sulfate, as the principal active ingredient.
   b. Baby shampoos have an amphoteric surfactant with both negatively charged oxygen and positively charged nitrogen.

4. Most components other than a detergent are in a shampoo only as the basis for advertising claims.
5. Hair is protein with both acidic and basic groups.
   a. Most shampoos have a pH between 4 and 7, which does not damage hair or scalp.
   b. Protein shampoos condition hair by coating the hair shaft with protein (glue).
   c. Shampoos for dry or oily hair differ only in the relative amounts of detergent.

L. Hair Coloring.
1. The color of hair and skin is determined by the relative amounts of two pigments: eumelanin and pheomelanin.
   a. Eumelanin is responsible for brown and black colors.
   b. Pheomelanin is the pigment in red hair.
   c. Blondes have little of either pigment; brunettes can become blondes by oxidizing the pigments with hydrogen peroxide.
2. Permanent hair dyes often are derivatives of *para*-phenylenediamine. These compounds penetrate the hair shaft and are oxidized to colored products (presumably quinones).
   a. para-Phenylenediamine produces a black color.
   b. para-Aminodiphenylamine sulfonic acid is used for blondes.
   c. Intermediate colors use other derivatives.
   d. Several of the hair-coloring diamines have been shown to be carcinogenic or mutagenic.
3. Hair treatments that restore color gradually use lead acetate solutions.
   a. The lead ions penetrate the hair and react with sulfur to form black, insoluble lead sulfide.

M. Permanent Waving: Chemistry to Curl Your Hair.
1. Adjacent protein molecules in hair are cross-linked by disulfide groups. To put curl in hair:
   a. A reducing agent such as thioglycolic acid is used to rupture the disulfide linkages.
   b. The hair is set on curlers; the protein chains slide in relation to one another.
   c. Disulfide linkages are formed in new positions.
   d. The same chemical process can be used to straighten hair.

N. Hair Sprays.
1. Hair can be held in place by resins (often polyvinylpyrrolidone, PVP, or its copolymers).
   a. The resin can be dissolved in a solvent and applied as a spray.
2. Holding resins can also be formulated as mousses (foams or froths).

O. Hair Removers (Depilatories).
1. These are strongly basic sulfur compounds such as sodium sulfide or calcium thioglycolate formulated into a lotion that destroys peptide bonds so that hair can be washed off.
2. Hair removers can damage skin.

P. Hair Restorers.
1. Minoxidil (Rogaine®) dilates blood vessels and produces growth of fine hair on skin containing hair follicles.
   a. To be effective, it must be used continuously.

Q. The Well-Informed Consumer.
1. Most cosmetics are formulated from inexpensive ingredients.
2. You don’t have to pay a lot for extra ingredients that contribute little to the performance of the product.

R. Green Chemistry: Practicing Green Chemistry at Home
1. We can’t reasonably design the products found in our homes, but we can use green principles to help control which ones we buy and how we use them.’
   a. Buy and use only what you need.
   b. Use safer products.
   c. Choose products that minimize resources depletion.

   d. Choose reusable, recyclable, or readily degradable products.
   e. Stay informed.
   f. Prevent accidents.
2. Some products are labeled or certified by the EPA Design for the Environment program to help consumers make greener choices. As a part of this program, the EPA works with product manufacturers to choose the safest and most sustainable chemicals using the principles of green chemistry.

Answers to Self-Assessment Questions

1. a Skin creams and lotions do not contain proteins.
2. a A manufacturer does not have to prove it safe and effective for its intended use to market a new cosmetic. Drugs must be tested before they are marketed, but cosmetics do not have to be tested.
3. b Ultraviolet-B rays in sunlight cause the skin to darken by triggering the production of melanin. UVA rays are less energetic and although they, too, cause tanning, they are less dangerous than UVB rays.
4. d Hair gets dirty when the sebaceous glands secrete oils that collect airborne dirt and dust.
5. a Bacteria in plaque metabolize sucrose and produce acids that attack the tooth enamel, causing decay.
6. d The top note of a perfume is made up of the smallest molecules which evaporate the most easily.
7. d Aftershave lotions do not contain proteins.
8. c Perfumes are responsible for many allergic reactions, so a cosmetic advertised as being hypoallergenic will likely not contain any perfume.
9. a When hair is wetted, hydrogen bonds are disrupted.
10. b Kohl is a colored pigment and has been used as eye makeup in many cultures for centuries.
11. c A detergent is the essential ingredient in a shampoo.
12. b Disulfide linkages are broken by reducing agents applied during a home permanent, so the hair can be wound around a curler and the disulfide linkages reformed.
13. a People with only a little melanin in their skin are more susceptible to sunburn and skin cancer because melanin is a protective chemical against ultraviolet radiation damage.

LEARNING OBJECTIVES

You should be able to . . .

1. Describe the structure of soap, and explain how it is made and how it works to remove greasy dirt. (21.1)

2. Explain the advantages and disadvantages of soap. (21.1)

3. List the advantages and disadvantages of synthetic detergents. (21.2)

4. Classify surfactants as amphoteric, anionic, cationic, and nonionic, and describe how they are used in various detergent formulations. (21.2)

5. Identify a fabric softener from its structure, and describe its action. (21.3)

6. Name two types of laundry bleaches, and describe how they work. (21.3)

7. List the major ingredient(s) (and their purposes) of some all-purpose cleaning products and some special-purpose cleaning products. (21.4)

8. Identify compounds used in solvents and paints, and explain their purposes. (21.5)

9. Identify waxes by their structure. (21.5)

10. Identify the principal ingredients in various cosmetic products. (21.6)

11. Describe the chemical nature of skin, hair, and teeth. (21.6)

12. Identify how green chemistry principles can be applied when selecting greener household products.

DISCUSSION

In this chapter we apply some of the principles learned earlier to a study of some of the chemicals used in and around the home. Our main focus is on chemicals used in cleaning, for these are among the most common—and often the most dangerous—of the household chemicals. Others are discussed elsewhere. Pesticides, fertilizers, and other "farm" chemicals (discussed in Chapter 20) are also used on lawns, gardens, and household plants.

In this chapter we also apply our knowledge of chemistry to the substances we put on our skin and hair to make us look or smell better. Some of the properties are desirable; often some are not. The properties are independent of our wishes. Often they fall far short of extravagant advertising claims. With this introduction to cosmetics you will be better equipped to judge for yourself the validity of some of the assertions of advertisers.

ANSWERS TO ODD-NUMBERED REVIEW QUESTIONS AND SOLUTIONS FOR ODD-NUMBERED END-OF-CHAPTER PROBLEMS

Review Questions

1. Soaps are excellent cleansers in soft water, relatively nontoxic, derived from renewable sources, and biodegradable. Their biggest disadvantage is their solubility in hard or acidic water. Because soaps are the anions of weak acids, they lose their charge when exposed to acidic water and precipitate out of solution. Also, the cations associated with hard water, Ca^{2+} and Mg^{2+}, form insoluble compounds with soaps, causing them to precipitate from solution where they are ineffective as cleansers.

3. (1) CH_3COOH, acetic acid, is an ingredient in vinegar (answer a).
   (2) $Na_2Al_2Si_2O_8$, an aluminosilicate, is a zeolite (answer f).
   (3) $NaHCO_3$, sodium bicarbonate, is a mild abrasive cleaner (answer b).
   (4) NaOCl, sodium hypochlorite, is used as bleach (answer c).
   (5) NaOH is used to make soap (answer d).
   (6) Na_3PO_4, trisodium phosphate, is a water softener (answer e).

5. Diluted ammonia solutions loosen baked-on grease and burned-on food, and clean mirrors, windows, and other surfaces. Ammonia vapors are highly irritating. Ammonia solutions should never be mixed with chlorine bleaches or used in a closed room.

7. Baking soda is a mild abrasive, and it absorbs odors.

Problems

9. Potassium soaps are softer and produce a finer lather than sodium soaps.

11. Most scouring soaps contain an abrasive such as silica or pumice.

13. Tripalmitin + lye $\rightarrow$ Sodium palmitate + glycerol

15. a. Potassium stearate: 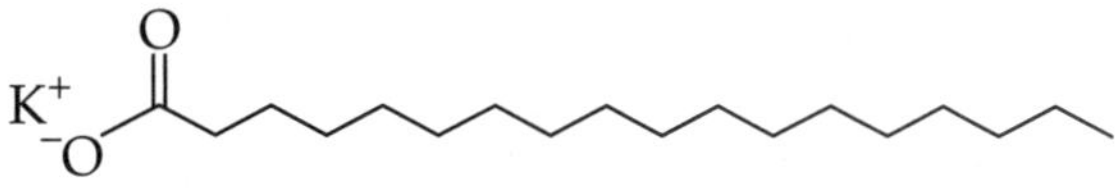

b. Sodium palmitate: 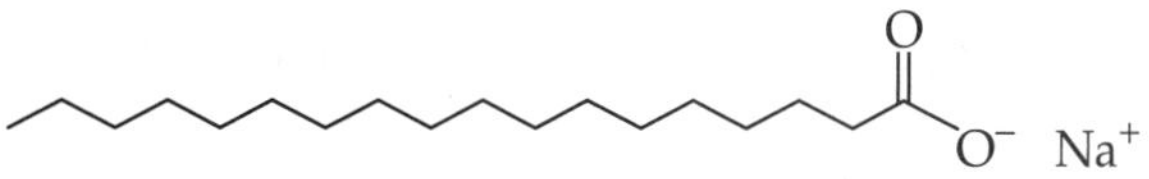

17. Trisodium phosphate softens water by precipitating calcium, magnesium, and iron(II) ions (hard-water ions) as insoluble calcium phosphate [$Ca_2(PO_4)_2$], magnesium phosphate [$Mg_3(PO_4)_2$], and iron(II) phosphate [$Fe_3(PO_4)_2$], and making the water more alkaline.

19. The ions that make water "hard" are Ca^{2+}, Mg^{2+}, and Fe^{2+}.

21. $3\ Ca^{2+}(aq) + 2\ PO_4^{3-}(aq) \rightarrow Ca_3(PO_4)_2(s)$

23. A detergent builder is any substance added to a substrate to increase its detergency.

25. Zeolites exchange calcium and magnesium ions for sodium ions, which allow the detergents to be more effective.

27. An amphoteric surfactant is a molecule that carries both a positive and a negative charge at different parts of its structure.

29. Cationic surfactants have a positive charge on the active part of the molecule, while anionic surfactants have a negative charge. When these two kinds of surfactants are used together, they attract one another and form clumps; thus, they are not effective in the solution.

31. Structure 1 is a linear alkyl sulfonate (LAS).

33. With no branching, all three structures, I, II, and III, are biodegradable.

35. Structures I and II are ionic surfactants. Structure III is a cationic surfactant.

37. A quaternary salt with two long carbon chains and two smaller groups in the nitrogen, such as hexadecyltrimethylammonium chloride, acts as a fabric softener.

39. Trichloroisocyanuric acid is the active ingredient in powdered chlorine-based bleach. Powdered oxygen-based bleaches usually contain sodium percarbonate or sodium perborate.

41. In hot water, sodium percarbonate and sodium perborate bleaches liberate hydrogen peroxide, an oxidizing agent that is responsible for the bleaching action.

43. Most organic solvents used around the home are volatile and flammable. Many have toxic fumes that are narcotics at high concentrations.

45. Gasoline should not be used as a cleaning solvent. It is too hazardous in many ways.

47. Emollients are creams and lotions that protect the skin by coating and softening it.

49. OMC, octyl methoxycinnamate, is not soluble in water and will not wash off as easily as PABA does when an individual sweats or is swimming.

51. Although kohl is mainly carbon, it includes finely ground lead(II) sulfide (galena) or, less commonly, antimony sulfide. Neither of these compounds is safe to use on the skin and near the eyes, especially for children. It is not safe to be taken internally.

53. As is the case with skin creams, lipsticks and lip balms are made up of an oil and a wax. Lipsticks and lip balms have a higher proportion of wax than do skin creams, which makes them firmer. They also contain pigments that skin creams do not contain.

55. Tooth decay is caused primarily by bacteria that convert sugars to plaque, a biofilm that is deposited on the teeth when they are not cleaned adequately. These bacteria convert sugars to acids such as lactic acid, and the acids dissolve tooth enamel. The best ways to prevent tooth decay are to brush and floss the teeth after eating.

57. a. Glycerol is used as a sweetener in toothpaste.
    b. Sodium fluoride is a protection against cavities.
    c. Sodium dodecyl sulfate is the detergent commonly found in toothpaste.
    d. Potassium nitrate is added to toothpastes used by individuals with highly sensitive teeth.
    e. Hydrogen peroxide functions as a whitening agent and helps prevent gum disease.
    f. Hydrated silica functions as an abrasive in toothpaste.

59. A perfume is a complex mixture of compounds used to make the wearer smell pleasant. A cologne is a perfume that has been diluted with ethyl alcohol.

61. Menthol is added to aftershave lotions because it creates a cooling effect on the skin.

63. Temporary dyes are water soluble and can be washed out. Permanent dyes penetrate the hair and remain there until the hair is cut or falls out. Because permanent dyes affect only the dead outer portion of the hair shaft, new hair will grow out in the original shade.

65. A reducing agent is used to break the disulfide linkages in hair. The hair is then wrapped around a rod. A mild oxidizing agent is then applied to reestablish disulfide linkages to reconfigure the hair in the shape of the rod.

67. Changing hair color from dark to lighter involves oxidizing the colored pigment to lighten it. This is often done with hydrogen peroxide. Changing gray hair to a darker color is more complicated as the process involves adding color. Permanent dyes affect only the dead outer layer of the hair, so new hair that grows in is the original gray color. Hair treatments that involve gradual shift from gray to a darker color often use lead acetate that forms lead sulfide. When it is rubbed into the hair, the Pb^{2+} ions react with sulfur to form lead sulfide, which is black. Repeated applications produce darker colors. The safety of using the lead compound is questionable.

69. Structure II, $CH_3(CH_2)_{10}COO^-$, is the anion of a fatty acid and is a soap.

71. Structure I, $CH_3(CH_2)_{15}N^+H_2CH_2COO^-$, is an amphoteric surfactant of the type used in baby shampoos.

73. The structure of isopropyl palmitate is $CH_3(CH_2)_{14}COOCH(CH_3)_2$.

75. Top-note components tend to be small molecules with greater volatility (lower boiling points) than middle-note components of perfumes. Although the sizes of phenylacetaldehyde and 2-phenylethanol are similar, 2-phenylethanol can form hydrogen bonds, while phenylacetaldehyde cannot. Therefore, the volatility of 2-phenylethanol is not as great as that of phenylacetaldehyde.

77. When combined with water, sodium hydroxide dissolves, producing considerable heat that works to melt any grease responsible for the clog. If the clog is caused by hair, the sodium hydroxide reacts with the grease, converting it to a soap that is more easily removed.

79. Both carnauba wax and spermaceti are similar because they are the esters of palmitic acid ($CH_3(CH_2)_{14}COOH$). They are different because the alcohol that reacts with palmitic acid to form carnauba wax has 34 carbon atoms, while the alcohol that reacts with palmitic acid to form spermaceti has 16 carbon atoms.

81. a. $CaCO_3(s) + 2\,HCl(aq) \rightarrow CaCl_2(aq) + CO_2(g) + H_2O(l)$
    b. $CaCO_3(s) + 2\,NaHSO_4(aq) \rightarrow Ca^{2+}(aq) + 2\,SO_4^{2-}(aq) + CO_2(g) + 2\,Na^+(aq)\;H_2O(l)$

83. The answer is c. Microorganisms can break down straight-chain hydrocarbon chains but not branched chains.

85. a. NaOCl is a bleaching agent.
    b. $CH_3(CH_2)_{14}COO^-Na^+$ is sodium palmitate, the sodium salt of palmitic acid and a soap .
    c. $Na_5P_3O_{10}$ is sodium tripolyphosphate, a detergent builder used in dishwasher detergents.
    d. $CH_3(CH_2)_{14}COOCH_2(CH_2)_{28}CH_3$ is an ester with two long hydrocarbon chains. Compounds with this kind of structure are waxes.

CHAPTER

22

Poisons

Toxicology: What Makes a Poison?

CHAPTER SUMMARY

22.1. Natural Poisons

Learning Objective: • Name some natural poisons and their sources.

A. A poison is a substance that causes injury, illness, or death of a living organism.
   1. A toxic dose may not be the same for every member of an affected species.
B. Toxicology is the study of the effects of poisons, their detection and identification, and their antidotes.
C. A toxin is a poison that is naturally produced by a plant or animal.
   1. Many insects and lots of microorganisms also produce poisons.
D. Poisonous Plants in the Garden and Home.
   1. Because many natural poisons are alkaloids that occur in plants, it is not surprising that poisons are found in gardens and on farms or ranches.

Answers to Self-Assessment Questions

1. d The study of poisons is called toxicology.
2. c A poison is best defined by dose. Even seemingly safe substances such as sugar and salt can be poisonous if taken in large quantities or by individuals who cannot tolerate them.
3. b Alkaloids are heterocyclic amines. Botulin is a bacterial toxin.
4. b The active ingredient in the poison given to Socrates was an alkaloid.

22.2 Poisons and How They Act

Learning Objectives: • Distinguish among corrosive poisons, metabolic poisons, and heavy metal poisons, and explain how each type acts. • Identify antidotes for some common metabolic poisons and heavy metal poisons.

A. Strong Acids and Bases as Poisons.
   1. Both acids and bases can hydrolyze proteins, destroying their function.

2. Acids break down lung tissue.

B. Oxidizing Agents as Poisons.
   1. Ozone and other oxidizing agents deactivate enzymes by oxidizing sulfur-containing groups.
   2. The amino acid tryptophan undergoes a ring-opening oxidation with ozone.
   3. Oxidizing agents can break bonds in many other chemical substances in a cell.

C. Metabolic Poisons.
   1. Carbon monoxide bonds tightly to the iron atom in hemoglobin and hinders the transport of oxygen in the bloodstream or interferes with oxidative processes in the cells.
   2. Nitrates are reduced to nitrites by microorganisms in the digestive tract.
      a. Nitrites oxidize the iron in hemoglobin from Fe^{2+} to Fe^{3+}. The resulting methemoglobin cannot transport oxygen.
         i. In infants, this condition is called the blue baby syndrome.
      b. Cyanides, compounds that contain a C≡N group, are among the most well-known poisons.
         i. Cyanide, as HCN or its salts, is lethal in amounts of 50 to 60 mg.
         ii. Cyanides act by blocking the oxidation of glucose inside cells; cyanide ions form a stable complex with iron(III) ions in oxidative enzymes called cytochrome oxidases.
         iii. Antidotes for cyanide poisoning can be in the form of providing 100% oxygen to support respiration, giving sodium nitrite intravenously to oxidize the iron atoms in enzymes back to the active 3+ form, followed by treatment with sodium thiosulfate ($Na_2S_2O_3$) which transfers sulfur atoms to the cyanide ions, converting them to thiocyanate ions.

D. Make your own poison: Fluoroacetic acid.
   1. Our cells use acetic acid to produce citric acid, which is then used as an energy source.
   2. When fluoroacetic acid is ingested, it is incorporated into fluorocitric acid.
      a. Fluorocitric acid blocks the energy-producing citric acid cycle by tying up the enzyme that acts on citric acid. Energy production ceases, and the cell dies.
   3. Sodium fluoroacetate (Compound 1080) was used both to poison rats and to poison predators such as coyotes.
      a. Its use is banned on federal land because of the devastating effect it had on the eagle population.
      b. Sodium fluoroacetate is found in the South African *gifblaar* plant and is used by the natives to poison arrow tips.

E. Heavy Metal Poisons.
   1. Metals with densities at least five times that of water are called heavy metals.
   2. Most metals and their compounds show some toxicity when ingested in large amounts.
      a. Even essential mineral nutrients can be toxic when taken in excessive amounts.
      b. In many cases, too little of a metal ion (a deficiency) can be as dangerous as too much (toxicity).
   3. Iron (as Fe^{2+} ions) is an essential nutrient.
      a. Too little iron leads to deficiency (anemia).
      b. Too much iron can be fatal, especially to a small child.
   4. Some heavy metals, including mercury, inactivate enzymes by reacting with sulfhydryl (SH) groups on enzymes.
      a. Mercury metal is a liquid at room temperature; its vapor is especially hazardous.
      b. Mercury is a cumulative poison with a half-life in the body of 70 days.
      c. British Anti-Lewisite (BAL) is an antidote for mercury poisoning.
         i. It acts by chelating (tying up) mercury ions, thus preventing them from attacking enzymes.

   ii. The BAL antidote is effective only when used right away.

5. Lead and its compounds are quite toxic.
   a. Metallic lead is generally converted to Pb^{2+} in the body.
   b. Lead, a soft, dense metal that is corrosion resistant, has many uses.
   c. We get lead in foods, drinking water, and the air.
   d. Large amounts of Pb^{2+} in a child's blood can cause mental retardation, behavior problems, anemia, hearing loss, developmental delays, and other physical and mental problems.
   e. Adults can excrete about 2 mg of lead per day.
   f. Lead poisoning in children causes mental retardation and neurological damage.
   g. Lead poisoning is treated with BAL and ethylenediaminetetraacetic acid (EDTA).
      i. The calcium salt of EDTA replaces the lead ions with calcium ions, and the lead–EDTA complex is then excreted.
   h. Damage to the nervous system is largely irreversible.
6. Cadmium is used in alloys, electronics, and rechargeable batteries.
   a. In cadmium poisoning, cadmium ions substitute for calcium ions; this leads to a loss of calcium from bones, leaving them brittle and easily broken.
7. Cadmium poisoning from drinking water contamination in Japan led to a strange, painful malady known as itai-itai, the "ouch-ouch" disease.

Answers to Self-Assessment Questions

1. a Of the compounds listed, HCN is the only one that is not corrosive.
2. d Carbon monoxide and cyanides are metabolic poisons. Carbon monoxide interferes with oxygen transport by hemoglobin and cyanides interfere with the oxidation of glucose in the cells.
3. d Nitrites poison by oxidizing the Fe^{2+} of hemoglobin, forming methemoglobin which can't carry oxygen to the cells.
4. d Cyanides poison by blocking the actions of cytochrome oxidase, an enzyme.
5. d The most important toxic heavy metals are lead, mercury, and cadmium.
6. c Excessive intake of iron can be fatal to children.
7. a One reason for the dramatic decrease in lead found in the environment is the banning of leaded gasoline.
8. a Lead poisoning can be treated by EDTA, a chelating agent.
9. c Cadmium compounds cause painful "ouch-ouch" disease.

22.3 More Chemistry of the Nervous System

Learning Objective: • Define *neurotransmitter*, and describe how various substances interfere with the action of neurotransmitters.

A. Acetylcholine (ACh) is a neurotransmitter; it carries messages across the synapse between cells.
   1. After carrying its message, acetylcholine is broken down by the enzyme cholinesterase to acetic acid and choline.
   2. Acetylase then converts acetic acid and choline back to acetylcholine, completing a cycle.

B. Nerve Poisons and the Acetylcholine Cycle.
   1. Various chemical substances affect the acetylcholine cycle at different points.
      a. Botulin blocks the synthesis of acetylcholine.
         i. No messenger is formed: no message is sent.
      b. Curare, atropine, and some local anesthetics block the receptor sites.

i. The message is sent but not received.

c. Anticholinesterase poisons block the action of cholinesterase. This blocks the breakdown of acetylcholine; the nerves fire wildly and repeatedly, causing convulsions and death.

C. Organophosphorus Compounds as Insecticides and Weapons of War.

1. Organic phosphorus insecticides are well-known nerve poisons.
2. Chemical warfare agents are Tabun (Agent GA), Sarin (Agent GB), and Soman (Agent GD); these are among the most toxic synthetic chemicals known (but still not as toxic as botulin).
   a. Nerve gases kill when inhaled or absorbed through the skin.
   b. The antidote for nerve gas is an atropine injection.
3. The insecticides malathion and parathion are closely related to the warfare agents but are far less toxic.
4. Nerve poisons have helped us gain a better understanding of the nervous system.

Answers to Self-Assessment Questions

1. a Receptor sites for the neurotransmitter acetylcholine (ACh) are blocked by atropine.
2. a Acetylcholinesterase is an enzyme (the names of enzymes end in *ase*) that breaks down acetylcholine.
3. c Curare acts by blocking ACh receptor sites, making it impossible for muscles to respond to motor nerve stimuli.
4. a Because it is an insecticide and the other listed compounds are nerve agents, malathion is much less toxic than the other compounds.
5. d The most hazardous nerve poisons are organophosphorus compounds.

22.4 The Lethal Dose

Learning Objectives: • Explain the concept of an LD_{50}, and list some of its limitations as a measure of toxicity. • Calculate lethal doses from LD_{50} values and body weights.

A. LD_{50}: A Measure of Toxicity.

1. The LD_{50} is the dose that kills 50% of a population of test animals.
2. LD_{50} values are usually given in mass of toxin per unit of body weight (for example, mg/kg body weight).
3. The larger the LD_{50} value, the less toxic the substance.
4. In general, substances with oral 50 LD values greater than 15,000 mg/kg are considered nearly nontoxic. Substances with 50 LD values from 7500 to 15,000 mg/kg are considered to be of low toxicity; those with values from 500 to 7500 mg/kg, moderately toxic; and those with values less than 500 mg/kg, highly toxic.

B. The Botox Enigma.

1. Botulin is the most toxic substance known, with a median lethal dose of only 0.2 ng/kg of body weight.
2. Botulin, in a commercial formulation called Botox®, is used medically to treat intractable muscle spasms and is widely used in cosmetic surgery to remove wrinkles.
   a. Botox is also used to treat afflictions such as uncontrollable blinking, crossed eyes, and Parkinson's disease.
   b. Botox curbs migraine headache pain in some people and is also used to treat excessive sweating.

Answers to Self-Assessment Questions

1. a LD_{50} is the dose that kills half a tested population.
2. b Of the substances listed, nicotine is the least toxic because it has the largest LD_{50} value.

22.5 The Liver As a Detox Facility

Learning Objective: • Describe how the liver is able to detoxify some substances.

A. The liver is able to detoxify some poisons by oxidation or reduction or by coupling them with amino acids or other normal body chemicals.
   1. The most common route of detoxification is oxidation.
      a. Ethanol is oxidized to acetaldehyde, then to acetic acid, and finally to carbon dioxide and water.
      b. Nicotine is oxidized to less toxic cotinine.
   2. The P-450 enzyme system in the liver can oxidize fat-soluble substances into water-soluble ones that can be excreted.
      a. They can also conjugate compounds with amino acids.
         i. Toluene is oxidized to benzoic acid, which is conjugated with glycine to form hippuric acid, which is excreted.
   3. Oxidation, reduction, and conjugation don't always detoxify.
      a. Methanol is oxidized to more toxic formaldehyde.
      b. Liver enzymes, built up through years of heavy ethanol use, deactivate the male hormone testosterone, leading to alcoholic impotence.
      c. Benzene is oxidized to an epoxide that attacks key proteins.
      d. Carbon tetrachloride is converted to trichloromethyl free radical, which can trigger cancer.

Answers to Self-Assessment Questions

1. a It is not true that benzene is inert and thus nontoxic.
2. b Methyl alcohol (methanol) is oxidized in the liver to formaldehyde, a compound that is even more toxic than methanol.

22.6 Carcinogens and Teratogens

Learning Objectives: • Describe how cancers develop. • Name and describe three ways to test for carcinogens.

A. Carcinogens cause the growth of tumors (abnormal growth of new tissue).
   1. Benign tumors are characterized by slow growth; they do not invade new tissue.
   2. Malignant tumors (cancers) generally show unrestricted fast or slow growth and invade and destroy other tissues.
B. What causes cancer?
   1. Most cancers are caused by lifestyle factors.
   2. Nearly two-thirds of all cancer deaths in the United States are linked to tobacco, diet, or a lack of exercise.
C. There are many natural carcinogens (for example, sunlight, radon, and safrole).
D. How Cancers Develop.

1. Some carcinogens modify the DNA, scrambling the code for replication and protein synthesis.
2. Carcinogens, radiation, and some viruses activate oncogenes that regulate cell growth and division. In addition, suppressor genes must be inactivated before a cancer can develop.

E. Chemical Carcinogens.
1. Best known are polycyclic aromatic hydrocarbons (PAHs), such as 3,4-benzopyrene.
    a. Found in grilled meat, coffee, and cigarette smoke.
2. Aromatic amines (ß-naphthylamine, benzidine) make up another class of carcinogens and are often compounds used in the dye industry.
3. Not all carcinogens are aromatic compounds. For example, dimethylnitrosoamine, vinyl chloride, and some lactone (cyclic esters) are carcinogens.

F. Anticarcinogens.
1. Antioxidant vitamins (vitamins C, E, and beta-carotene, a precursor to vitamin A) are believed to protect against some forms of cancer; flavonoids and the food additive butylated hydroxytoluene (BHT) may protect against stomach cancer.

G. Three Ways to Test for Carcinogens.
1. The Ames test is used to test chemicals to see if they cause mutations in bacteria. Mutagens are often carcinogens.
2. Animal tests employ large doses of the suspected substance on small numbers of laboratory animals such as rats.
    a. Such tests have limited relevance; humans are exposed to lower doses for longer periods, and human metabolism can differ from that of the experimental animals.
    b. However, good correlation exists between known human carcinogens and cancer in laboratory animals.
3. Epidemiological studies correlate the incidence of various cancers in human populations with various occupations, dietary practices, and exposures to chemicals.

H. Birth Defects: Teratogens.
1. A teratogen is a substance that causes birth defects.
    a. The most notorious is the tranquilizer thalidomide.
    b. The drug was used in Germany and Great Britain, but not in the United States, thanks to Frances Kelsey of the FDA who saw evidence to doubt the drug's safety.
2. Other chemicals that act as teratogens include isotretinoin (Accutane®), a prescription medication approved for treating severe acne.
3. By far the most hazardous teratogen, in terms of the number of babies born with birth defects, is ethyl alcohol, which causes fetal alcohol syndrome.

Answers to Self-Assessment Questions

1. b One type of gene when switched on that is involved in cancer development is oncogene.
2. d Suppressor genes are involved in preventing cancer, and when they are switched off, cancers can develop.
3. b Ethylenediamine is not an aromatic compound.
4. b A source of polycyclic aromatic hydrocarbons (PAHs) is automobile exhaust.
5. d Vinyl chloride is a prominent aliphatic (nonaromatic) carcinogen.
6. b Bis(epoxy)butane is not an anticarcinogen. This compound is a heterocyclic carcinogen (see Figure 22.8).
7. a The fastest way to determine whether a substance is likely to be a carcinogen is by performing an Ames test.
8. d The Ames test is a test for mutagenicity used as a screen for potential carcinogens.

9. a A teratogen is a substance that causes birth defects.
10. d Thalidomide was discovered to be a teratogen after women in Europe using it in the first trimester of their pregnancies gave birth to infants with shortened or absent arms and legs.

22.7 Hazardous Wastes

Learning Objective: • Define hazardous waste, and list and give an example of the four types of such wastes

A. Hazardous wastes are those that can cause or contribute to death or illness or that threaten human health or the environment when improperly managed.
   1. There are four types of hazardous wastes:
      a. Reactive wastes react spontaneously or vigorously with air or water.
         i. They explode when exposed to shock or heat.
         ii. Examples of reactive wastes are sodium metal, TNT, and nitroglycerin.
      b. Flammable wastes are those that burn readily upon ignition, presenting a fire hazard.
         i. Hexane is an example of a flammable waste.
      c. Toxic wastes contain or release toxic substances in quantities that pose a hazard to human health or the environment.
         i. Examples of toxic wastes include polychlorinated biphenyls (PCBs).
      d. Corrosive wastes corrode their containers.
         i. Examples of corrosive wastes include strong acids and bases.

B. The best way to handle hazardous wastes is to not produce them in the first place, which is the major focus of green chemistry.
   1. Many industries have modified manufacturing processes to prevent wastes (green chemistry principle 1) and maximize atom economy (principle 2); some wastes can be reprocessed to recover energy or materials.
   2. The best current technology for treating organic wastes including chlorinated compounds is incineration.
   3. If a hazardous waste cannot be used or incinerated or treated to make it less hazardous, it must be stored in a secure landfill.
   4. Bacterial biodegradation is another method that shows great promise.
      a. Through genetic engineering, scientists are developing new strains of bacteria that can decompose a great variety of wastes.

C. What Price Poisons?
   1. We have to decide whether the benefits we gain from hazardous substances are worth the risks we assume by using them.

Green Chemistry: Designing Safer Chemicals with Green Chemistry

Learning Objectives: **• Explain how green chemistry can lead to both hazardous properties and safer chemicals. • Trace the process for designing safer chemical structures.**

A. The public and regulators are becoming more concerned about risk, as shown by the European Union's Registration, Evaluation, Authorisation, and Restriction of Chemicals (REACH) legislation and proposed changes to the Toxic Substances Control Act in the United States.
   1. REACH requires companies to submit human health and environmental safety test data and an assessment of risks before manufacturing or importing chemicals.

2. Designing safer chemicals relies on understanding what makes a molecule hazardous.
3. Scientific advances are being sought to more effectively identify hazards based on structure during chemical design because traditional current testing methods are slow and expensive, rely on the use of lots of test animals, and do not provide information about how molecules are toxic.
4. For the future, researchers are trying to develop chemical design tools that can quickly screen large groups of structurally similar or diverse chemicals and build the knowledge base to support the design of safer and commercially successful chemicals.

Answers to Self-Assessment Questions

1. c Reactive wastes tend to react vigorously with water.
2. a Common strong acids and bases are corrosive wastes.
3. d Discarded unused pesticides are toxic wastes.
4. b Charcoal lighter fluid is flammable.
5. d The best way to handle hazardous wastes is to minimize or eliminate their use.

LEARNING OBJECTIVES

You should be able to . . .

1. Name some natural poisons and their sources. (22.1)
2. Distinguish among corrosive poisons, metabolic poisons, and heavy metal poisons, and explain how each type acts. (22.2)
3. Identify antidotes for some common metabolic poisons and heavy metal poisons. (22.2)
4. Define *neurotransmitter*, and describe how various substances interfere with the action of neurotransmitters. (22.3)
5. Explain the concept of an LD_{50} value, and list some of its limitations as a measure of toxicity. (22.4)
6. Calculate lethal doses from LD_{50} values and body weights. (22.4)
7. Describe how the liver is able to detoxify some substances. (22.5)
8. Describe how cancers develop. (22.6)
9. Name and describe three ways to test for carcinogens. (22.6)
10. Define *hazardous waste*, and list and give an example of the four types of such wastes. (22.7)
11. Explain how green chemistry can lead to both hazardous properties and safer chemicals.
12. Trace the process for designing safer chemical structures.

DISCUSSION

It has been said that "all things are poison." which isn't far from the truth if you take dosage into consideration. However, some things are much more toxic than others. Also, some substances present an immediate hazard (acute poisons), whereas others exhibit their harmful effect only over a much longer period of time (chronic poisons). The public often gets contradictory, misleading, and erroneous information about toxic chemicals. A few chemical principles will help you to gain a better understanding of what toxic substances are, how they act, and what can be done about toxic wastes.

ANSWERS TO ODD-NUMBERED REVIEW QUESTIONS AND SOLUTIONS FOR ODD-NUMBERED END-OF-CHAPTER PROBLEMS

Review Questions

1. Taken in small amounts, ethanol, the alcohol present in alcoholic drinks, is not poisonous and can be tolerated by many individuals. Larger amounts can cause problems, for example, when liver enzymes that oxidize alcohols also deactivate the male hormone testosterone, accounting for alcoholic impotence. Taken in abnormally large amounts, the same substance can be poisonous for anyone.

3. Examples of the toxicity depending on the route of administration are the toxicity of water when it is inhaled and the fact that nicotine is 50 times more toxic when taken intravenously than when taken orally. The route of administration affects the speed of the poison's action and the rate at which it can be detoxified.

5. Some sources of mercury are broken thermometers, dental fillings, fish consumption, and agricultural uses. Some sources of lead are the lead that leaches from lead pipes and lead solders, the lead in paint, and the lead in the glazes on some food containers.

7. Sunlight, radon, and safrole in sassafras are examples of natural carcinogens. In addition, plants produce compounds to protect themselves from fungi, insects, and higher animals, some of which are carcinogenic. For example, natural carcinogens are found in mushrooms, basil, celery, figs, mustard, pepper, fennel, parsnips, and citrus oils.

Problems

9. Sulfuric acid (a) and sodium hydroxide (b) are corrosive poisons.

11. The hydrolysis of the amide (peptide) linkage between amino acids in proteins disrupts the protein's ability to function in the way in which it was intended.

13. Metabolic poisons prevent cellular oxidation of metabolites by blocking the transport of oxygen in the bloodstream or by interfering with the oxidative processes in the cells. Carbon monoxide and nitrates are examples of metabolic poisons.

15. Nitrate ions act by oxidizing the Fe^{2+} in hemoglobin to Fe^{3+}, forming methemoglobin which is incapable of carrying oxygen.

17. The answer is c and d. Heavy metals have densities at least five times that of water. Note: To answer this question, you will have to look up the densities of the elements listed and realize from the question

itself that there is more than one correct answer. The densities are: beryllium, 1.85 g/cm^3; lithium, 0.54 g/cm^3; molybdenum, 10.28 g/cm^3; and tellurium, 6.54 g/cm^3. Only molybdenum (answer c) and tellurium (answer d) have densities at least five times greater than that of water (density of water = 1.00 g/cm^3), so, of this list, Mo and Te are heavy metals.

19. The mercury in fish is more toxic than that ingested in other ways. Fish convert ingested mercury into methyl mercury, which is extremely toxic.

21. EDTA ties up, or forms chelates with, lead ions into complexes that can be excreted.

23. Three nerve gases that were developed for use in warfare are tabun, sarin, and soman.

25. Various substances can block acetylcholine. For example, botulin blocks the synthesis of acetylcholine, with curare, atropine, and some local anesthetics acting by blocking acetylcholine receptor sites.

27. The liver detoxifies nicotine by oxidizing it to less toxic cotinine. The added oxygen atom makes cotinine more water soluble and therefore more readily excreted than nicotine in the urine.

29. No, P-450 enzymes do not always detoxify foreign substances. The enzymes just catalyze oxidation. Some substances become more toxic once oxidized than they were when they were more reduced (methanol is an example).

31. The liver enzymes that oxidize alcohols also deactivate the male hormone testosterone. Buildup of these enzymes in a chronic alcoholic leads to a more rapid destruction of testosterone, which leads to alcoholic impotence, one of the well-known characteristics of alcoholism.

33. Lethal dose = (125 lb)(0.454 kg/lb)(140 mg/kg) = 7950 mg or 7.95 g of methyl isocyanate

35. Mutagens are compounds that alter the genetic material in a cell. The Ames test would be an effective tool to use to make this determination.

37. Polycyclic hydrocarbons are formed during the incomplete burning of nearly any organic material. For example, they have been found in charcoal-grilled meats, cigarette smoke, and automobile exhausts.

39. Animal tests involving low dosages and millions of rats would cost too much, so tests are usually done by using large doses and a few dozen rats, with an equal number of rats serving as controls. The tests are not conclusive because humans are not usually exposed to comparable doses and there may be a threshold below which a compound is not carcinogenic. Further, human metabolism is somewhat different from that of the test animals, so a carcinogen might be active in rats but not in humans (or vice versa). There is only a 70% correlation between the carcinogenesis of a chemical in rats and that in mice, and the correlation between carcinogenesis in either rodent and that in humans is probably less.

41. Oncogenes are genes that trigger or sustain the processes that convert normal cells to cancerous cells.

43. A toxic waste contains or releases toxic substances in quantities sufficient to pose a hazard to human health or to the environment. Examples of household toxic wastes include paint, pesticides, motor oil, medicines, and cleansers.

45. Number of molecules in lethal dose = (66 kg)(200 pg botulin/kg)(1×10^{-12} g/pg)
(1 mol botulin/1.50×10^5 g)(6.02×10^{23} molecules botulin/mol botulin)
= 5.3×10^{10} molecules

47. a. The molecular formula of thujone is $C_{10}H_{16}O$.
b. lethal dose = (55 kg)(87.5 mg thujone/kg) = 4800 mg or 4.8 g
c. mass of thujone = (10 mg/L)(0.50 L) = 5 mg

49. a. % smokers in study = (1350 smokers/1357 studied)(100%) = 99.48%
% smokers in control group = (1296 smokers/1357 studied)(100%) = 95.50%
The control group contained a few percent fewer smokers. The proportion of smokers in both groups was roughly equivalent.
b. The odds of smoking for cancer sufferers were 0.9948:1. The odds of smoking for the control group were 0.9550:1.
c. The odds ratio is 0.9550/0.9948 = 0.96:1, which indicates that the control group was a good representation of the cancer sufferer group.

51. The data in Table 22.1 indicate that one would have to ingest relatively large masses of some compounds (for example, sucrose) before the dose reached the LD_{50} level, but only extremely small amounts of other substances (ricin and botulin toxin, for example) to surpass the LD_{50} level.

53. Lethal dose = (55 kg female)(200 μg/kg)(1 mg/1×10^3 μg) = 11 mg
a. A 1.5-mg dose is considerably less than 11 mg, so it is not considered likely to be lethal.
b. A 20-mg dose is almost twice the LD_{50} for plutonium, so this dose is lethal.

55. Benzene is generally inert in the body. If it is ingested, it does not react until it reaches the liver where it is slowly oxidized to an epoxide. The epoxide is a highly reactive molecule that can attack certain key proteins, sometimes resulting in leukemia, a form of cancer.

57. a. mass of ethanol that would likely kill the rat, g = (395 g rat)(1 kg/1000 g)
(10.3 g ethanol/kg rat)
= 4.07 g ethanol
volume of ethanol, mL = (4.07 g ethanol)(1 mL ethanol/0.79 g ethanol) = 5.15 mL ethanol
volume of vodka, mL = (5.15 mL ethanol)(100 mL vodka/50 mL ethanol) = 10.3 mL vodka
b. volume of water in rat, mL = (395 g rat)(0.67 g water/g rat)(1.0 mL water/1.0 g water)
= 265 mL water
concentration of alcohol in rat, mg/mL = (4.07 g ethanol)(1000 mg/g)(1/265 mL water)
= 15.4 mg/mL

59. The answer is b, use of rats and mice for testing.

61. Saving millions of dollars, improving consumer safety, and reducing the load of poisons in the environment are all benefits of applying green chemistry principles on a large scale.